Megha Sud
Political Ecology of the Ridge

MEGACITIES AND GLOBAL CHANGE

MEGASTÄDTE UND GLOBALER WANDEL

herausgegeben von

Frauke Kraas, Martin Coy, Peter Herrle und Volker Kreibich

Band 23

Megha Sud

Political Ecology of the Ridge

The Establishment and Contestation of Urban Forest Conservation in Delhi

Umschlagabbildung: © Mohd Salman

Bibliografische Information der Deutschen Nationalbibliothek:
Die Deutsche Nationalbibliothek verzeichnet diese Publikation in der Deutschen Nationalbibliografie; detaillierte bibliografische Daten sind im Internet über <http://dnb.d-nb.de> abrufbar.

Angenommen als Dissertation durch die Mathematisch-Naturwissenschaftliche Fakultät der Universität zu Köln. Verteidigung am 20.04.2015, die Gutachter waren Prof. Dr. Frauke Kraas und Prof. Dr. Boris Braun.

Druck: Hubert & Co, Göttingen
Gedruckt auf säurefreiem, alterungsbeständigem Papier.
Printed in Germany.
ISBN 978-3-515-11714-2 (Print)
ISBN 978-3-515-11715-9 (E-Book)

CONTENTS

LIST OF TABLES

LIST OF FIGURES

LIST OF MAPS

ABSTRACT

The remains of the timeworn Aravalli Mountains that extend through Delhi are known as the Ridge. The range exists in four main segments today and these are legally protected as conserved forests. This book presents an examination of the conflicts and negotiations that led to the establishment of conservation as the dominant discourse in the Ridge, highlighting the role of different actors and the unequal outcomes of these deliberations.

The framework used here is based in post-structural political ecology and draws on concepts of hybrids and metabolism developed in urban political ecology. An actor approach was taken to analyse discursive interactions regarding the conservation of forests in Delhi. In particular argumentative discourse analysis with its concepts of storylines and discourse coalitions was used to understand the conflicting, changing and overlapping components of various discourses and the unequal power relations embedded in these negotiations. This analysis is contextualised in the broader processes of urban development that conservation conflicts are mediated by.

The demands for conservation of the Ridge have been temporally and spatially fragmented. Therefore, a historical overview of conservation of parts of the landscape and an examination of beginnings of environmental activism surrounding the issue, are followed by a detailed analysis of trajectories of the three specific parcels where active conservation and restoration projects are being undertaken. It was found that the conflict of interests and variation of outlooks between state agencies and the demands of middle class environmentalists have enabled the establishment of specific legal and administrative arrangements in each of the three areas. These conservation projects have negative implications for marginalised populations in the vicinity in terms of restriction of access and eviction. However these groups have not been able to participate in debates that lead to policy formulation due to their positioning in the dominant discourse as illegal and detrimental to conservation and due the fragmentation in political participation in Delhi.

The study contends that the forested spaces of Delhi must be understood as implicated in the socio-political structures of the city rather than attempting to create pristine wilderness areas. This would be a necessary first step towards holistic decision-making regarding socio-environmental problems, which is crucial if urban conservation policies are to be sustainable both socially and environmentally.

ZUSAMMENFASSUNG

Die geologischen Überreste des präkambrischen Faltengebirges der Aravalli Mountains durchziehen Delhi als Bergrücken und werden als Ridge bezeichnet. Heute sind vier Teilbereiche dieses Bergrückens als Wald gesetzlich geschützt. Die vorliegende Studie untersucht die Konflikte und Aushandlungsprozesse innerhalb der Naturschutzdiskurse, die zum Schutz dieser Waldbestände geführt haben. Hierbei wird besonders auf die Rolle der verschiedenen Akteure und deren ungleichen Einflüsse eingegangen.

Der Untersuchungsrahmen basiert auf dem Ansatz der post-strukturellen politischen Ökologie und bezieht Hybrid- und Metabolismuskonzepte mit ein, die in der urbanen politischen Ökologie entwickelt wurden. Ein Akteursansatz wurde für die Analyse der diskursiven Interaktionen im Zusammenhang mit dem Schutz der Wälder in Delhi gewählt. Hierbei werden die unterschiedlichen Handlungsstränge und Diskurskoalitionen mit Hilfe einer argumentativen Diskursanalyse herausgearbeitet, um die unterschiedlichen, sich verändernden und überlappenden Bestandteile der verschiedenen Diskurse und die zugrundeliegende ungleichen Machtverhältnisse zu verstehen. Diese Analyse der Naturschutzdiskurse ist eingebettet in die übergeordneten Stadtentwicklungsprozesse.

Die Forderungen im Hinblick auf den Schutz der Ridge variieren sowohl zeitlich als auch räumlich. Aufbauend auf einen historischen Überblick über die Diskurse zur Ridge sowie einer Untersuchung der Anfänge des Umweltaktivismus erfolgt eine detaillierte Analyse der Entwicklungspfade von drei ausgewählten Fallstudien, die sich auf Teilflächen beziehen bei denen aktiv Naturschutz- und Wiederaufforstungsprojekte durchgeführt werden. Die Ergebnisse der Untersuchung zeigen, dass Interessenskonflikte und unterschiedliche Auffassungen zwischen den Behörden und Naturschützern aus der Mittelklasse zu spezifischen gesetzlichen und verwaltungstechnischen Regelungen für die jeweiligen Teilflächen geführt haben. Diese unterschiedlichen Naturschutzprojekte haben jedoch negative Auswirkungen für marginalisierte Bevölkerungsgruppen in der Nachbarschaft der Gebiete und haben u.a. zu Zugangsbeschränkungen und Räumungen von Siedlungen geführt. Da die Bedürfnisse der marginalisierten Gruppen jedoch dem vorherrschenden Naturschutz-Diskurs (conservation discourse) widersprachen, und sie als illegal und ihre Praktiken als unvereinbar mit dem Naturschutz eingestuft wurden, konnten sie nicht an den Debatten, die zur Formulierung der gesetzlichen Richtlinien führten, teilhaben. Dieses ist Ausdruck der Fragmentierung der politischen Beteiligung in Delhi.

Die Arbeit argumentiert, dass der Schutz der Ridge nicht als Versuch verstanden werden darf ursprüngliche Wildnis wieder herzustellen, sondern die Entwicklungen sind vielmehr Ausdruck der sozio-politischen Strukturen innerhalb der Stadt. Dieses anzuerkennen erscheint als ein erster notwendiger Schritt hin zu einem holistischen Verständnis von sozial-ökologischen Problemen und Grundvoraussetzung für sozial sowie ökologisch nachhaltige Entscheidungsfindung im Hinblick auf den Schutz der Ridge.

ACKNOWLEDGEMENTS

I would like to extend my gratitude Prof. Dr. F. Kraas for accepting me as a doctoral student after our meeting at a very stimulating summer school in Bombay and making this study possible. I hope she will find the journey I have made as a researcher since then reflected in this work. Prof. Dr. B. Braun also provided comments at the beginning of the study which were very helpful in formulating the initial direction of research.

I am grateful to all the interviewees who gave me their time, views and documents, often fitting me into their busy schedules and inviting me into their homes and offices.

Several people have read parts of this book and provided feedback and/or proofreading: Tine Trumpp, Pamela Hartmann, Stephani Leder, Gerrit Peters, Hugues Vincent-Genod, Mhd. Salman, Alexander Follman. Thank you so much for your time and efforts that have greatly enhanced the quality of this document.

The AG Kraas has been indispensable to my work in Cologne. From stimulating discussions to fielding innumerable queries, they have provided the support base which anchored me in the institute.

Salman accompanied me on several trips to the Ridge, took many pictures, including the one on the cover of this book and provided constant encouragement on almost a daily basis during the writing phase. I am grateful for his enduring friendship.

A scholarship from the DAAD made this research possible. The staff at the DAAD as well as the International Students Office of the University of Cologne has always been extremely helpful and have kept things smooth so I never had to worry about finances or official formalities. Harald Schmitt and Sussane Henkel from Franz Steiner Verlag tirelessly assisted me in preparing the manuscript for publication. I take this opportunity to thank them all for their support and assistance.

My parents have been a constant source of encouragement and have made possible all these opportunities and freedoms the make my life what it is. They also provided room and board and great company while I was in Delhi for field work. As always, I am thankful to them. Thanks also to Nina for her late night company and for occasionally obliging me with much needed cups of chai.

I now understand why people thank their partners at the end of the acknowledgements. Hugues Vincent-Genod has seen the best and worst of me in the process of this research and managed to be always patient and supportive. I would like to thank him for holding fort while I was absent for large periods of time either physically or otherwise and for all the ways in which he encourages me to do better.

LIST OF ACRONYMS

ABWLS	Asola Bhatti Wildlife Sanctuary
CEC	Centrally Empowered Committee
CEMDE	Centre for Environmental Management of Degraded Ecosystems
CGWA	Central Ground Water Authority
CPQLW	Citizens for Protection of Lakes and Quarry Wilderness
DCF	Deputy Conservator of Forests
DDA	Delhi Development Authority
DPTA	Delhi Preservation of Trees Act
DSIDC	Delhi State Industrial Development Corporation
DSMDC	Delhi State Mineral Development Corporation
EIA	Environmental Impact Assessment
EIAA	Environmental Impact Assessment Authority
EPCA	Environment Pollution (Prevention and Control) Authority
ETF	Environmental Task Force
GNCT	Government of National Capital Territory
GSI	Geological Survey of India
IFA	Indian Forest Act
INTACH	Indian National Trust for Art and Cultural Heritage
JJ	Jhuggi Jhopri (refers to shanties/slums)
JNGO	Joint NGO Forum to Save the Ridge
LG	Lieutenant Governor
MLA	Member of Legislative Assembly
MoEF	Ministry of Environment and Forest
MP	Member of Parliament
MPD	Master Plan Delhi
MPISG	Master plan Implementation Support Group
RBA	Ridge Bachao Andolan
RMB	Ridge Management Board
RWA	Resident Welfare Associations
MCD	Municipal Corporation of Delhi
NCR	National Capital Region
NCT	National Capital Territory
NCTD	National Capital Territory of Delhi
NDMC	New Delhi Municipal Corporation
NGO	Non-governmental Organisation
NGT	National Green Tribunal
PIL	Public Interest Litigation
STF	Special Task Force
TCPO	Town and Country Planning Organisation
TSM	Total Station Method

UPE	Urban Political Ecology
WLPA	Wildlife Protection Act
WII	Wildlife Institute of India
WWF (I)	Worldwide Fund for Nature (India)

1 INTRODUCTION

1.1 SOCIAL CONSTRUCTION OF URBAN FORESTS IN DELHI

Having walked dusty tracts among shrubs and stunted trees for about an hour chasing a sighting of the rare and elusive sandgrouse, we climb a small hillock so the Nature Education Officer can point out where leopard pug marks were found a few weeks ago. Standing on this slight elevation within the Asola Bhatti Wildlife Sanctuary, one is at once affronted by the sights and sounds of the bustling metropolis that is Delhi; the constant growling and honking of distant traffic and the humming and hammering of construction sites are carried by the wind, puncturing the quiet of the forest. From this height, the multi-story apartments that are quickly rising in the horizon on the other side of the state boundary and Asia's largest 'unauthorised colony' bursting at the seams at the edge of the sanctuary, are clearly visible. One is prone to the feeling that the city is poised to swallow this last refuge of nature within it. It was perhaps similar sentiments that have driven calls to 'save the Ridge' by citizens in different parts of the city through the years. These various demands for conservation and protection of the forests have led to the establishment of the aim of maintaining the Ridge in its 'pristine glory' in the master plan of the city (DDA, 1990, p. 3).

Such sentiments of nostalgia and the fear of losing 'Eden' characterise preservationist thinking that presents spaces set aside for nature conservation as representing a primordial, undisturbed nature (Neumann, 2003). Certain areas have been historically invested with romantic ideas of being places of contemplation and beauty as opposed to areas meant for production and profit, obscuring the power relations and conflicts embedded in the creation and sustenance of these areas (Cosgrove, 1984; Neumann, 1998; Williams, 1973). An overview of the history of the Ridge disproves any idea of it being a pristine forest, and a deeper analysis of its trajectory reveals that rather than being an antithesis of the city, the forest is implicated in the material, socio-political, economic and historical fabric of Delhi.

It is argued in the following chapters that the Ridge has been socially constructed. Social construction of nature refers to the material and physical construction of environment, as well as construction of concepts and ideas regarding the environment which are dynamic and geographically and historically situated (Castree, 2003; Demeritt, 2002; Escobar, 1999). These two elements are intimately linked as it is through the social construction of the concepts of nature and the negotiation between these concepts that society comes to shape the material world and to interpret the changes it affects (Demeritt, 2002). Smith uses the term 'ideology of nature' to refer to the false belief in the ontological fixity of natural environment and presents the idea of social construction of nature as a refutation of this ideology, he suggests:

> "What jars us so much about this idea of the production of nature is that it defies the conventional, sacrosanct separation of nature and society, and it does so with such abandon and without shame. We are used to conceiving of nature as external to society, pristine and pre-human, or else a grand universal in which human beings are but small and simple cogs" (Smith, 1984, p. xiv).

The aim of re-framing natural environment as socially produced is to provide a political critique that exposes the power relations embedded in the formation and implications of such environments. This research is based in the fields of post-structuralist political ecology and urban political ecology which focus on the debates on socio-political conditions regarding the definition, experience and solutions to environmental problems (Blaikie & Brookefield, 1987; Forsyth, 2008a; Robbins, 2012; Zimmerer, 1993). Through the political ecology framework, this study presents spheres of politics, society, state, legality and that of environment as interactive and interdependent and argues that they need to be seen as such while making policy decisions regarding any landscape. So while political economy studies economic distribution of conflict, political ecology studies 'ecological distribution of conflict' i.e. "social, spatial and temporal asymmetries or inequalities in the use by humans of environmental resources and services" (Guha & Martinez-Alier, 1997, p. 31). Rather than seeing urban and nature as separate entities, political ecology engages in a consideration of urban natures as embodiments of the socio-political elements that produce them (Swyngedouw, 1996).

1.2 AN INTRODUCTION TO THE DELHI RIDGE

Delhi has long been a centre of strategic and political importance in Northern India (Spear, 2002). As a capital city to several ruling governments of the region over the decades (Frykenberg, 1994), the city has undergone multiple phases of urban re-structuring. The seat of the British colonial government was shifted from Calcutta to Delhi in 1911, giving rise to 'New Delhi' which was to be planned and built to reflect imperial power (Legg, 2006). Delhi continued to be the capital of independent India and since 1957 has been subjected to modern state-led zoning plans (Batra, 2010; Baviskar, 2003a). Recent metropolitan transformation in Delhi is linked to a shift towards neo-liberalism that accelerated in the early 1990s, due to structural adjustments made in India's economic policy (Dupont, 2011). The present governance structure of the city is complex, due to its status as National Capital Territory (NCT). India has a federal system of government and the central government of India is based in Delhi, but the city also has an elected state legislature headed by a Chief Minister. The civic administration (Delhi Administration) is headed by the Lieutenant Governor who represents the President of India at the centre and not the Chief Minister.

The city occupies an area of 1483 square kilometres with a population density of 11,029 persons per square kilometre and a floating population of 0.3 to 0.4 million people per day (Census of India, 2011). With an expanding population within city limits and a rapidly densifying urban agglomeration in the surrounding areas, the pressure on land is high and increasing (GNCTD, 2014). Despite this, 20.08 per cent of the total geographical area of the city consists of forests and tree cover while in 1993, this figure stood at merely 1.48 per cent (FSI, 2013). Legally recorded forests (under forest conservation laws) cover 5.73 per cent of the land area, of this around 91 per cent is Reserved Forest area and the rest is Protected Forest land[1] (FSI, 2013), all of which lie on the Ridge. The conflicts and deliberations that led to the establishment of conservation areas, which are part of the reason behind the increase in forest cover in Delhi in the last two decades, are traced in the following chapters.

The Ridge is one end of the Mewat branch of the Aravalli Range that extends through the neighbouring states of Gujarat, Rajasthan, Haryana and Delhi in the Northwest of India. The Aravalli is the oldest mountain range in the country that has eroded over time, now existing as low hills at a height of 2.5 to 90 meters in Delhi (Singh, 2006). The range extends for about 35 kilometres in the city, from Wazirabad in the north and curving around Bhatti Mines in the south, with some intermittent, scattered outcrops (Sinha, 2014), evidence to a larger area covered in the past of which four distinct patches are legally protected as forest land today. It is suggested that the Ridge once covered as much as 15 per cent of the geographical area of the city (Agarwal, 2010). However by the 1980s, an administrative note stated that 40 per cent of this had already ceased to exist as a distinctive landscape[2]. An idea of the extensiveness of the original range can be gleaned from the sketch of Delhi's environs in (circa) 1807, showing an unbroken hilly outcrop extending through the city (see Map 1). Further evidence towards the fact that the Ridge was once a continuous landscape is contained in the toponomy of the city. Historian Narayani Gupta traces the erstwhile extent of the Ridge thought the names of localities that signify their hilly origins:

> "The Ridge […] was once a distinct range of hills. Muradabad Pahadi[3] in Vasant Vihar, Pahad Ganj and Pahadi Dhiraj in Central Delhi, Raisina Pahadi on which the President's house stands, Bhojla Pahadi which is the base of Jama Masjid, Anand Parbat in West Delhi, are reminders of the hills we have lost" (Gupta, 2010, p. 96).

The flora of the Ridge is classified as tropical thorn forests (Champion & Seth, 1968), consisting of a semi-arid open scrub type of vegetation. The dominant tree groups here are Acacia and the related Mimosa (Krishen, 2006). These have been

1 'Reserved Forest' is a legal category of conserved forests where all uses are prohibited except those specifically sanctioned. 'Protected Forests' are those in which certain uses are specifically banned

2 Town and Country Planning Organisation, 1982, 'Note on the Status of the Ridge in and around National Capital'. The note refers to a study conducted by the Delhi School of Planning and Architecture which provides this figure (Ganguli, 1975)

3 The words 'pahad', 'pahadi' and 'parbat' mean hill/mountain in Hindi

mixed with exotic species, especially Prosopis juliflora (Ekta, 2014). In 1883-84, a large variety of fauna was noted to be found in the Ridge including foxes, wolves, jackals, mongoose, wild pigs, nilgai (blue bulls), blackbucks, chinkaras and leopards (Gazetteers Organisation, 1999). While most of the large mammals are rarely sighted today,the nilgai is common, foxes, jackals, porcupines can be seen occasionally and the leopard is known to stray intermittently into the Southern Ridge from the other side of the Haryana border (Sinha, 2014). The city has a rather high bird population of 110 resident species and 200 migrant species of birds (a third of the total bird count of the subcontinent) (Sinha, 2014).

The Ridge and the river Yamuna, the two main environmental landmarks of Delhi, formed the boundaries of the city historically (Dhawale, 2010; Mandal & Sinha, 2008). The Delhi Districts Gazetteer of 1883–84, mentions in its description of Delhi:

> "The tract thus limited [by the river and the Aravallis], though exhibiting none of the beauties of mountainous districts, possesses a considerable diversity of physical features, and in parts is not wanting in picturesqueness. This it owes to the hills and to the river" (Gazetteers Organisation, 1999, pp. 1–2).

Reasons for the location of the city in the triangle between the river and the Ridge included the availability of water and the strategic defence provided by the hills (Shokoohy & Shokoohy, 2003). While the city may have grown beyond its original natural boundaries, the Ridge continues to function as a protective barrier from the dust and hot desert winds from Rajasthan, as Delhi lies in the leeward side of the Aravallis (Agarwal, 2010; Dhawale, 2010). Other noted advantages of the forests in Delhi include lowering of temperatures, provision of a noise buffer and reduction of pollution levels (Kothari & Rao, 1997; Srishti, 1994). The Ridge was also a source of several water channels that fed the Yamuna in the past. Most of these are now lost or turned to sewers (Misra, 2010). Quartzite rock that is found in abundance in the Aravallis is fairly porous and enables the Ridge to function as a ground water recharge zone; according to one estimate, 80 per cent of the rain falling in this area is recharged (Soni, 2007). For a city that lies in the semi-arid climatic zone, with a fast depleting water table (Shekhar, Purohit, & Kaushik, 2009), this is an important function.

There is no clear definition or delimitation of what exactly qualifies as the Ridge in Delhi today. Different authorities have argued according to their own viewpoints. Moreover contestations between various actors (including local residents, environmentalists, planning and forest bureaucracies and courts) have resulted in a montage of legal and administrative and arrangements across the Ridge. The Delhi Development Authority (DDA), the planning body of the city, in the Master Plan for Delhi, 2001 offers this vague definition:

> "The Ridge in Delhi is defined as a rocky outcrop of the Aravalli range stretching from the University in the North to the Union Territory boundary to the South and beyond" (DDA, 1990, p. 53).

Administratively, the DDA defines the Ridge as the four distinct segments that received legal notification as Reserved Forests in 1994 (see Map 2). These segments are: The Northern Ridge which is the smallest section of around 87 hectares lying in old Delhi in the northern campus of the University of Delhi. The Central Ridge[4] spread across 864 hectares lies in the administrative centre of Delhi. These parts are largely fenced-in and have well defined boundaries. The South Central Ridge or the Mehrauli Ridge is highly fragmented by built-up area and contains the Sanjay Van (around 633 hectares). This segment is dotted with medieval monuments including the Qutab Minar and the Mehrauli Archaeological Park as well as university campuses of the Jawaharlal Nehru University and the Indian Institute of Technology. The Southern Ridge or the Tughlaqabad Ridge is the largest segment of around 6,200 hectares and contains the Asola Bhatti Wildlife Sanctuary. This segment lies on the periphery of Delhi coinciding with the southern state border (DDA, 2007; Sinha, 2014; Srishti, 1994).

These four segments however, do not cover the entirety of the Ridge. The Forest Department has defined the Ridge as all areas displaying the morphological characteristics of the Aravalli hills according to Geological Survey of India (Sinha, 2014), which is a much larger area spread across the city. In a segment of the South Central Ridge, not included in the DDA's administrative definition, a long drawn contestation by environmentalists and local residents resulted in the establishment of the Aravalli Biodiversity Park for the conservation and active restoration of forests in the area. It can be summarised that though city planning itself is an outcome of the interaction between actors, contesting visions of the Ridge as a landscape have led to the establishment of fragmented and varied forest conservation spaces in Delhi that the plan does not capture.

To one unacquainted with the Ridge, it may seem a surprising fact that within a congested megacity lie a wildlife sanctuary, a biodiversity park, a citizens' forest restoration project and large areas of landscaped parks, all under the label of the 'Ridge'. Each of these conservation units represents an aspiration to preserve the city's forests for different reasons. There is little interaction between these various models of conservation as each bounded segment is administered separately and has resulted from different contextual histories. This research examines the discursive construction of these spaces as worthy of conservation and the outcomes of such ideas in practice. In other words, the following chapters examine how the need to 'save the Ridge' was established as the dominant discourse but also raise and answer questions regarding how, why, by whom, for whom and against whom is the Ridge being saved.

1.3 OUTLINE OF THE BOOK

A review of the vast and contested field of political ecology is presented in the next chapter in order to locate this study within the field and define key concepts and ideas. This study synthesises the aims and concepts of two distinctive branch-

4 The present day Central Ridge was labelled the Southern Ridge in earlier notifications. The current nomenclature is used throughout the book

es of political ecological traditions which have developed from different roots, namely post-structural political ecology and urban political ecology.

Chapter 3 presents the research questions, methodology and methods used to study the Ridge. Actors were analysed using argumentative discourse analysis that provides the concepts of storylines and discourse coalitions to capture how complex environmental issues are deliberated upon (Hajer & Versteeg, 2005; Hajer, 1993, 1995, 2006). The second half of the chapter details the methods of data collection.

A historical background to the Ridge as a space of degradation and a site of afforestation projects is provided in chapter 4, along with a review of various orders and notifications that have established conservation spaces in the Ridge since 1913. The aim of the chapter is to provide a background of the legal and administrative status of the Ridge as well as to introduce the idea of the forest as socio-politically shaped.

Chapter 5 is dedicated to unpacking the variety of aims and motivations of the state, a task which many political ecology studies have been accused of ignoring (Robbins, 2012). Three main state agencies involved in shaping the Ridge have been examined to argue that the state is not one but many actors and these must be studied in their elements rather than as a given whole. This chapter also serves as a background to the detailed discussion on the establishment of conservation units in the Ridge, as state agencies are central actors and state policies provide a context for socio-political struggles (Bailey & Bryant, 2005; Walker, 1989).

Beginning in 1979, student groups brought the issue of conserving the Ridge to the attention of the highest executives in the city. This phase of environmentalism, examined in chapter 6, presents an early example of a sustained citizens' movement to preserve urban green spaces in India. Though it did not continue past the mid-1990s, it led to the solidification of conservation in official discourse as reflected in policy documents and administrative boundaries.

Chapter 7 provides a localised analysis of the three distinct conservation projects located in the Ridge. Apart from compiling micro-histories of these spaces to elucidate the evolution of these three parcels of land as being based in the historical development of the city, the chapter examines the various discourses around these and their interaction to locate reasons for discursive domination and subordination of the claims of certain groups.

The question of why certain claims are effective in the formation of conservation spaces while other discourses remain unrepresented in the current situation is further deliberated upon in chapter 8. The chapter presents the struggles regarding the Ridge as related to wider socio-political structures by engaging with literature related to environmentalism, conservation, urbanism and political participation in India to identify patterns of power, injustice and inequality.

The penultimate chapter re-constructs the discursive struggles related to the conservation of the Ridge through the concepts of storylines and discourse coalitions to illuminate the motivations, practices and outcomes related to various actors. The metaphor of metabolism is used to illustrate this unequal flow of discourses and benefits.

The conclusion briefly summarises the arguments made. Secondly, it deliberates on the strengths of using methodologies and concepts related to post-structural political ecology and urban political ecology in conjunction, to counter the weakness of both in analysing urban environmental issues before providing some closing reflections.

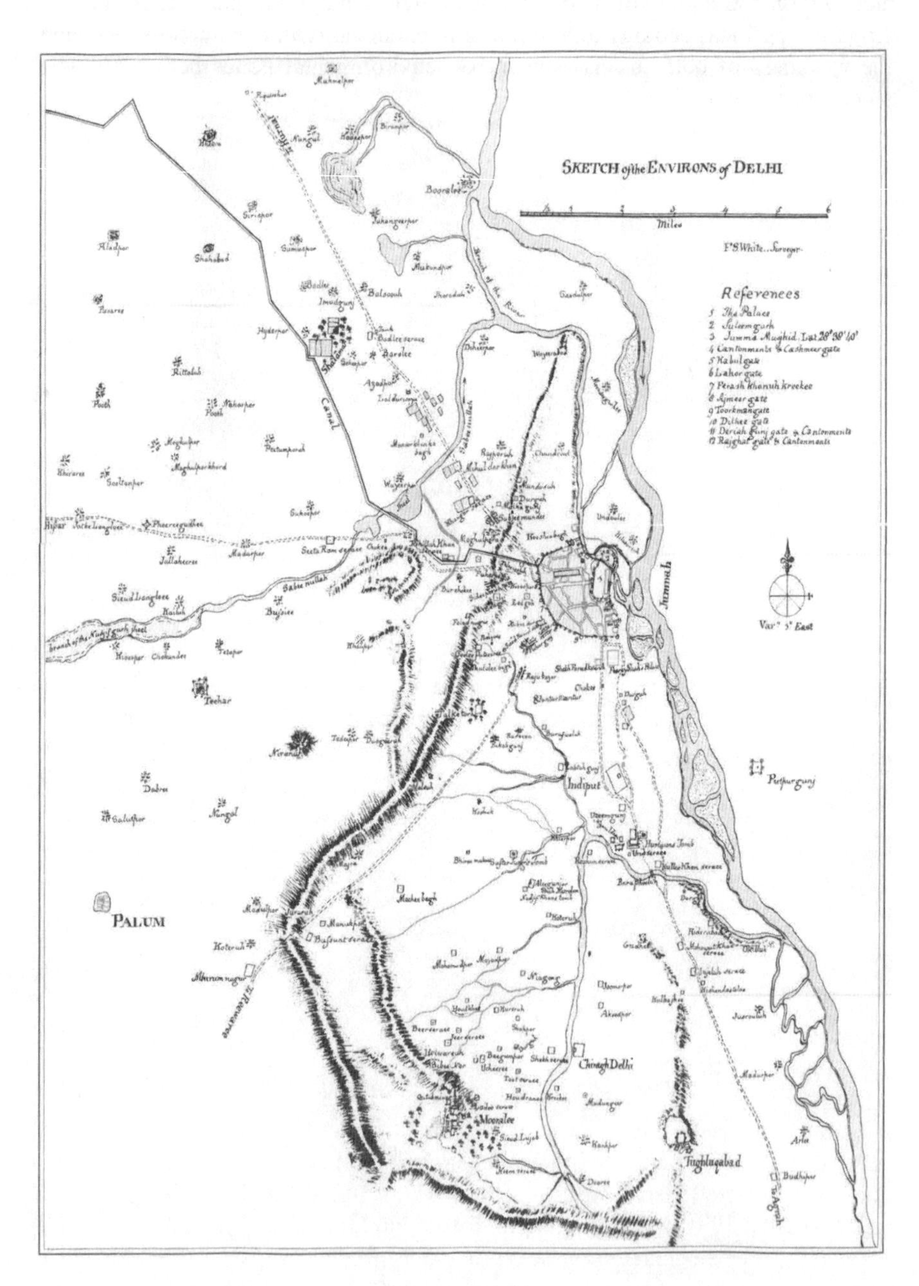

Map 1: Sketch of the Environs of Delhi, 1807. Scale 1:79,200
(Source: Survey General of India, 1989)

Map 2: Delhi Ridge, Current Administrative Segments
(Source of Base Map: DDA, 2007. Modified by the author with inputs from DDA 2007)

2 STATE OF THE ART: LOCATING THE STUDY IN THE FIELD OF POLITICAL ECOLOGY

The traditional environmental realist school worked with an epistemological and ontological separation between science and nature (Heynen, Kaika et al., 2006). The bulk of scientific studies on the environment treated it as a purely physical phenomena that can be diagnosed and rectified with scientific means and is separate from the conundrum that is society (Macnaghten & Urry, 1998). Following critique of such scientific rationality from various quarters, a more complex view of environmental problems and solutions gained ground. It is increasingly recognised that 'real world' policy questions, such as the one addressed in this research, demand a more pragmatic and interdisciplinary handling (Beck, 1999; Bryant, 2000; Eden, 2000). Political ecology studies have more often than not used such a grounded approach in order to understand the "underlying contexts and processes" (Basset, 1998, p. 453) in a variety of biophysical and cultural settings. This study is situated in a landscape that has developed in diverse ways spatially and temporally and is surrounded by a variety of human interests and demands. Thus, a rather flexible approach was needed to enable a grasp of the nuances of the situation. It is argued in this chapter that political ecology provides a suitable social theoretical framework for such a study.

The chapter begins with a discussion on the development of political ecology as a framework for studying socio-environmental relationships. This is important as political ecology is not a theoretically bound, methodologically clear approach (Bailey & Bryant, 2005; Blaikie, 1999; Robbins, 2012), therefore a look at the influences, development, main ideas and debates within it give us an understanding of its content and tools. Next, the field of urban political ecology is discussed along with its main concepts. This is followed by a summarisation of the central elements of political ecology and its urban counterpart that provide a framework for building a political ecology study. In the final section, it is argued that while urban political ecology has contributed some useful concepts, it could benefit from an application of methods espoused by post-structural political ecology a well as contextualisation through literature regarding local socio-environmental politics. Through this synthesis, the study aims at a wider urban political ecology that acknowledges at structural aspects without being deterministic.

2.1 POLITICAL ECOLOGY: THE PROBLEM OF DEFINITION

The origins and development of the field of political ecology remain poorly understood (Bailey & Bryant, 2005). In part, this is because much of the earlier work that can now be seen as situated within the field did not allude to the term 'political ecology'. The term was first defined in the 1970s by Eric Wolf (1972), but it was not until the 1980s that it became more common (Bailey & Bryant, 2005; Blaikie, 1994).and began to appear as a "self-conscious reference point" (Blaikie, 2008, p. 766) by the end of the 1980s. The beginnings of political ecology coincide with the politicisation of the environment when the modern environmental movement gained form in the Global North (Blaikie, 1999; Robbins, 2012), as well as, with the epistemological shift in social science through the arrival of post-structuralism, post-modernism and post-colonial studies (Blaikie 2008). Various definitions have tried to capture what is implied by 'political ecology'. The central idea is that politics should be foregrounded in an attempt to understand how human-environment interactions could lead to environmental degradation (Bryant, 1998).

The areas of enquiry that political ecologists have engaged in are only broadly similar (Robbins, 2012) and scholars have used different approaches to look at environmental change and conflict. The key elements of such approaches have been a consideration of specific environmental problems (e.g. soil erosion), concepts (e.g. sustainable development), actors (e.g. the state), regions (e.g. South-East Asia), or various combinations of these approaches (Bailey & Bryant, 2005; Robbins, 2012). Political ecologists have largely favoured empirical analysis rather than focussing on theory building (Bailey & Bryant, 2005; Blaikie, 2008; Peet & Watts, 1996). This led to a research field "grounded less in a coherent theory but in similar areas of enquiry" (Peet & Watts, 1993, p. 239). It has been referred to as "something people do" or a "community of practice united around a certain text" revolving around certain themes, adopting a critical attitude (Robbins, 2012, p. 20) to address socio-environmental systems with explicit considerations of relations of power.

Although political ecology encompasses a largely eclectic body of work and theoretical heterogeneity, such that those who have tried to review and conceptualise it retrospectively have declared a "bibliographic overload" (Blaikie, 2008, p. 767), it is argued that this also enabled a robustness in the face of rapid permutation of environmental research in general and provided a freeing up[5] of methodological and conceptual frameworks to make them suitable for studying complex socio-natures (Bailey & Bryant, 2005; Blaikie, 1994, 1999; Zimmerer & Basset, 2003).

The underpinning coherence in this field was the understanding of political ecology being predicated upon the assumptions and ideas of political economy (Bailey & Bryant, 2005; Blaikie, 1999). Blaikie & Brookefield (1987), in their seminal text *Land Degradation and Society,* provided a much cited definition of

5 From traditional disciplinary methodological frameworks and epistemologies

political ecology, especially Third World political ecology[6], as being about the combined concerns of ecology and political economy. However, It has been pointed out that this would only be a useful definition if political economy and ecology were clear notions in themselves (Zimmerer, 1994). Political economy for example, is a very wide field that ranges from political right (neo-classical thought) to political left (neo-Marxism). As elaborated later in this chapter (section 2.1.4), the stress on political economic structures was expanded to broader socio-political analysis by later developments in the field.
It has been suggested that the most effective way to address the problem is

> "not through a grand theoretical exposition but through a selective engagement with the political economy literature as, and when, the literature is appropriate to the argument. The aim is to explore political ecology's possible multiple interpretations of 'political economy' rather than to assert in a dogmatic fashion a single 'correct' interpretation" (Bailey & Bryant, 2005, p. 13).

This rather opportunistic statement highlights the ambiguous terrain that political ecology studies must place themselves within.

Reflecting on this problem of lack of coherence, political ecologists have tried to argue, depending on their own disciplinary affiliation, what the goals and means (in terms of theoretical lens and methodological frameworks) of political ecology should be (Blaikie, 1994, 1999, 2008; Vayda & Walters, 1999; Walker, 2005; Zimmerer & Basset, 2003). Two main aspects of argument are regarding the role of politics and the question of how to bring nature into the picture in a largely social science dominated field. These are discussed in some detail below. Political ecology remains a dynamic and reflexive field with much discussion and debate among scholars. It can be seen as a "creative idea that brings together contradictory and complementary notions from its origins, rather than expect it to provide a readymade single and coherent framework" (Blaikie, 1994, p. 3). It would therefore be useful to outline these 'notions' and 'origins' in order to understand political ecology and its application, which is done in the next section.

2.1.1 Disciplinary Location and Genealogy

Political ecology has been shaped by various disciplinary influences over its evolution. In practice it continues to be based in several disciplines or combinations of disciplines, such that, political ecology is seen as making regular "disciplinary transgressions" (Blaikie, 2008, p. 767). The approach taken here is rooted in geography. Zimmerer (1993) placed political ecology as a common way in which human geography approaches ecological concepts. Bailey and Bryant (2005) characterise it as being a geography-based research field that nonetheless maintains strong links to other fields like anthropology and sociology. A quick overview of the genealogy of political ecology helps elucidate its current agenda, methodology

6 Third World political ecology' is a term often used in political ecology studies based in the Global South (for example Bailey & Bryant, 2005; Bryant, 1992)

and areas of interests and provides a better understanding of the field as it stands today.

Neo-Marxist analysis was the mainstay of political ecological research of the late 1970s and early 1980s, contributing to it a concern for relations of production, commodification, access and control (Paulson et al., 2003). It provided a way for linking local oppression and environmental degradation to wider socio-political structures (Blaikie 1985a; Bunker 1985). There have been some concerns with the studies of this period regarding their implicit structural determinism and the brushing over of the nuances of local politics and its role in mediating resource access as well as of simplistic views of actors involved, especially the state (Bailey & Bryant, 2005). The most widely cited definition for political ecology in this phase was the one provided by Blaikie and Brookfield:

> "The phrase "political ecology" combines the concerns of ecology and a broadly defined political economy. Together this encompasses the constantly shifting dialectic between society and land-based resources, and also within classes and groups within society itself" (Blaikie & Brookefield, 1987, p. 17)

Other strands of work also influenced political ecology. Anthropologists contributed with a body of work in the 1960s and 1970s regarding culture and environmental management practices. The work of cultural ecologists challenged the dominant paradigm where human and environmental geographies remained separate (Kasarda & Crenshaw, 1991). This combination of anthropological-style local research along with a political-economic structural analysis became the key concern of political ecology (Bailey & Bryant, 2005). Cybernetics and systems theory in the 1970s, lent a vocabulary and set of concepts to human-environment systems, such as flows of energy and materials and metabolism (Heynen, Kaika et al., 2006). Radical geographers contributed to the research agenda by highlighting the interaction between political-economic structures and ecological processes through their work on disasters and hazards in the early 1980s (Susman, O'Keefe, & Wisner, 1983; Watts, 1983). In geography and anthropology, peasant studies and a critique of colonialism highlighted the questions of control, marginalisation and power (Paulson et al., 2003).

The 'second phase' in beginning the late 1980s saw the introduction of a variety of theoretical sources exploring the complex human-environment interaction (Guha, 1989; Peet & Watts, 1996; Peluso, 1992). Post-structuralism[7] and discourse theory added the dimensions of discursive interactions and power as mediators in the political-ecological outcomes (Escobar 1995; Blaikie 1999; Moore 1998b; Bryant 2000). This phase is discussed in detail in the sub-section on 'post-structuralist and discursive political ecology' as it is within this sub-field that the present study is placed.

Given this plethora of influences, methodologies and theoretical backgrounds, political ecology seems to be an overwhelmingly diverse field of study. It was

7 Called thus because it goes beyond the Marxist analysis of political economic structures such as class and relations of production

selected it as an umbrella area for this study, since its core concerns address the research questions within a framework that allows politics and power relations to be at the centre of the analysis.

2.1.2 Role of Politics in Political Ecology: Beyond Sustainable Development

Politics holds a key place in the world view of political ecology and has roots in its association with peasant and development studies, social movements theory, and studies of indigenous knowledge and symbolic and discursive struggles over resources (Walker, 2007). This focus on politics and the substantiation of what that term entails has also been one of the areas of constant debate within the field. There has been some accusation of political ecology as being insufficiently politicised due to its structuralist legacy[8] (Moore, 1996; Walker, 2007). It is seen as a failing on the part of such rigidly structured studies that they glossed over details that cannot be captured by airtight containers of their conception of state, international organisations, peasant organisations etc. As Peet and Watts (1996, p. 8) described it: "there is very little politics in political ecology as it is concerned increasingly with institutions and organisations and the shifting configurations of state and market roles". These political ecologists were pointing out that not enough attention was paid to the 'micro-politics' of environmental conflict in the Global South and criticised the monolithic portrayal of the actors involved especially, the state, obscuring the complex and diverse interests and actions involved. This was a direct critique of the political ecology studies of the 1970s to mid-1980s and some effort has been expended in rectifying the economic determinism prevalent in those times. The term politics is now expected to encompass a "panoply of processes and mechanisms through which power is circulated and wielded" (Walker, 2007, p. 364) or more specifically, the ways in which claims are made and negotiated (Peet & Watts, 1996) rather than simply the formal political process. This could include movements of various kinds, everyday resistances by marginalised resource users (Scott, 1985), contestations and interactions between various state authorities, various kinds of civil society activism etc. It is argued that to understand the existing socio-environmental conditions, it is of pivotal importance to acknowledge the existence of 'politicised environments' i.e., political interests and struggles are key to understanding and therefore to solving the current environmental problems (Bailey & Bryant, 2005; Bryant, 1998; Escobar, 1996; Robbins, 2012).

8 See for example Blaikie and Brookfield's (1987) Chains of explanation model in which they try to capture the actors at various scales and their influence on local land degradation. Though a useful early effort in including political concerns into socio-environmental studies, it has been much criticized for its rather simplistic presentation of politics. Blaikie himself later admitted this shortcoming and advocated more nuanced, ground-level work (Blaikie, 1995, 2008)

On the other hand, too narrow a focus on local land users may be insufficient as well, given the weight of the effect of other actors especially the state (Bailey & Bryant 2005). The idea is to move towards a broader political ecology addressing the politics of environmental change and at various scales to have an impact. Blaikie also ruminates over this question of scale and scope and accepts that while a broader view is needed, a comprehensive scale linking the international, national, regional and local is seldom, if ever, possible. This is due to problems of data and thin evidence of linkages and causality (Blaikie 2008; Blaike 1994). This study attempts to work upwards from the landscape and get an idea of the relationship of the local residents and users as well as planners, activists, scientists, bureaucrats, political leaders and others in order to build a fuller picture of the broad politics around the forest.

This research looks at local users and residents (especially economically marginalised ones) along with other actors as one of the distinguishing features of the research field of political ecology is that it tends to have a 'radical perspective' (Bailey & Bryant, 2005; Blaikie, 1999). It is radical in the sense that it seeks to uncover local resistance and alternative knowledges of the environment. Two unifying points of agreement, despite different approaches that political ecology studies have taken, have been: Firstly, that the environmental problems societies are facing today are not simply due to policy or market failure but rather a manifestation of broader political and economic forces. Therefore, any quick fix technological solutions will not suffice to deal with a complex and deep rooted issue. Secondly, and consequently, far reaching changes to the political-economic process will be required to address these issues (Peet & Watts, 1996) and an exposition of the unequal power relations inherent in these systems is necessary. It is in the service of this much broader social project of raising the emancipatory potential of environmental ideas and to engage directly with the larger landscape of debates over modernity, its institutions, and its knowledges (Peet & Watts, 1996) that political ecologists have found common ground. Political ecologists have more often than not suggested a stronger role in decision making for local, place-based actors and a mere coordinating role for traditionally powerful non-place based actors with a view to achieving social justice and environmental conservation (Bailey & Bryant 2005; Blaikie 1985).

The concept of sustainable development has been criticised as being under-politicised as it often sees technocratic fixes and not a change in socio-political systems as the solution (Escobar, 1998). The sustainable development discourse is also seen as having been co-opted by the mainstream and therefore being powerless to affect real change (Bailey & Bryant, 2005; Bryant, 1991; Escobar, 1998). Political ecology critically engages in the very areas which the sustainable development discourse glosses over. For example, while sustainable development studies often work with the assumption that the state (which is often taken as omnipotent and undifferentiated), through sound implementation of techno-managerial innovations can deal with environmental issues to bring about a 'balance of nature' (Forsyth, 2003, p. 6); political ecology focusses on complex socio-political forces and their relationship with environmental change to reveal a much more contested state of affairs where everything from the definition of the problem to the nature of the solution contains several claims and counterclaims.

2.1.3 Politics vs. Ecology: In Favour of a Social Science Study of Environmental Change

Given the growing body of literature in the field of political ecology, there has been a healthy debate among researchers on its direction (rather directions) since the 1990s (Paulson et al. 2003; Blaikie 1994). At its inception, it served as a platform for bringing together of political and biophysical ecological studies, but by the 1990s an influx of studies by social scientists is seen (Blaikie, 1999, 2008; Robbins, 2012; Walker, 2005). While some found this a disturbing trend and demanded a re-focussing on natural science methodology (Vayda & Walters 1999), others have seen this as a welcome broadening of the field as captured in this statement defending social science studies in political ecology:

> "A key question is, of course, what passes for the environment and what form nature takes as an object of scrutiny. And here Vayda and Walters display their own parochialism [...] For Vayda and Walters (1999) the only expression of environment can be the biophysical events of environmental change […] But political ecology rests on the dialectical and non-linear relations between nature and society in which environment can be approached in a number of ways […] what political ecology has done obviously is to open up the category of the environment itself and explore its multiform representations […] Another way to approach the environment is to examine knowledge of the environment and why and how particular forms of knowledge predominate [...]" (Watts 2003, p. 8-9, in 'For political Ecology,' unpublished manuscript, quoted in Walker 2005).

Given this difference in view on the basic formulation of political ecology, some scholars have tried to recount the main approaches political ecologists have developed in order to demonstrate the variety of ways in which the central research concerns of political ecology are sought to be fulfilled (Blaikie, 2000; Zimmerer, 2000). These range from the ones that employ natural science to a larger extent and try to examine the environmental changes that occur due to various levels of social and political influence, to those social science studies that assess the effect of environmental change on socio-economic and political relationships i.e. how environmental changes affect socio-economic inequalities (Bryant, 1991).

Latour (1993) tried to work out the tensions between 'social' and 'natural' by forwarding the concept of socio-nature hybrids which are nature and culture at once with no clear margins or separations. Here, the agency of nature is acknowledged as a co-producer of hybrid environments. Forsyth (2008) also refers to this deep connection when he points out that such a debate is misplaced as it depends on a priori definitions of politics or ecology, which overlook how the two are linked.

This study makes no claim of containing any primary, natural science based examination of the biophysical environment. It places itself in the spectrum of political ecology studies (as contested and vibrant as this spectrum may be) among studies that seek to understand the environment in its various representations to see how environmental changes reflect and modify socio-political processes at various levels. Socio-environmental outcomes unfold in heterogeneous and diverse ways in different contexts (Bailey & Bryant, 2005) and this unevenness is primed for socio-environmental analysis (Zimmerer & Basset, 2003). As Harvey (1993, p. 25) points out:

> "All ecological projects (and arguments) are simultaneously political-economic and vice-versa. Ecological arguments are never socially neutral any more than socio-political arguments are ecologically neutral. Looking more closely at how ecology and politics inter-relate, then becomes imperative if we are to get a better handle on how to approach environmental/ecological questions".

Such an analysis falls cleanly into the category that Zimmer and Basset (2003) describe as one of the main approaches that constitute political ecology, where the primary emphasis is on environmental politics (politics is broadly defined as social and institutional practices) and a discursive turn[9]. Why is the current study placed under the aegis of political ecological concepts and not simply titled a 'social science/humanities study of environmental politics' (Walker, 2005, p. 73) if it makes no claims to add any purely ecological (in the sense of biophysical ecology) insight about the Delhi Ridge? Firstly, because the central concerns developed and espoused by some political ecology scholars[10] were found relevant to this study which aims to contribute to the body of similar work in the field of urban political ecology by analysing a site for conservation in a megacity in the Global South. Secondly, political ecologists have developed a useful set of tools and concepts to tackle such a study. To use the analogy of political ecology being a "critical toolbox" (Robbins, 2012, p. 72) for studying nature society relationships, it can be argued that one need not use all the tools available in the box to build one's case.

Science itself is a socially constructed discourse (Forsyth, 2003; Lele & Norgaard, 2003). Scientific means remain necessary to document and compare environmental change and though they are powerful discourses, they are not the only ones involved in knowing and seeking to change environments. Scientific fact does not capture the multiple reasons, means and contexts of human agency and action which are aimed at shaping nature (Forsyth, 2008a). Political ecology is as much a question of the social and the political as it is of the biophysical (Rademacher & Sivaramakrishnan, 2013b). While there is weight in the idea of research bringing together biophysical ecological study and socio-political analysis, this does not negate the need or viability of studies that focus on one of these aspects as long as they are not deterministic and do not discount the influence of the other factors and the use of other methodologies. Showing the social construction of a concept or the object which it refers to does not necessarily disprove it, it only

9 The other main approaches being: a) a primary emphasis on political economy and commodification of nature in particular; b) an emphasis on politicised biogeophysical changes and c) feminist political ecology

10 See section 2.3 of this chapter

subjects them to critical analysis in terms of aims and means (Forsyth, 2008a). Indeed, it is this expansiveness that has allowed political ecology to evolve in several different but useful directions.

2.1.4 Post-structuralist, Discursive Political Ecology

From its roots in Marxist inspired analysis of resource control, through the influence of wide ranging developments in social theory in general, a post-structuralist phase of political ecology can be discerned. This has enabled the political ecology of current times to address a much broader (in terms of actors) and deeper (in terms of various discourses rather than simply an overt control of resources) set of concerns and "eschew theoretical baggage like 'structures' and 'systems' that mystify the concrete relationship between things'' (Robbins, 2012, p. 23).

The political ecology of the 1980s and early 1990s is now considered the structuralist phase of political ecology due to its focus on political economic container structures. Post structuralist political ecology thus focuses on "local level studies of environmental movements, discursive and symbolic politics and the institutional nexus of power, knowledge and practice" (Walker, 2005, p. 75). A more robust political ecology resulted from an incorporation of theoretical insights from post-structuralist, interactionist[11] philosophy (Bailey & Bryant, 2005; Blaikie, 1999; Escobar, 1996). The discursive turn in political ecology allowed for "richly textured empirical work (a sort of political-ecological thick description) which matches the nuanced beliefs and practices of the world" (Peet & Watts, 1996, p. 38). Natural processes are experienced and interpreted socially (Blaikie, 1994) and competing understandings play a formative role in material and discursive struggles that shape the distribution of resource rights (Moore, 1998b) therefore accounts of the past, interpretations of the present and aspirations for the future are the key to human action and agency.

One of the celebrated gains of such a political ecology has been its assault on normalised ideas regarding environmental change, conservation and the social causes including powerful scientific and state sponsored versions of 'given truths' (Forsyth, 2003; Robbins, 2003a). Environmental 'truths' have been revealed as narratives through an investigation of how they are produced and normalised[12]. This scepticism has further spurned an investigation into alternative knowledge of the environment (Basset, 1998; Robbins, 2003a). Therefore studies began, with insights form post-structural debates especially critical science studies, to investigate the political nature of environmental knowledge and its institutionalisation.

11 Interactionist approaches are called interactionist as they elucidate that the establishment of "truths" about the environment is inevitably shaped by the relationship or interaction between the subject and object of research, such that different actors may interpret the same environment in very different ways (Blaikie, 1999)

12 Examples of such 'truths' that can be read as narratives include the Malthusian stress on population growth in the Global South rather than consumption patterns as a source of environmental degradation (Bryant, 1997)

This was done through turning attention to discourse and narrative (Forsyth 2008). Environmental narratives have been described as convenient but simplistic ideas and beliefs about the character, causes and impacts of environmental problems. These in turn influence subsequent generation of environmental research and proposed solutions (Leach & Mearns, 1996). It is not with the aim of testing their scientific validity, but to uncover their origin in social settings with incumbent political backgrounds and objectives, that post-structural analysis concerns itself with environmental narratives (Forsyth 2008; Blaikie 1995; Castree et al. 2009).

These studies and their post-modern epistemology enabled them to emphasize politics rather than purely economic differences, varied accounts of reality, agency and resistance, rather than structural inequality alone (Blaikie, 1999). It was felt that the economically deterministic approach of the earlier studies failed to account for the other causes of environmental change such as factions within the state and inter-state forces, weaker sections of society as well as ecological/biophysical factors (Bryant, 1992). A critique of dominant environmental policies and views is not entirely new, structuralist studies attempted the same end but post-structuralism brought in a new epistemology as well as a new empirical focus as it does not aim to disprove and replace one established 'truth' with an alternative 'truth' but focuses on how environmental interpretation are produced, represented, contested (Blaikie, 1995). Thus there is a focus on social construction of nature; this implies both the construction of ideas about nature as well as the production of material nature mediated through these ideas (Demeritt, 2002).

There is a range of positions that social scientists have argued from, from 'hard' to 'soft' constructionist approaches (Robbins, 2012, p. 127). On one end is the approach which denies any ontological existence of nature outside that which is presented in discourse, while others are critical of unlimited relativism and insist that social construction plays an important part in political ecology analysis but not of the sort that negates the existence of a material nature (Castree, 2002; Demeritt, 2002; Zimmer, 2010). Escobar advocates an 'antiessentialist'[13] conception of nature which implies that nature is always constructed by discursive processes and therefore the natural is also social and cultural (Escobar, 1999). Latour (2004, p. 9) in arguing for a more complex understanding which leaves no space for the separation of nature and culture argues that political ecology has to "let go of nature" as a concept in itself and embrace nature-culture hybrids. On the other hand scholars like Zimmerer (1994, 2000), Blaikie (1999), and Robbins (2003) work with an understanding of nature as ontologically given but transformed and interpreted socially.

The work of some post-structuralist scholars has been criticised on grounds of excessive emphasis on language, to the detriment of practices and social effects of historical processes. This carries the danger of a "morally hollow relativism" (Davis, 2003, p. 162); if nothing can be concretely read, then how do we change it

13 Essentialising refers to reducing an individual or group's recognisable traits to fixed characteristics which are naturalised through discourse i.e. the characteristic is made inseparable from its identity (Fuss 1989)

positively? Normative inputs for environmental policy are extremely hard to propose based in such analysis as these must account for aspects that cannot be explained with meaning and language alone (Gandy, 1996). Therefore linguistic analysis has to be placed within socio-political context.

As regards the denial of an ontological basis of nature by some post structuralist scholars, it must be pointed out that biophysical nature is not infinitely changeable. The biophysical context also dictates the limits and directions of human action and thus we cannot assume it is entirely a matter of social construction, it must be seen a matter of interaction between the two (Blaikie, 1999; Rademacher & Sivaramakrishnan, 2013b). Sivaramakrishnan (1999, p. 282) argues:

> "Human agency in the environment, mediated by social institutions, may flow from cultural representations of processes in 'nature' but we cannot forget the ways in which representations are formed in lived experience of social relations and environmental change".

Gandy (1996) also points out the problems of overly relativist positions which deny any external reality or foundation that forms the base of human knowledge. He points out that if an environmental problem was strictly a construct, we could make it go away by simply not indulging in a discourse about it. It has been argued out that realist studies are carried out and are embedded in social structures while more relativist work must be based on interpretations of locally bound experiences and 'truths' held by various groups, thus it is not possible to be entirely realist or relativist in either case (Forsyth, 2008b).

The present study takes the view that while the environment is socially constructed and its understanding is also socially constructed, some elements are more subjected to this construction than others (Blaikie, 1999; Blaike, 1994; Demeritt, 2002). Blaikie (1999, p.114) also argues for a pragmatic form of constructivism in order for research to be effective:

> "By adopting an epistemology which avoids relativism and unreconstructed pluralism[14], it may be possible to address specific audiences in languages they recognise to identify real and feasible choices. This epistemology may be characterised as a form of 'weak social constructivism', which claims that environment is socially constructed but that provisional truths may be shared for a while between the actors involved"

Take for example the fact that the invasive tree species Prosopis juliflora reduces the biodiversity of the Ridge by curtailing undergrowth and other tree species, is rather uncontested. However, how it affects different groups of actors and what actions are seen as desirable to deal with this, are not clear uncontested 'facts'. Like some other political ecologists, this research borrows from critical realists[15]

14 Here Blaikie refers to the kind of research that indulges in deconstructing plural knowledges without 'reconstructing' them by placing them in relevant socio economic structures to make sense of power relations through them

15 Critical realism by Bhaskar (1975) speaks of understanding 'externally real' items, such as soil erosion or tree growth, as 'peeling the layers of an onion' because it distinguishes between 'actual' observations (day- to-day experiences), 'empirical' measurements (scientific

the idea that nature has an ontological basis and is an agent in its own right, coupled with our understanding of nature's agency as socially mediated and it is this interaction that created the socio-nature hybrid which is the concern of this study (Forsyth, 2008a; Zimmerer & Basset, 2003).

2.2 URBAN POLITICAL ECOLOGY

While political ecology contributed insights to nature-society relationships in their various forms, for a large part it neglected to extend its gaze to cities. The urban is often considered 'unnatural' (Heynen, Kaika, et al., 2006) or even the "opposite of nature, ecology and environment" (Keil, 2003, p. 729) and therefore beyond the concern of research that looks at environmental questions[16]. In trying to answer the question of why political ecologists ignored the urban context, Angelo and Wachsmut (2014, p. 3) postulate that one of the central aims of political ecology is the politics of environmental degradation and rehabilitation. Thus the urban was left out as nature in cities is considered already subjugated to humans and seen as impossible to rehabilitate as there was "no environment left to rehabilitate". Given this perceived dichotomy between city and nature it was remarked that "urban ecology makes as much sense as dry land hydrology" (Trepl, 1996, p. 85 cited in Kiel 2003, p.729). This neglect, if left uncorrected, would have excluded large swaths of areas, people and issues that concern them as more than half the world's population is now urbanised (UNFPA, 2007). Thus, Keil (2003, p. 728) declared that: "UPE (urban political ecology) is well overdue" as the urban is now central to any discussion on human futures. After all, "it is on the terrain of the urban that accelerating metabolic transformation of nature becomes most visible, both in its physical form and its socio-ecological consequences" (Swyngedouw & Heynen, 2003, p. 907).

Urban environments are increasingly gaining the attention of scholars in various fields who have come to develop a useful vocabulary to encapsulate and problematise environmental issues in the urban context. Contemporary environmentalism, environmental change and conservation are no longer the concerns of the countryside alone (Zimmerer & Basset, 2003). Urban political ecology has developed as an offshoot of political ecology since the late 1990s (Keil, 2003; Zimmer, 2010)[17]. On the one hand, it has led to the extension of political ecological analy-

research) and the insights these give about 'real' structures (underlying causes) (Forsyth, 2008a, p. 760)

16 The question of the urban environment has not always been missing from the imaginary of the city. It has been documented that residents of the early modern cities lamented that cities were so unnatural and ways to bring about a more 'wholesome' urban life have been sought all through (Swyngedouw & Heynen, 2003, p. 900). Engels in describing the industrial city in The condition of the working class in England also pointed out the terrible ecological conditions for example (Engels, 1845)

17 The term urban political ecology was first used by Eric Swyngedouw in 1996 in his article 'The city as a hybrid: on nature, society and cyborg urbanism' (Angelo & Wachsmuth, 2014;

sis to the urban space and on the other it has also provided a way of looking at the city as socio-naturally co-produced (Angelo & Wachsmuth, 2014; Heynen, 2013). While urban political ecology and political ecology have developed quite separately (as elaborated later in the chapter), this research attempts to draw on the strengths of both.

The antecedents of urban political ecology are placed in the arguments by Lefebvre (Lefebvre, 1991, 1996) and Castells (Castells, 1977, 1978). Lefebvre's formulation that urban societies are the currently relevant forms of historical capitalist socialisation and Castel's problematisation of the urban as a privileged historical and social place for the survival of capitalism coupled with a structural critique of capitalism left an impression on the field (Keil, 2003). More recently, a debt is owed to the large area of urban sustainability studies spurned on by the global efforts as embodied in the Brundtland report and its successors, and the work of the UN among other influential organisations (Low et al., 2000). There has been a push for urban planning and policy to improve urban environments in a bid for a more sustainable future. Urban political ecology is suspicious of such (largely) positivist, policy oriented studies but also leans on them, not least in the fact that urban sustainability as a concept created a political climate for actors to intervene in the ways cities are built, planned and lived in (Keil, 2003). The field of urban political ecology (just as that of political ecology) differs from the urban sustainability studies in that, while the later focuses on policy oriented and problem solving agendas, urban political ecology is more 'critical' in that it links environmental problems to larger socio-ecological solutions. Moreover, it is pointed out by studies in the field, that a sustainable city is a slippery concept as any environmental change that may benefit some groups may be detrimental to others (Peet & Watts, 1996; Swyngedouw & Heynen, 2003).

The conceptual approach of urban political ecology is largely grounded in structuralist, neo-Marxist traditions with its emphasis on political economic aspects broadly in line with the work of Smith and Harvey (Harvey, 1993; Smith, 1984). The major contribution of Marxist urban political ecology was its assertion that environmental inequalities must be looked at as products, or at least reflections of social power relations (Holifield, 2009). Therefore the city is a site for unpacking social relations that produce it (Lawhon et al., 2014).

There have been additions to the traditional (if they can be called that, given how recent they are) political economic approaches of the field, these have been summarised as follows; Firstly, from viewing environmental inequities purely in terms of class differences a more nuanced view evolved, especially in the U.S. where race and ethnicity became a lens of viewing the question of environmental justice (Heynen, Perkins et al., 2006). The field also gained from the environmental justice movement in highlighting inequalities inherent in urban environments, though environmental justice studies are mostly advocate driven (rather than aca-

Heynen, 2013; Zimmer, 2010) thus 'hybrids' and 'metabolism' the two concepts elaborated later in this chapter, have been part of urban political ecology vocabulary from the very inception

demic), praxis driven (rather than theoretical), small scale, hazard focused and informed by a liberal idea or distributive solutions (rather than more radical societal change) as opposed to urban political ecology studies (Heynen, Kaika, et al., 2006; Lawhon et al., 2014; Swyngedouw & Heynen, 2003). Secondly, there was a push for democratisation as the environment movement in the Global North pushed environment into the public arena and democratic deliberation and pluralist negotiation came to the centre stage. Finally, the works of Latour (1993) and Haraway (1991) made space for non-human actors by conceptualising environment as a 'combined socio-physical construction' which is produced both in terms of social content and physical qualities, actively and historically (Keil, 2003, p. 726).

While the 'first wave' of urban political ecology was Marxist, the 'second wave' (Heynen, 2013) inspired by post-humanist works like that of Escobar and Latour has begun to explore broader social dynamics intrinsic to nature as explicitly political (Gandy 2012) leading to a re-examination of culture, politics and nature (Heynen, 2013). In these studies, cities are understood to be made and remade through metabolism embedded in "discursively scripted, culturally imagined understandings" and therefore the transformation of nature is ripe with certain understandings and ideas about nature (Swyngedouw & Heynen, 2003, p. 903). Historically specific social and physical natures are thus embedded in power relations that can be read in discourses and practices. A study of discourses has been used as a way of understanding the power politics involved in the manufacturing of legitimacy of certain ideas and practices. Kaika (2006), for example, uses discourse analysis to uncover dominant, normalised views on water scarcity in Athens to show that the nature of water as an unpredictable resource rather than faulty distribution policies were presented as the problem. If natures are produced, then the aim of exposing the socio-political process behind this is to provide an opportunity for democratising this production in order to move towards an environment more in line with the aspirations and needs of those who inhabit it (Swyngedouw & Heynen, 2003). In other words, aspirations of urban political ecology stay true to the radical ideals of post-structuralist political ecology or 'liberation ecology'[18]. Thus, while Marxist structuralism remains the mainstay of the field, it has incorporated an analysis of discourse and a 'weak' form of Latour's actor network theory which allows it to consider the agency of non-humans while insisting that the power originates in capitalist social relations (Castree, 2002; Lawhon et al., 2014).

Urban political ecology has maintained an indispensable interest in political concerns that foreground the issue of environmental justice and questions of access and control (Keil & Boudreau, 2006). Kiel (2003, p.726) summarises that:

> "The political in UPE is largely constituted by distributional and systemic inequities in the socioeconomic order that underlies the societal relationships with nature both world-wide (North/South) and locally (environmental justice)".

18 The term 'Liberation Ecologies' is from the title of Peet & Watts (1996)

This emphasis on politics demands an analysis of the procedural aspect of co-production of urban environments within unequal social relations and different relations of social groups with nature. Thus:

> "Urban political ecology describes a political project intending to investigate dynamics which (re)produce certain socio-economic conditions within the city. Asking who produces what kinds of conditions in whose interest, allows formulating claims towards a more democratic handling of environmental problems" (Zimmer 2010, p. 348).

Social relations as well as societies' relations with nature are in constant negotiation and reinforce each other (Keil & Boudreau, 2006). While an actor could be empowered by professing a dominant discourse or performing an environmental activity seen as legitimate, this also changes or reproduces a certain interaction with nature (Zimmer, 2010). Thus an examination of why certain discourses and practises are considered legitimate and desirable while others are not, would be a step towards understanding the opportunities and needs for democratisation.

Along with focussing attention on power relations, urban political ecology studies have provided a space for the consideration of everyday socio-natures, in that they look not only at environmental conflicts and events but relationships within existing political, economic and social settings by aiming at an "engaged query of the status quo" (Mueller 1999, p. 430; cited in translation in Zimmer 2010, p. 345). A study of cases where conflict is not very apparent and no resistance or criticism of environmental conditions or changes has been broadcast, is important if we are to attempt a political ecology that is not simply reactive, but proactive in exposing the politicised natures that people 'live and work' in (Miller & Hobbs, 2002, p. 331).

A further contribution of scholars in the field is that they have looked at incorporating the agency of nature within their analytical models in response to the criticism that ecology is lost in political ecology (Vayda & Walters, 1999; Walker, 2005). This is done through the key concepts of 'hybrids' and 'metabolism' which are discussed in the following sections.

2.2.1 Hybrids

Urbanisation, as presented by studies in urban political ecology, is not a linear separation of humans from nature which occurs through the taking over of nature by cities, but a process in which particular and complex relationships of society and nature are altered and created (Keil, 2003). To conceptualise this, studies in the field have often worked with the concept of hybrids as mixtures of nature and culture as suggested by Latour (Latour, 1993). Socio-nature hybrids can be seen as tightly intertwined social and political aspects and environmental transformations (Zimmerer & Basset, 2003). An example could be avenue trees planted by city administrations in line with scientific knowledge (Zimmer, 2010) or a lawn planted with certain species of plants, watered, weeded and fertilised (Robbins & Sharp, 2006) or indeed the Ridge forest, deforested and reforested in line with the views and compulsions of the administration with certain species considered appropriate and certain land uses deemed legitimate at particular historical points. For Latour the concepts of 'nature' and 'culture' are merely convenient reference points. A useful metaphor in this regard has been the 'cyborg city' which is cultural and biophysical with no delimitations in between.

According to Latour, modernity has on one hand accelerated the formation of hybrids and their ubiquity and on the other hand increased the aspiration for 'purity', i.e. the prevalent powerful discourses try to present environments that are actually hybrids as either purely nature or purely culture. These two poles of nature and culture are not seen as equal, rather nature is seen as inferior (Latour, 1993). Zierhofer calls for an examination of power relations embedded in these 'purification' ideas and terms the discursive dichotomy of nature and culture the "caste system of modernity" (Zierhoffer 2002, p. 210; cited in Zimmer 2010, p. 346) which see the natural and cultural as separate then valorises one over other.

It is the processes of formation of hybrids that urban political ecology has drawn attention to. This formation includes physical and material processes as well as discursive and cultural ones steered by actors and the social relations between them (Swyngedouw, 2004). Thus hybrids have multiple dimensions: physical and material, commodified and a discursive constructionist dimension. Social power relations as well as societal relationships with nature meditate the processes belonging to these dimensions (Zimmer, 2010). These human and non-human relationships and processes that produce hybrids include ecological conditions (such as geology, topography), economic conditions (such as the particular economic system in place and the flow of goods and value), political conditions (such as the nature of the state, organisation of marginalised groups) etc. (Heynen, Kaika, et al., 2006; Kaika, 2006; Robbins & Sharp, 2006; Swyngedouw, 1996). Further, cultural frames provide the cognitive patterns and acceptance of legitimacy for discursive and material practices which influence social relations and relationships between human and non-human elements even as they are based on them (Zimmer, 2010).

Swyngeduow (2004) also theorises that social relations on the one hand and discourses and practices on the other are in a dialectical relationship; while the latter is embedded in the former, it can also modify social relations. Hybridisation is a power laden process and its result (the hybrids) is also unequally shared by different actors and therefore they are contested and negotiated constantly. It is thus important to look at who the winners and losers of hybridisation and 'purification' are (Latour, 1993; Zimmer, 2010). Hybrids therefore provide a way of looking at spatially, temporally, and culturally specific discourses and practices as well as the power relations and the societal relationships with nature that are negotiated on a daily basis in micro-politics.

This idea of interwoven nature-cultures provides a way to engage with such complex matters without falling into counterproductive dichotomies. It also replies to the need to include the agency of nature in social constructivist analysis

(Gandy, 1996; Heynen, Kaika, et al., 2006). Furthermore, the nature-cultures perspective enables us to highlight the processes that lead to uneven urban environments. Rather than separating nature and society, it forces us to look at the questions such as: How are hybrids produces? Who benefits from them and who doesn't?

2.2.2 Metabolism

Metabolism is another concept that is used to politicise the analysis of urban natures. It has been used in various ways and its meaning is contested (Zimmer, 2010). As a metaphor, metabolism in organisms conjures the image of input of material and energy, their transformation and storage, and the removal of waste in set patterns (Warren-Rhodes & Koenig, 2001). The use of the term in socio-natural theory goes back to Marx[19] in the tradition of historical geographical materialism in which all organisms are understood as needing to transform 'nature' and in the process are changed themselves as well. This concept is well illustrated by this quote that urban political ecology studies have frequently referenced (Heynen, 2013; Swyngedouw & Heynen, 2003):

> "Labour is, first of all, a process between man and nature, a process by which man, through his own actions, mediates, regulates, and controls the metabolism between himself and nature. He confronts the materials of nature as a force of nature. He sets in motion the natural forces which belong to his own body, his arms, legs, head, and hands, in order to appropriate the materials of nature in a form adapted to his own needs. Through this movement he acts upon external nature and changes it, and in this way he simultaneously changes his own nature […] (Labouring) is the purposeful activity aimed at the production of use-values. It is an appropriation of what exists in nature for the requirements of man. It is the universal condition for the metabolic interaction between man and nature, the ever-lasting nature-imposed condition of human existence, and it is therefore independent of every form of that existence, or rather it is common to all forms of society in which human beings live" (Marx, 1990, pp. 283–290).

Abel Wolman used the term metabolism early on, in the article titled 'The Metabolism of Cities' (1996). Here, metabolism implied material flow and aimed at technical and behavioural solutions, i.e. it stopped at a systems-theoretical awareness of the material streams that sustain a city and its residents and did not apply a critical political analysis. Though this conception of the term does not account for the political, symbolic and discursive flows, it did enter the vocabulary of urban thinkers and has been used and developed upon since (Keil, 2005). To this early conception, urban political ecology deliberations on Metabolism, based on its Marxist legacy and leaning, reformulated, expanded and politicised the concept. In his seminal article where Swygedouw first called for an urban political ecolo-

19 The original German term 'Stoffwechsel' simultaneously denotes exchange, circulation and transformation of material elements and is seen as more illustrative than 'metabolism' as it implies a creativity rather than a repetitive pattern in set paths (Heynen, Kaika, et al., 2006; Heynen, 2013; Swyngedouw & Heynen, 2003; Swyngedouw, 2006)

gy, he stated: "[…]this hybrid socio-natural 'thing' called the city is full of contradictions, tensions and conflicts [...]" (Swyngedouw, 1996, pp. 65–66). To capture these contradictions, he attempted a synergy between the representational, discursive, ideological and material aspects of uneven power relations through the notion of urban metabolism (Heynen, 2013).

Thus in its current usage in the field of urban political ecology, metabolism can be described as a flow of material and energy as well as discourses and ideas, which is a historical product and therefore humans can control the input to satisfy their respective needs (Zimmer, 2010). Metabolism takes place within existing social relations and, therefore, certain groups have a greater control on its process and results (Swyngedouw & Heynen, 2003). This firmly added a social and political dimension (for example modes of regulation, cultures of consumption) as well as (and as a result of the recognition of these factors) a creative aspect. Rather than seeing metabolism as static, repetitive and habitual flows in set paths, urban metabolic flows are seen as negotiated, contested and therefore dynamic (Heynen, 2013; Keil, 2005). In other words, since unequal power relations are the context and product of urban environments, creative metabolism embodies the ever changing interplay between humans and non-humans and among humans resulting in constant dialectical spatial reformation of urban environments (Heynen, 2013).

In terms of discourses and practices, hegemonic, dominant ones overlay the subaltern ones. Non-human dynamics are also acted upon by social relations as natural process are governed and used for human aims within the social setting (Heynen, 2006; Zimmer, 2010). Metabolism is thus politicised in that even though it is co-produced, it is dominated and mediated by social relations and humans (Swyngedouw, 2006).

Thus, both metabolism and hybridity are concepts that seek to illuminate the process and historicity behind cities and especially 'nature' in cities (Heynen, Kaika, et al., 2006; Swyngedouw & Heynen, 2003; Zimmer, 2010). They also bring in a concrete political agenda of situating environmental change in larger processes of uneven development. On the basis of this, it can be argued the answer to environmental problems must be linked to social aspects rather than relying merely on technocratic solutions. In reviewing urban political ecology and its directions, Zimmer (2010) draws up the points a research in the field would address as follows: an examination of hybrids and their co-production by humans and non-humans through processes that are socially embedded and historically specific (i.e. a sensitivity to power relations) and also that there be attention paid to the dynamics instabilities, discontinuities, conflicts, in and around practices, discourses and social relations as well as relations of society with nature.

Angelo and Wachsmut (2014) have summarised the contributions of the urban political ecology framework as (1) providing arguments against the false dualisms such as town/country, urban/rural and nature/culture as prescriptions for social problems which leads to creations of certain kinds of nature and access structures (2) exposing economic and social injustices inherent in uneven urban landscapes and their (re)creation (3) a vocabulary to build alliances among social and environmental activists.

Though ecological modernisation and sustainability discourses dominate urban environmental research in general (Angelo & Wachsmuth, 2014; Bryant, 1997; Escobar, 1996), urban political ecology scholars have argued that the question of urban sustainability is a political one in its essence (Swyngedouw & Heynen, 2003) since it is intertwined with the question of power relations and injustice. Therefore, the way in which we regulate our relationship with nature in cities, is ultimately a question of "democracy, governance and the politics of everyday lives" (Keil, 2003, p. 729).

2.3 CENTRAL ELEMENTS OF POLITICAL ECOLOGY AND URBAN POLITICAL ECOLOGY

To make what is meant by political ecology more tangible, given the lack of theoretical and methodological coherence, scholars have tried to draw up lists of common themes, intentions, theses and assumptions (Bailey & Bryant, 2005; Blaikie, 1994, 1999, 2008; Robbins, 2012). These threads that unify political ecology studies delineate the substance of political ecology more clearly and help illustrate its central role in guiding this study. To this end, the central elements or research manifesto drawn up by researchers of the field are summarised below:
Reviewing the political ecology framework in the 1990s, a leading figure of the field, Blaikie (1994) put forward the following points that a political ecology study must address:

1. Epistemological view that ideas about nature are socially constructed and must be looked at critically.
2. Complex socio-environmental relations are to be seen in light of local historical development.
3. Different scales of analysis must be used to elaborate a study. Local place-based actors and their interactions as well as larger non-place-based pervasive actors (like the state or international organisations) must be linked.
4. Critical analysis of international policy and global environment must remain in sight.
5. Local, micro studies must be part of the research as it is often the first link in the chain of explanation. More importantly, they provide an opportunity for political ecology to contribute meaningful insights to the tension between theories and disciplines regarding the roles of structure and agency.
6. The role of the state must be scrutinised as the state in most countries (in varied ways) plays a role in mediating and shaping human-environment interactions.
7. Conflict and contestation is a central element of political ecology, both real contestation (for access to resources) and in terms of ideas.
8. The contestation of knowledge systems must also be addressed.

Bryant (1997) suggests a framework for understanding environmental change in the Global South (though it is unclear why he restricts the applicability of this framework to this context) through studying: Firstly, the contextual sources of change like the state policy and interstate relations, and local capitalism. Secondly, a study of location specific conflict or struggle to capture the historical and contemporary dynamics and, finally, a look at the socio-political ramifications of environmental change (though not in an environmentally deterministic way).

Robins (2012) summaries that political ecologies: 1) track winners and losers of environmental change, to understand persistent structures of winning and losing; 2) are narrated using human and non-human dialectics (in that they look at processes and relationships rather than things in themselves); 3) start from or end in a contradiction (i.e. they start with a problem in the human-nature relationship and end up scrutinising taken for granted approaches, techniques or ideas); 4) simultaneously make claims about the state of nature and claims about the claims about the state of nature; i.e., while they must describe environmental change in realist terms, political ecology studies must also critically analyse and juxtapose various ways of knowing nature.

A 'manifesto' was drawn up for urban political ecology in a collection of studies of the field (Heynen, Kaika, et al., 2006). It focuses on the idea of environmental and social changes co-determining each other through metabolism in a particular historic-geographical setting and leading to different positive and negative outcomes for various actors-based on social power relations. It adds a point regarding the agency of nature which Blaikie certainly considers in later studies of political ecology. The core concerns of political ecology and urban political ecology remain the same: to describe how society and environments co-produce each other within unequal power relations. Both see nature as socially constructed and this construction as fraught with contestation.

In sum, the political ecology framework (including that of urban political ecology) demands an analysis of the social construction of nature. This would include all actors from local level micro-politics to broader policy formulations. This must be done within a historical and geographical context and must account for the various relationships within society and of society with nature. The unequal power relations in society must be exposed in the shaping of environments and in bearing the outcome of environmental change.

These summarisations serve a researcher well in the absence of a clear definition as it outlines the core of the research concerns and options for useful approaches, while leaving open space for theoretical and methodological innovation. An effort was made to use these elements as a guideline in the research conceptualisation and throughout the analysis. In keeping with one of the main lines of enquiry of the vast field, this research analyses the environmental politics of a contested landscape in an urbanised context. Politics here is meant in the broader sense and encompasses "the practices and processes through which power, in its multiple forms, is wielded and negotiated" (Paulson et al., 2003).

2.4 POST-STRUCTURALIST URBAN POLITICAL ECOLOGY IN THE GLOBAL SOUTH

As mentioned above, urban political studies continue to be based mostly in the cities in developed countries while the larger field of political ecology has immersed itself in the rural, agricultural context of the Global South (Angelo & Wachsmuth, 2014; Heynen, 2013; Keil, 2003; Robbins, 2012; Zimmer, 2010; Zimmerer & Basset, 2003). Each has been restricted both in terms of geographical location and analytical frameworks, so much so, that they have been stated to exist as "two solitudes" (Angelo & Wachsmuth, 2014, p. 7). The separation of rural and urban in the agendas of the two fields seems to be an ironic reflection of the separation in the concepts of nature and culture.

Given the agendas of political-environmental justice of both these fields, cities of the Global South demand attention. Firstly, these cities are affected by aggravating environmental problems (Véron, 2006; Zimmer, 2010) in a context of high demographic growth and high levels of poverty (Zérah & Landy, 2012). Secondly, their complex and diverse histories have resulted in specific structures of heterogeneity and inequality (Benjamin, 2008; Rademacher & Sivaramakrishnan, 2013b; Roy, 2009, 2011a). For example, some of these urban spaces can be charecterised as containing

> "land tenure systems that combine colonial landholding arrangements with national imperatives of planning and zoning; where squatter settlements and sharecropper agriculture proliferate alongside gated zones of residence and industrial factories and where ties of patronage and clientalism thrive in the interstices of the electoral regimes of liberal democracy" (Roy, 2011a, p. 310).

Themes across political ecology, address relevant concerns of these regions by paying attention to scale of poverty and inequality, state control of resources and issues of access and livelihood (Adams & Hutton, 2007; Bailey & Bryant, 2005; Bryant, 1997; Forsyth, 2008a; Robbins, 2012). Moreover political ecology studies have drawn attention to colonial histories focussed on natural resource exploitation that gave rise to a state that controlled the human-environmental interactions in a certain way and invented tools and techniques for the purpose, many of which continue to function today (Vandergeest & Peluso, 1995; Williams & Mawdsley, 2006b).

A political ecology study based urban areas in the Global South, would have

> "similar concern for social stratification, power, and unequal access to natural resources and ecological services. It would likewise confront the precarious lives of the poor in conditions of environmental degradation. [...] (yet) contemporary urbanism raises new kinds of questions about networks, neighbourhoods, enclaves, and communities, highlighting the different ways that city dwellers are embedded in dense urban webs [...]" (Rademacher & Sivaramakrishnan, 2013b, p. 10).

However, while a political ecology approach applied in the urban context may direct focus to the problems of the cities of the Global South, it also leads to the construction and reinforcement of these areas as historically exceptional

(Rademacher & Sivaramakrishnan, 2013b). The solution suggested, is a careful uncovering of the historically and politically situated actions of various groups. This has to be done without falling into traps of generalities, such as the assumption that poverty and population growth are the causes environmental degradation (Bryant, 1997; Rademacher & Sivaramakrishnan, 2013b). Contextual literature on urban socio-environmental relations provides a means for situating political ecology in the cities of the South.

Similarly, contextualisation is called for in order to apply concepts of urban political ecology to cities in the South. Lawhon et al. (2014) have recently argued for a theoretical heterogeneity in urban political ecology studies in order to allow for a 'provincialised UPE'. Here they borrow the term 'provincialising' from Chakarbarty (2007, p. 114), which rather than rejecting established European concepts, adds that thought is related to place and universal concepts like justice, democracy etc. "encounter pre-existing concepts, categories, institutions and practices through which they get translated and configured differently". Thus they call for a widening of the subject, ideas and experiences of urban political ecology to make it more suitable and explanatory in contexts wider than the Global North. This would be done through place-based studies of micro politics and a widening of the political economic framework to include other factors which may be encountered. Through such an expansion, it is hoped to create a more vigorous urban political ecology which can see the cities of the South as an "epistemological location" rather than a "geographical container" (Lawhon et al., 2014, p. 505). Rademacher & Sivaramakrishnan (2013) also point to a similar insight when they mention that while urbanisation itself is a ubiquitous phenomenon, its experience in terms of environmental and social processes has no single pattern. An exploration of the politics of place is seen as a way to illuminate the interface between environmental change and urban trajectories

Another way of expanding the enquiry of urban political ecology to include multiple ways in which urban natures are created and used, is through post-structural methodology. It has been pointed out that roots in structuralist Marxist geography have restricted the analytical frame of the field (Lawhon et al., 2014; Zimmer, 2010). Hybrids provide a useful concept to reveal a socio-ecological ontology; a new way of seeing the city, but more must be done to reveal the political opportunities and implications of these (Keil, 2005; Zimmer, 2010). What is needed is

> "a political methodology of social nature sensitive to multiple social ecologies of city life and is able to recognise the different political opportunities offered by different fragments of metropolitan ecology" (Keil, 2005, pp. 647–648).

The structuralist focus on political economic aspects (Keil, 2003; Swyngedouw & Heynen, 2003) has also led to an exclusion of factors other than class, such as region, caste, education etc.; cultural and social resources which also influence matters of access and discursive power. Thus urban political ecology studies often miss out on stakes, meanings, contributions to and effect of environmental transformations on the everyday lived experiences of various groups of society

(Rademacher & Sivaramakrishnan, 2013b). More attention needs to be paid to the diversity of social relations with nature (Keil, 2003, 2005; Zimmer, 2010) in order to demonstrate that various actors have differential relations with nature and between each other, rather than limiting the analysis to formal actors such as the state, industries etc.

Studies rarely take the actor oriented approach in urban political ecology (Zimmer, 2010), which is one way to illustrate the multiple pathways of societal expression in creation of hybrids. Further, there has been an under-exploration of the constructionist perspective which could lead to new insights into the contestations and effects of hybrids (Lawhon et al., 2014). The post-structuralist approaches to political ecology (as discussed in section 2.1.4), have covered much ground in providing and testing the methodologies and frameworks for capturing such situated politics. These are applicable in the urban context as well and are used in this study to expand urban political ecology theorisation.

To summarise, urban political ecology provides a useful entry point with its concepts of hybrids and flows and highlights the unevenness of production of environments while post-structuralist political ecology contributes to exploring deeper and more varied explanations on why certain urban environments develop, who benefits and who loses from the resulting situation. In this study, a strand of post-structural political ecology is used following Blaikie (1994, 1995, 2008), Forsyth (2003, 2008b), Gandy (1996, 2002) an others, that does not deny an ontological basis of environment but points out that environment and environmental change is experienced and interpreted through the cultural, experiential, technical and value laden means of socio-political relations (Blaikie, 1995).

3 RESEARCH QUESTIONS, METHODOLOGY AND METHODS

3.1 RESEARCH QUESTIONS

Political Ecology holds that since ideas about the environment are inevitably shaped by interaction between nature and society and between various sections of society, different actors; scientists, government officials, local residents, environmental activists, may hold different ideas and aspirations for the environment (Blaikie, 1999; Hajer, 1995; Macnaghten & Urry, 1998; Robbins, 2012). This is of consequence as it is the day to day interaction, conflicts and alliances that ultimately structure resource use (Sutton & Anderson, 2004). To explore these multiple standpoints and their dynamics, this research aims at combining what are perceived as the strengths of political ecology through an exploration of the access and use practises in the conservation units in the Ridge, with an attempt to unlock the 'black box of the state', an aspect political ecology studies are accused of often ignoring (Robbins, 2003b). For this purpose, a methodology was developed for a case study integrating an actor-based approach and an argumentative discourse analysis to answer the following basic research questions:

- Which are the actors involved in the creation and contestation of the conservation spaces in the Ridge forest in Delhi?
- What are the discourses surrounding the conservation of the Ridge? What are the overlapping, contradictory or changing components of the various discourses?
- How are competing interests and ideas negotiated to shape the environment?
- How do these changes affect various actors?

To answer these questions, methods described below were used to investigate the discourse, and practices that contest and change the urban landscapes under study. This case study looks at the role of actors in the case of the Delhi Ridge by combining methods such as interviews of varying lengths, participatory and non-participatory observation, focus group discussions and textual analysis including newspaper reports, grey literature, official plans, judicial records, legislation, archival records and photographs. Using the data elicited from these methods, a critical discourse analysis was conducted through the argumentative approach which resolves the contradiction between Marxist and Foucauldian discourse concepts and deliberates on the issue of structure and agency.

3.2 RESEARCH DESIGN

3.2.1 Case Study

To investigate the research questions, a case study method was employed. There are specific and well documented advantages to studying human-environmental relationships through case studies and therefore it is no surprise that the review of literature (Blaikie & Brookefield, 1987; Heynen, Kaika, et al., 2006; Rademacher & Sivaramakrishnan, 2013a) shows that political ecologists more often than not favour case studies.

It has been noted that case studies emphasise the role of 'geographical difference' which encompasses the biogeophysical and social particularities associated with a place (Zimmerer & Basset, 2003). This is a concept of much interest in political ecology as well as geography. Case studies are therefore common forms of political ecology studies as they reveal the "idiosyncrasies, contextual outcomes and local surprises'' that defy the constraints of a rigid theory building (Robbins, 2012, p. 84) and highlight the location specific aspect of environmental change. The implementation of three different models of conservation in the Ridge[20] highlight the specificities that are at play in deciding socio-environmental outcomes. It has been seen that human-environment relationships are formed in specific socio-political contexts within time and space (Sundberg, 2003) and therefore case studies are most suitable to capturing these differences and providing insights for how similar situations can play out differently based on various factors.

Biophysical and social components of urban environments are dynamic and therefore a case study will inevitably capture a snapshot of these dynamics; however case studies remain necessary to understand the means and ends forwarded by various actors and how some of these are rendered powerful and effective as against the others through an exploration of politics of place (Rademacher & Sivaramakrishnan, 2013b). The study of conflict over access for example, affords insights into how contextual actors impinge on specific socio-ecological conditions and relationships in different ways (Bryant, 1991). For example, the marginalised sections may comply, find alternatives or resist, the state may act coercively, ignore transgressions or change the rules in different cases (Long & Long, 1992). Case studies provide an opportunity to investigate what particular factors make possible the existing situation.

It has also been pointed out that detailed case studies are important to dispel the idea of the monolithic state and undifferentiated community that ignore the complex and changing political alliances and contradictions within and among them that reveal different cultural constructions of the land itself (Moore, 1998a).

To summarise, the case study method was chosen as it helps define topics broadly rather than narrowly, to cover complex contextual subjects and not just isolated variables and to rely on multiple rather than single or few sources of evi-

20 See chapter 7

dence (Yin, 2003). Moreover, this research is based on an actor oriented discourse analysis, both the actor approach and discourse analysis are case study oriented.

3.2.2 Actor Approach

An actor approach is well suited to provide an understanding of local level micro-politics including how various actors interact with each other and the environment on a particular issue and how this affects them (and the environment). The uncovering of multiple interpretations and effects of environmental transformations is necessary to expose the social and political construction of 'environmental orthodoxies' held and promoted by powerful actors and voices opposing them (Batterbury et al., 1997; Forsyth, 2008a). In their seminal political ecology research, Blaikie and Brookfield (1987) pointed out the importance of capturing the plurality of perceptions and definitions of problems regarding the environment. Actor-oriented discursive approaches can be seen as responding to this very demand. A look at the roles of the actors also makes the larger socio-environmental processes tangible and meaningful in political terms (Blaikie, 1995) i.e. it exposes power dynamics involved in the formation of socio-natures.

Analysing actors enables an examination of other main political ecological elements through their interests and actions and helps maintain a well rounded line of enquiry. Bailey and Bryant (2005) list the approaches taken by studies that share a political ecology perspective including the more 'traditional geographic approach' of basing a research around a particular socio-environmental problem (deforestation and loss of biodiversity in this case). This approach captures the human impact on a particular aspect of the environment, explaining it through interlinked socio-political and economic forces[21]. Another approach takes socio-economic characteristics like class, ethnicity and gender as the cornerstone of analysis[22]. A third entry point for studies is concepts related to political ecology, sustainable development being a common one, conservation and restoration are examples relevant to this case[23]. This focus on concepts is akin to looking at major discourses to understand the ways in which "ideas are developed and understood by different actors, and how attendant discourses are developed to facilitate or block promotion of a specific actor's interest" (Bailey & Bryant, 2005, p. 21). This is an important arena of analysis since influential discourses carry dominant assumptions about nature and society and the socio-political and economic conditions that make these assumptions possible (Escobar, 1996). It was found that an actor based approach made it possible to capture the key concerns of other approaches especially that of tracing the major discourses as a means to understand the positions and interrelations of the actors.

21 For example Blaikie (1985) on soil erosion

22 For example Heynen, Perkins, et al. (2006) on the correlation of race and ethnicity factors and distribution of green space

23 A specific example on a concept based study is Hajer (1995) on Ecological modernisation

An actor approach would also deal with criticism of rigid structuralism some political ecology (especially urban political ecology) studies face, by analysing the motivations and actions of various actors including those considered 'weak', injecting an acceptance of agency of actors in socio-environmental affairs (Bailey & Bryant, 2005). That is to say that it is not only state policy and capitalist interest that determine environmental outcomes; various actors do have certain agency within the given structures. As noted earlier, urban political ecology largely stresses structural basis for explaining urban problems and yet within the same structural conditions, there could be different responses. The actor approach can help uncover the complexity of the many claims, meanings and practices that surround environmental problems. Scott, in arguing against overly structuralist explanations also endorses such an approach that looks at the micro-politics and asserts that

> "only by capturing the experience in something like its fullness will we be able to say anything meaningful about how a given economic system influences those who constitute and maintain or supersede it" (Scott, 1985, p. 42).

Thus, the role of agency and agent is illuminated better by looking at actors, their discourses and practices.

While this agency is situated within the structures, structures cannot be understood without allowing for human agency (Giddens, 1984). Giddens elaborates that structures can be both enabling and constraining and points out the 'duality of structure'. According to him, "the structural properties of social systems are both medium and outcome of the practices they recursively organize" (Giddens, 1984, p. 25). Thus his structuration theory encapsulates how structure and agency interact in a way that the subject has agency, but this agency is embedded within, and is a constitutive element of, social structures. While structures can enable and constrain social behaviour, these actions reproduce and transform structure as well (Giddens, 1984, 1987). He explains this duality as follows:

> "In following the routines of my day to day life I help reproduce social institutions that I played no part in bringing into being. They are more than merely the environment of my actions [...] they enter constitutively into what I do as an agent. Similarly, my actions constitute and reconstitute conditions of actions of others, just as their actions do mine" (Giddens, 1987, p. 11).

Structural forces certainly have an impact on the 'life worlds' of the individuals and social groups. Once they enter these, they are mediated and transformed by the actors who are part of these structures (N. Long, 1992; Seur, 1992). Thus the question of agency is accounted for by the actor-oriented approach which takes off from the point that similar structures can and do give rise to different social formations and therefore it is important to look at the role of actors and how they deal with situations to illuminate the reasons for this (N. Long, 1992).

According to the actor-oriented approach developed by Norman Long and other scholars of the Manchester school (Long & Long, 1992), an actor is not just a person (or an organisation/institution) making free rational choices or even an abstract social category like a particular class. An actor is a social construct in that his agency and relationship with other actors is formed within his context. In this approach, individual choices are made within the given frames of meaning and practises but different patterns of social organisation result from the actors' agency. An actor thus interacts, struggles, and negotiates with other actors in order to strategize his interests within the particular context. A multiplicity of discourses is available to an actor and however limited the choice may be, the actor draws from

> "a stock of available discourses that are to some degree shared with other actors, contemporary and predecessors[...] (in his) search for order and meaning and in which they themselves play (wittingly or unwittingly) a part in affirming or restructuring" (N. Long, 1992, p. 25).

Agency therefore would imply that an actor has the means to process social situations and find ways of dealing with them; he has 'knowledgeability' and 'capability' implying that within any structures, actors have the "possibility of doing otherwise" though they have certain limits like uncertainty and access to information and resources (Giddens, 1982, p. 9). Agents actively and reflexively monitor the impact of their own actions, the flow of society and contingent circumstances (Giddens, 1984). Giddens explains agency in the terms of individuals but it can apply also to various actors. In the actor oriented approach, only those social entities qualify as actors, which may be attributed with agency, i.e., their actions must have the capacity to make a difference in a pre-existing situation (N. Long, 1992) in the sense that they must be actively involved in attempting to enrol other actors into their scheme (Latour, 1986).

The definition of power (intimately related to that of agency) in such a scheme is also based on interaction of between actors. Power is not a given, it is not a concrete thing that is possessed in perpetuity by its holder and exerted on others. This actor approach holds a Foucauldian idea of power which implies that power is only formed relationally as a product of social interaction and must be constantly negotiated (Foucault, 1980). Foucault writes that each society has a 'regime of truth' based on economic power and political hegemony but this 'truth' is diffused in society (power is in interaction, not just in the power centres) and is based on a process infused with social struggles. As power is formed of social relations and can only be effective through them, each actor must try to use the means at his disposal to win over other actors, enrol other actors in a given social and political scheme. Power then is a result and not a cause of action (Latour, 1986). Thus, discourses and practices are the key to looking at how power is exerted (Verschoor, 1992) as power relations are communicated, reproduced and also challenged through discourses (Barnes & Duncan, 1992). Further, as Giddens (1984) points out, power is never unidirectional. All agents, no matter how subordinate, engage in the construction of their social worlds. All forms of dependence also carry resources to influence activities of the dominant. Thus this study considers the meanings, struggles, negotiations and micro-histories of various actors, even the seemingly 'powerless' ones. These are often left out of the picture as they are not considered legitimate authorities with regard to the environmental question. Policy formulation and at times contestation by NGOs are the more visible forms of this struggle over meaning and resources but these are not the only actors involved in interpreting and shaping the environment. In short, the actor oriented approach looks at

> "social phenomena from the point of view of social practice and perception, which implies giving recognition to individual strategy and understanding. But it also requires analysis of social forms that result from a mix of intended and unintended actions, as well as understanding how macro-representations and phenomenon shape social behaviour and individual choice" (Long & Long, 1992, p. 277).

Therefore, this approach does not ignore structure but also does not limit its analysis to it, in keeping with the arguments of post-structuralist political ecology. Further, this actor approach focusses on reading the power relations between actors through discourses and practices as sites of struggles to define the social outcomes (Verschoor, 1992).

3.2.3 Discourse Analysis

Enlightenment philosophy considered 'truth' to be singular and objective and understood by people by their experience of a concrete reality (Barnes & Duncan, 1992). Post-structuralist studies account for the difference in this experience. In the words of Gregory and Walford (1989, p. 2):

> "our texts are not mirrors which we hold to the world, reflecting its shapes and structures immediately and without distortion. They are, instead creatures of our own making though their making is not entirely of our own choosing".

These 'creatures' that make reality legible and constitute it are discourses.

There are many definitions and descriptions presented for the term discourse[24]. A discourse can be seen as an area of language expressing a particular standpoint, located socially and politically. These may conflict with others and also uneasily co-exist with the hegemonic discourse (Peet & Watts, 1996). Barnes and Duncan (1992, p. 8) describe them as "frameworks that embrace particular combinations of narratives, concepts, ideologies and significant practices, each relevant to a particular realm of social action" these are important because they have a "naturalising power which is largely unseen" (ibid., p. 9). Wodak and Meyer (2008, p. 6), define discourse as "relatively stable uses of language serving the organisation and structuring of social life". Each definition puts forward that discourse is a linguistic area of analysis that is located in, and demonstrative of, social facets. Bové, suggests a way around simply defining discourse by concentrating on its process and effect. He states

> "we can no longer easily ask such questions as, 'What is discourse?' or, 'What does discourse mean?' […] We should, then, ask another set of questions: How does discourse function?

24 For the purpose of this study, I distinguish between discourse and text (oral and written) as the concrete manifestation of discourse which are more abstract structured forms of knowledge (Reisigl & Wodak, 2009)

Where is it to be found? How does it get produced and regulated? What are its social effects?" (Bove, 1995, pp. 53–54).

Discourse can therefore be a key to understanding how social actors construct meaning in a given context by noting the bargaining and accommodation that take place in the interactions between actors (A. Long, 1992).

Lees (2004) reminds us that there are in fact two strands of discourse. The first is based in empiricist traditions in sociology and Marxist ideology. This form takes the positions of the actors as stable and based in the material conditions. Here, discourse is analysed to expose hegemonic ways of thinking that support vested interest. A more current form of discourse analysis within this line looks at how discourse coalitions are formed and how they promote certain interests and actors and shut off others. The basis of coalitions, in this framework is not only material interest (as with the earlier Marxist versions) but persuasion and subscription to similar terms (Kaika, 2006). Therefore rhetoric used in policy debates influences the relationship between actors as much as it reflects them. The second strand is that of Foucauldian discourse analysis that is post-positivist and draws on feminist, cultural and postcolonial studies among fields (Escobar, 1996). This view defines discourse as constitutive, i.e., language constructs actors and their relationships and power is diffused in society rather than held by certain actors.

In practise, these to strands have long been mixed (Lees, 2004) and one can see the benefit of using these strands together; one suits well the study of the interests based in the material context and the other serves to understand the effects of discourse and power structures that shape 'truths' about human-environment relationships and inform social justice (Waitt, 2005). Together these are important elements to understand the politics of environment. However a tension exists in seeing discourse as constitutive as held by Foucauldian approaches and ascribing agency to the actors in using discourse to fulfil their interests.

Foucauldian discourse analysis provides the basis for looking at discourse beyond purely linguistic composition by turning attention to the broader context in which the discourse is produced and the institutional practises in which it is based, paying attention to the ways in which certain discourses are rendered powerful and meaningful (Foucault, 1969, 1979). Discourse cannot be reified and yet it can be read and described as discursive practices take the form of cultural statements expressed in language, material objects and practice (Foucault, 1980).

An approach to discourse based on Foucault's work can be characterised by certain basic assertions. Firstly, Foucault sees discourse as constitutive and not merely descriptive. For Foucault discourse is social structure and discursive practice represents a social practice (Diaz-Bone et al., 2008). Not only are they based in society, they effect the reality which they seek to depict, they are a process of creation of social reality (Escobar, 1998, 1999; Norman Fairclough, 2012; Wodak & Meyer, 2008) i.e., discourses are not only constituted *by* the social world, but are constitutive *of* the social world (Fairclough, 1992). Secondly, all forms of social interactions are based in and reflect power relations (Feindt & Oels, 2005). In as much as discourses are rooted in the power structures of society, they are also an articulation of knowledge and power (Escobar, 1996). Discourses decide what questions can be asked and who can participate. Power is repressive in that it limits what is possible in a certain discursive field but it is also enabling. Therefore discourses can be constricting as well as empowering (Foucault, 1981).

Discursive practices refer to historically specific, mutually shared rules and norms through which discourse is developed, upheld and changed[25] (O'Farrell, 2005). Since discourses and practices are constitutive of reality, they are essential in understanding environmental change and its effects. Our values and beliefs related to the environment are shaped by a variety of discourses that we have access to. These influence the actor in deciding "what exists, what is good, and what is possible" (Herndl & Stuart, 1996, p. 3). This has a direct impact on how various actors seek to influence the shaping of the natural environment as

> "the environment […] is the product of the discourse about nature established in powerful scientific disciplines such as biology and ecology, in government agencies […] and its regulations, and in nonfiction books and essays […]" (Herndl & Brown, 1996, p. 3).

Applying Foucauldian discourse analysis however, is a task that presents some difficulty. Foucault provided no concrete means of operationalising his concepts and his main concepts are often used in a confusing manner with no stable definitions and different usages (Diaz-bone et al., 2008). The second challenge is in combining Foucauldian discourse with the actor oriented approach which seeks to retain a space for agency of the actor. In Foucault's earlier work, discourse is a powerful social entity of stability and change. Individuals function as an instrument of discourse there is little space for ideas, interests or choice (Foucault, 1969). It is noted that "conclusions about the interests or cognitions of actors are ignored by the interpretative analytics from Foucault" (Diaz-Bone, 2006, p. 76, cited in translation in Winkel, 2012, p. 83). The role of the individual actor is not in focus and discourse is seen as a "kind of practice that belongs to collectives" (Diaz-Bone et al., 2008, p. 10). In later work, Foucault does deliberate on discursive strategies and relations between individuals and discourse. His concept of subjects "points at the same time to an actor capable of initiating action and to a being subjected by power, so that actors are never fully determined by a strategic situation" (Feindt & Oels, 2005).

The difficulties of operationalising discourse as a concept for analysis and the ambiguities of the agency of the individual in Foucauldian discourse were deliberated upon by Maarten Hajer in his study of the discourse of ecological modernisation in the context of acid rain in Britain and the Netherlands (Hajer, 1993, 1995). Hajer (1995) presents a framework that helps that makes a Foucauldian discourse perspective graspable to analyse environmental politics, as well as, provides a space for strategic action by actors. He also provides the useful concepts of storylines and discourse coalitions that illuminate the ways in which discursive interactions between actors takes place to affect environmental changes.

25 For example, judicial proceedings, electoral procedures, NGO Reports, legal Notifications

3.2.3 (a). Argumentative Approach: Storylines and Discourse Coalitions[26]

A variety discourses can be traced regarding a certain environmental problem. Along with multiplicity of discourses competing for space, each of these discourses is not always coherent but fragmented and contradictory containing a multiplicity of claims and concerns. Yet policy seems to define coherent problems. What consequence does this have? According to Hajer (1995), discourse analysis would draw on all those ways in which environmental problems are understood and influenced to explain how specific social constructions of the problem at hand eventually define environmental outcomes. To analyse how discourses interact to produce the problem definition and solution and how this affects different actors, is a central question of this study. Hajer (1995) provides a relevant and useful framework to engage in such an enquiry drawing on the work of Foucault and social psychologists like Harré (1993) and Billig (1989) and also provides a conceptual tool to operationalise discourse analysis through 'storylines'.

Firstly, Foucault (1979) asserted that looking at institutions is not enough to explain societal change. He broke larger discourses (related to sexuality and discipline and punishment in his work[27]) into smaller component discourses. These 'micro powers' or smaller, less obvious practises, mechanisms and techniques that are at play below the institutional level, determine how large institutional systems work. These interactions are crucial to societal change according to him and must be analysed. Secondly, he contended that discourses are not merely inert tools used by actors with pre-determined interests to manipulate reality to their liking. There exists a set of rules or discursive order (Foucault, 1973, 1968, 1981). Subjects must act within the context of a group of rules according to which his ideas are formed. The subject's role and position is inscribed in the discursive field. Therefore, discourse is not simply a random collection of signs, words etc., with which a subject changes reality towards an a priori interest, the discourse also shapes the subject. Interests, then have to be reproduced and changed through discourse. Further, power is not a given feature of a certain actor or institution but is defined relationally, i.e. it depends on the way that actor or institution is implicated in the discourse (Foucault, 1976). The internal rules of discourses or the discursive order makes discourses behave as structures within which actors (subjects of discourse for Foucault) must act (Foucault, 1976). These structures of the discursive field can be enabling as well as constraining. Discourses are constraining in that they do not allow certain ideas, arguments to be brought up or even certain actors to participate but they can also be enabling by providing the right rules and resources (Barnes & Duncan, 1992).

Although Foucault demanded we see the role of the subject as conditional upon his position and function as inscribed in the discursive field, what exactly this role is, remains unclear (Hajer, 1995). In order to develop the role of the subject and to operationalise Foucault's abstract ideas in the study of concrete political

26 This section draws largely on Hajer (1995)
27 Foucault (1976, 1979)

events (i.e. to explain the effect of discourse on practise, the interaction of discourses and individuals role in bringing together discursive elements from different discourses available to them), Hajer turns to theories developed in social psychology which he terms 'social-interactive discourse theory' (Hajer, 1995, p. 53). This is a theory concerned with inter-personal interaction that holds that persons are constituted by discursive practises[28] and the interaction between them is an exchange of arguments or contradictory perceptions of reality. The object of research in this framework would be to capture the idea of reality or status quo as upheld by key actors through discourse and practises through which actors seek to persuade others to see reality in their way. Therefore it would be important to examine what positions are being criticised by each actor (Billig, 1989). In short, actors not only try to make others see reality in their way but also position other actors through their discourse. Discourse analysis must look at how a particular framing presents certain elements as desirable or undesirable, moral or amoral, what is considered fixed or appropriate and what elements are seen as problematic (Hajer, 1995).

Thus, the argumentative interaction of discourses is central to discourse formation and reproduction in this framework and reasons for the prevalence of certain arguments can be found in the study of this interaction[29]. Discourse is often presented as a standpoint defined in relation with other discourses (Barnes & Duncan, 1992). What are the crucial claims in an issue? What is seen as a persuasive argument? What form of presentation or historical positioning justifies a particular action? These are the relevant questions to understanding a socio-political discursive interaction (Hajer, 1995).

This framework fills the gap in Foucault's theory by treating the subject as actively involved in creating and transforming discourse according to Hajer. Discourse analysis does not only reveal the subject positions but structure positioning (that is, how the subject positions himself in relation to the others and which structural elements are seen as changeable and which are seen as fixed). The subject is not totally free as he sees reality from his position, from the concepts, storylines and metaphors it entails (Davies & Harre, 1990). These routinised understandings inherent in the position do not deny the subjects ability to make (at least in theory) choices in terms of various practices available to them. However there is considerable power in structures and subject may not always consciously recognise the moment of positioning (Hajer, 1995).

28 This disagrees with Foucault's 'roles' which assumes that actors can be separated from their roles. This theory put forward the concept of subject positions as actors can only make sense of the world with the terms of discourse available to them

29 This idea that discourses develop through argument and persuasion is different from Foucault. For Foucault the change comes when discourses developed for some other purposes are brought together for a new purpose: He shows this through the example of the Plague epidemic in Europe. Technologies developed to observe, examine and categorise people during the disruption of the plague, later became the way in which governments controlled populations in everyday governance (Foucault, 1979)

From these traditions, Hajer develops his argumentative approach (Hajer & Versteeg, 2005; Hajer, 1993, 1995, 2006) with its central concept of 'storylines', which are "generative sort of narratives that allow actors to draw upon various discursive categories to give meaning to specific physical or social phenomenon" (Hajer, 1995, p. 57). They present a symbolic reference point that suggests a common understanding. Storylines could include a historic reference repeated by many actors, a sense of collective fear, ideas about roles of other actors etc. Through them, ideas of 'blame' 'responsibility', 'urgency' etc. are established and certain actors positioned as victims and others as legitimate decision makers etc. (ibid., p. 66).

The central function of storylines is to present some sort of unity in a vast array of discourse components that surround a problem and therefore provide discursive closure[30]. It is assumed that people draw on storylines rather than engage with all the complexities of the comprehensive components of a discourse. Storylines therefore function as metaphors; by uttering one aspect, the entire storyline is evoked. As they are used by more and more actors, they are 'ritualised' (used with regularity) and give permanence to the debate and various actors find their position according to it. Storylines play a role both in subject positions as well as the structure and therefore new storylines may lead to new understandings. One role of agency then, is to find an appropriate storyline. However, though individuals pick certain storylines, they do so due to their 'discursive affinities' (ibid., p. 67) i.e. they may not seek understand the detail of the argument but it appeals to them due to their place in the discursive field and therefore 'it sounds right' (ibid., p. 64) to them. This may be because of their trust[31] in the actor that the storyline came from, the practice in which it is based or what it means for their own discursive identity (though this may not be an active cognitive choice). This concept of storylines therefore operates as a 'middle ground' between agency and structure (and between episteme and individual construction) (ibid., p. 67). The role of the storyline is in "clustering knowledge, the positioning of actors and ultimately the creation of coalitions among actors in a given domain" (ibid, p. 65).

Actors form 'discourse coalitions' around a storyline in the struggle for dominance (Hajer, 1993, 1995, 2006). These are not coherent coalitions that coordinate their action like political or advocacy coalitions, rather a set of actors loosely grouped around a particular storyline. Previously Independent practises can get actively related to each other in a common discourse to form a coalition with common understandings. In sum, these coalitions are ensembles of sets of storylines, actors who utter these storylines and practises related to these discursive areas (Hajer, 1993, 1995, 2006). Each actor in the coalition may carry their own interest and have different interpretations of the storylines involved. The discur-

30 Hajer describes 'discursive closure' as reaching a definition of the problem so that deliberations and solution finding can ensue (Hajer, 1995)

31 'Trust' in Hajer's understanding refers to the suppression of doubt through various means such as the discourse holders past record or his access to certain knowledge resources (Hajer, 1995)

sive order is reproduced when the cognitive commitments implicit in the storylines are routinised. Storylines may at times seem vague or contradictory, but they are important in connecting actors with different or sometimes overlapping understandings. Thus,

> "what people in an environmental discourse coalition support, is an interpretation of threat or crisis, not a core set of facts and values that can be teased out through content or factor analysis" (Fischer, 2003, p. 103).

The enrolment of new actors may change elements of the discourse since each actor draws on a variety of storylines, making discourse coalitions a site of change and development of discourses (Hajer, 1993, 1995).

A storyline may achieve political dominance once it has gathered enough socio-political resonance and enrolled central actors into a discourse coalition (Hajer, 1995). A discourse (and its associated storylines) is seen as dominant in a particular domain if it attains 'discourse structuration' i.e. central actors refer to it and 'discourse institutionalisation' or solidification in institutional practices for example official policy documents (Hajer, 1995, pp. 61–62, 2006, p. 70).

Discourse coalitions are therefore similar to the concepts of interpretative communities (Yanow, 2000) and issue networks (White, 1994). The main import of this concept is that discourses are created through interaction and interpretation and are therefore part of social and political processes. Therefore discourse coalition theory fulfills the need to

> "appreciate that agreements and communication will reflect local circumstance of language, shared interest and perceived purpose between actors and not an absolute transfer of defined concepts from one group to another" (Forsyth, 2003, p. 161).

Hajer's framework therefore provides a means to politicise discourse analysis, by making space to consider actors and providing a

> "way to combine the analysis of the discursive production of meaning with analysis of socio-political practice from which social constructs emerge and in which actors that make these statements engage" (Hajer, 2006, p. 67)

3.3 ENTERING THE FIELD AND FRAMING THE STDY

The Ridge was an ideal case study to examine the socio-political aspects of forest conservation in an urban setting, as it is currently undergoing a transformation with the implementation of various models of conservation and restoration. There were also some practical reasons for choosing this case. The researcher is well acquainted with the city of Delhi, having lived, studied and worked there for seven years and could engage immediately in fieldwork rather than having to familiarise with the language, cultural norms and other contextual elements of the setting of the study. This was a valuable advantage given the relatively short time available in the field.

The research was conducted in three phases, each ending with around three months of field work. The first phase lasted from October 2011 to April 2012, the second phase stretched over May 2012 to May 2013 and the third and final phase from June 2013 to March 2014. A summarisation of the activities carried out in these phases is provided in Figure 1 and is based on Hajer's (2006) ten steps of carrying out discourse analysis.

Phase		
Phase I	1.	Desk Research- A general survey of relevant documents leading to a first chronology and a first reading of events and positions of main actors.
	2.	'Helicopter interviews'- With main actors to provide an overview of the field from different perspectives
	3.	Document analysis- For structuring concepts, ideas, categorisations to identify storylines, metaphors and sites of discursive struggle and a basic notion of process of events.
	4.	Interviews with key players- To collect information on causal chains, interviewees discourse and positions and cognitive shifts.
	5.	Sites of Argumentation- Search data for argumentative exchange
Phase II	6.	Analyse for positioning effects- to show how actors get caught up in an interplay
	7.	Identify key incidents- To understand discursive dynamics and outcomes.
	8.	Analyse practices in particular cases of argumentation- To see if what is said in data can be related to practices in which it is said.
Phase III	9.	Interpretation- Come up with an account of discursive structures, practices and sites of production in a particulate domain and time.
	10.	Revisit some key respondents- Respondents should recognise some of the hidden structures of language.

Figure 1: Research Phases (Based on Hajer 2006, pp. 73–74)

The first exploratory field work in 2012 was conducted to check the viability of the case for the intended study of the complex ways in which biophysical environments are shaped and how this affects various sections of society in an urban setting. The first step was to identify the relevant actors and self-identifications pertinent to the issue, and then to develop a methodology capturing the nuances involved. The idea was to use this first experience to select the theory and methods, rather than use a priori explanations for complex interwoven networks. The first interviews were conducted with officials of different authorities involved with the management of the Ridge and some known environmental activists in Delhi. This initial information enabled "studying up" from these interviews (Paulson et al., 2003, p. 211) and the contacts they provided, to map actors and events surrounding the forest. Following the first phase of fieldwork, a review of literature was carried out, which led the researcher to the vast, varied field of political ecology, providing the basis to plan the rest of the framework for the study. Specific segments of the Ridge were identified for detailed study during future field trips and a list of interviewees was prepared based on previous interviews, news clippings, judicial documents and NGO reports.

The second phase of fieldwork, in early 2013, consisted of further interviews with officials of state agencies involved in administering and managing the Ridge, including the Delhi Development Authority, Ministry of Environment and Forest (GNTD), Officials of the Forest Department, field staff of government agencies like the DDA and Forest Department, scientists working with the Centre for Management of Degraded Ecosystems, retired bureaucrats and city planners, politicians etc. Interviews were also carried out with civil society members and individuals involved in the past or present in various forms of activism on issues surrounding the forest. Finally, participant and non-participant observation and informal interviews were carried out with users of various kinds and residents of varied settlements around the forest segments isolated for detailed framework.

The third phase of fieldwork, completed in March 2013, enabled interviews with some key respondents for the second or third time in order to understand recent developments and to clear earlier lacunae. Sites for observation were revisited for new information and to check for any information that may have been missed during the previous visits. Some new respondents were also included in this phase (these had earlier been unable to speak given the paucity of time or unavailability during the period of earlier fieldwork). A few key respondents were presented preliminary findings and their responses were invited regarding gaps or faults in the data or understanding. The Delhi State Archives were also accessed during this time to get a clearer historical picture of the legal background and discursive formation of conservation landscapes in the Ridge over time.

During the second phase, specific areas for detailed fieldwork were selected based on observational visits in the first phase. Two segments of the Ridge, the South Central Ridge and the Southern Ridge, were chosen. Within these broad segments, three bounded spaces were selected to be examined closely. This selection was carried out in order to concentrate on the various aspects of these areas as well as in keeping with considerations of access to the researcher. Each of these segments is a site of recent and ongoing contestations and each is presented a specific model for the conservation and restoration of the Ridge by the authority involved. There exist, around these three parcels of land (referred to as conservation units in this study as they contain active conservation projects, managed separately from each other), a variety of settlements ranging from informal settlements to formal housing for the middle class and elite groups. The other two segments; the Northern and Central Ridge are both located in more institutionalised areas. The former is in the campus of University of Delhi and the latter, largely in the administrative centre of Delhi. These areas were legally protected much earlier than the rest of the Ridge[32]. The major problems in these areas (according to the Forest Department and the Delhi Development Authority) are horticultural practises converting forest to park in the Northern Ridge and dumping of construction debris in the Central Ridge. Both these areas also suffer from encroachment by government bodies and religious institutions (Kalpavriksh, 1991; Srishti, 1994). These two parts have been included in the overall study through expert interviews

32 See chapter 4

and textual analysis, but have not been subject to location-specific informal interviews with local residents and users.

The South Central ridge is about 633 hectares, it is highly fragmented and has a high amount of built-up area. The segments of this part of the Ridge include the Mehrauli Ridge, Mahipalpur Ridge, Nanakpura Ridge and the campuses of Indian Institute of Technology and Jawaharlal Nehru University (Sinha, 2014). These areas exist as ecological islands with built-up spaces between them. Within the South Central Ridge the following two bounded spaces were chosen as different models of conservation are implemented here (the other areas are not under active conservation and restoration projects):

- The Aravalli Biodiversity Park: The park is the larger of the two existing biodiversity parks in the city and was partially opened to the public in 2011. Five more such parks are proposed in different parts of the city which makes it important to analyse the aims, implementation and socio-political dynamics of the model. The Delhi Development Authority is responsible for the area, but the scientists of the Centre for Environmental Management of Degraded Ecosystems of the Delhi University make key decisions as the technical partner. The area has a conflict ridden history. It has transformed from an extensively mined area, to a site for land contestation for commercial purposes, to a heavily fenced green space. Over the last decade, the miners' settlements were cleared and the landscape, pockmarked with mine-pits was converted to heavily guarded lakes and plantations. In the early 1990s, proposals for a building first a road and then a hotel complex were met with severe opposition and court battles. Further citizen movements were spurned by the building of two malls on the same parcel of land leading to the formation of the Ridge Bachao Andolan (Save the Ridge Movement) and the Citizens for Protection of Quarry and Lake Wilderness (CPQLW). In 2004 an informal settlement, Lal Khet was demolished and the area was designated for the biodiversity park. Today two major informal settlements exist at the borders of the park, and conflicts regarding access to the space for firewood, grazing, through fares, sanitation and recreation are common. Apart from these, the residents of middle to high income residential areas of Vasant Vihar and Vasant Kunj as well as residents of Mahipalpur village use the forest for recreational purposes.
- Sanjay Van: This patch of forest included farmland and pasture up to the 1980s after which it was designated a state owned green space (DDA, 2013) and was notified as a Reserved Forest in 1994. The Working With Nature (WWN) group has been coordinating with the DDA since 2010 with the aim of reforesting the area. The attempt here, as in most other parts of the Ridge today, is to re-introduce native species of Aravalli vegetation. This is a different model from the biodiversity park as a group of volunteer citizens are working with the administration to manifest their vision of what a forest in the city should be. Tensions exist between the managing authorities and existing 'encroachments' including temples and dargahs (mausoleums of Muslim saints) in the area that are under constant threat of eviction but have managed

to stay so far through political support. The surroundings areas include Mehrauli (ward number two) and Kishangarh villages, the residents of which collect wood, graze pigs and use the area for recreation.

The Southern Ridge is by far the largest part of the Ridge (around 6,200 hectares[33]) and lies in the peri-urban fringe of Delhi; it contains the Asola-Bhatti Wildlife Sanctuary which was the third study site, as well as village land and privately owned farms. The Forest Department is in charge of the area and afforestation in some parts has been carried out with the help of the paramilitary Eco-task Force. Delhi's largest unauthorized settlement, Sangam Vihar, is at the border of the sanctuary, and Sanjay Colony which self-characterises as a village lies within the sanctuary. The locals from the area access the sanctuary for illegal mining, pasture, sanitation and fuelwood. Two informal settlements were removed on account of falling within the area designated for the sanctuary, however, the large estates of the elite (called farmhouses but seldom containing any farms) in the area continued to exist. Currently, the settlement of land claims is underway, under pressure from the National Green Tribunal (Sinha, 2014).

3.4 METHODS OF DATA ELICITATION

This inductive study attempts to derive new insights by selecting a problem of interest and conducting a detailed case study with multiple forms of data collection (Wodak & Meyer, 2008). Grounded, inductive research methods have been found to fit well with the current mould of social science (Denzin & Lincoln, 1994), where the stress is not on positivism but on relevance to the questions (Guba & Lincoln, 1994). Heavily structured quantitative methods were unsuitable for this study as this would have involved imposing the researcher's understanding on the interviewees, and would foreclose the possibility of exploring multilayered views and "the sheer density of feeling [...] and complex relationships between ideas of nature and wider critiques of progress and societal change" (Macnaghten & Urry, 1998). Instead, methods that allowed observation of various settings and the participants, helped gaining in-depth or 'rich' information from respondents and left space for respondents to provide their own interpretations and various kinds of information, were needed. Therefore, qualitative social science methods such as semi-structured interviews, field observations, focus group discussions and secondary sources described below, were employed in this study.

33 These figures are provided by the Forest Department and the DDA (DDA, 2007; Sinha, 2014). However, the demarcation of the Southern Ridge area is yet to be completed and several different claims to land exist within the protected area

3.4.1 Semi-structured Interviews

Interviews could well be the most widely used qualitative tool (Gaskell, 2000). A qualitative research interview can be described as an interview, whose purpose is to gather descriptions of the life-world of the interviewee with respect to interpretation of the meaning of the described phenomena" (Kvale, 1983, p. 174). A semi-structured, open-ended format was considered appropriate to capture the complexity of the issues and the wide range of information sought. Semi-structured interviews also provide the respondent an opportunity to provide their own understanding and experience rather than imposing pre-defined categories (King & Horrocks, 2010).

For the in-depth interviews, purposive sampling and snowball methods (Creswell, 2009) were used by asking interviewees to identify other actors and individuals relevant to the study area. This quickly led to saturation with regard to possible interviewees as the area of the study was well defined and most experts and activists were well acquainted with individuals and organisations that have been active in the various capacities, regarding the conservation of the Ridge. As regards the local residents and users, a purposive method was used for key respondents (for example the head of the village council or an active member of the Residents Welfare Association). Other interviewees in the localities were selected through random sampling and snowball methods; people were drawn into conversations during participant and non-participant observations and neighbourhood meetings and through referral from other interview participants.

The questions in the interviews aimed at eliciting information about the participant's background, opinion, experience as well as non-verbal details. Respondents were asked about their association and involvement with the forest, their relationship with and opinions about other actors, their ideas on how the forest has changed and what this means for them and other actors, adequacy and suitability of current policy, aspirations for the future etc. The more general questions helped cross-check information and sometimes add additional aspects to the information provided by others. There were also several specific questions for each respondent apart from these general themes. The interviews were in English, Hindi or both, depending on the interviewee. The longer interviews were recorded on a digital recorder whenever permission was granted and the setting allowed for clear recordings[34]. The interviews of local residents and users and certain government officials were recorded by taking notes as it was noted that many respondents were ill at ease with a digital recording and were less inclined to offer specific information. Detailed notes were taken during and immediately after all interviews to aid transcriptions and add observational details.

A total of 42 in-depth interviews were conducted. These were pre-scheduled in almost all cases and respondents included activists associated with the Ridge, officials of various government departments, a few residents in the vicinity of the

34 In a few cases, for example while walking with respondents in the forest or in a room with a ceiling fan, the recording was unclear and handwritten notes proved an important backup

conservation units and academic experts. Additionally, shorter informal interviews were carried out with local residents and users of the forests. These were mostly unscheduled, impromptu and conversational (though mostly they turned into semi structured interviews) and were recorded by note-taking as part of the field notes. These interviews aimed at ascertaining the ways in which these areas are used and by whom and document ties of work, affection, responsibility, interest and memory (Relph, 1985). Amongst other things, respondents were asked since when, for what purpose and how often they came to the forest, if they remembered the time the current arrangement (depending on where the was conducted, this could mean the biodiversity park, the wildlife sanctuary or the afforestation project in Sanjay Van) did not exist, what was their opinion on the change to the said model of conservation, their relationship with other actors and what they would like in this area in the future.

These interviews were very helpful in grounding the study in local realities and gaining access to alternative meanings, contestations and demands. The uncovering of various interpretations and effects of environmental transformations, helped counter-pose the social and political construction of the forest as a space for conservation of pristine wilderness held and promoted by powerful actors.

3.4.2 Focus Group Discussions

Focus Group Discussions were not planned in the research design but they came up organically during interviews with people in informal settlements and inside the forest. These took place in more structured settings like in a meeting of the women's committee of an informal settlement, to which the researcher was invited by one of the members who was found collecting fuelwood in the forest. Similarly one was held in a vocational class in another settlement adjacent to the forest. In other cases these were completely informal; while interviewing an individual, the neighbours and passers-by would join into the discussion. Given that the streets in the informal settlements are narrow and spacing between houses minimal or non-existent, almost all interviews in these areas were carried out in a rather public setting since "much of slum life is lived outside, in the open, and on display" (Ghertner, 2011b, p. 299).

These focus groups were useful in generating data on different perspectives of social and environmental issues. It has been suggested that focus groups may be characterized as an approximation of Habermas' public sphere (Gaskell, 2000). Focus group discussions were particularly helpful in revealing new information as more than a few times, the group brought up new elements in the course of the conversations among themselves ("don't you remember when they built the wall a few years ago? Tell her about it" or "its ok I already told her that we collect firewood, you can tell her about what the guard said to us"). A group seemed certainly more than its parts as the interaction among them produced new insights. As the members of the group were familiar with each other, they were more informal and open than they would have been in a one on one interview with a stranger.

3.4.3 Participant and Non-participant Observation

Apart from detailed interviews with various actors, participant observation was carried out in the areas selected for field work as and when possible. Obervation served as was "a way to collect data in naturalistic settings [...] (and to) observe and/or take part in the common and uncommon activities of the people being studied" (DeWalt & DeWalt, 2002, p. 260). This played out, more often than not as a 'go-along' (Kusenbach, 2003) with people in their activities and in a few cases as 'friendship as method'[35] (Tillman-Healy, 2003). It involved, among other things, walking, jogging or playing cricket with recreational users, joining meetings and discussions, joining plantation drives for school children, hikes with experts, guided walks, organised nature walks for groups etc.

Non-participant observation was used to make notes especially with regard to forest access and usage to corroborate with the interviews conducted. For example, observation were made bout women collecting firewood, use of the breaches in the boundary walls that were meant to keep residents of the informal settlements out of the forest, houses that stocked firewood in their courtyard or had animals to graze.

These methods provided a chance for the researcher to experience day-to-day realities within which the politics and representation are embedded and to record various experiences of the environmental transformation in a specific context. It allowed a recording within a naturalistic setting of activities, exchanges, gestures, conversations, and generally learn from people about their reality (Agar, 1996).

Observation also allowed for identifying new respondents who access the forest for various reasons, and some of them were drawn into conversations about the forest and their opinions, memories, patterns of access and use, agreement or disagreement with the current situation among other things. It also helped respondents gain a certain comfort level in an informal setting and increased their willingness to share sensitive information (for example related to illegal forest use and uses considered both illegal and anti-social like using the forest for alcohol consumption). Participant and non-participant observation helped improve the quality of data and interpretation by providing a better understanding of the context and helped formulate more contextualised questions and lines of enquiry for interviews.

3.4.4 Document Analysis

A large range of textual sources were consulted and systematically analysed. These include documents produced by various actors under study, such as government reports, NGO reports and campaign materials, planning documents, legal proceedings and court orders, archival documents, newspaper articles, broader

35 This was the case with members of the bird watching community that I knew well and NGO members that I have worked with in the past

cultural and historical sources (memoirs, letters, and notes of actors regarding the Ridge), official websites of government agencies and NGOs. Other grey material like conference presentations, pamphlets, brochures were also used though these were largely used as information points to verify during interviews or cited as illustrations to an argument.

If "a text corpus is the representation and expression of a community that writes" (Bauer, 2000, p. 133), then one can expect to uncover discourses in text just as in an interview. To analyse the text, categories, ideas and concepts based in Hajer's (2006) methodology for discourse analysis described in Figure 1 were identified and examined. Texts were also helpful in building timelines of events and inventories of actors to interview. And finally textual information was used wherever possible and relevant to verify interview data.

Photographs were taken in each site and maps were collected (though these were surprisingly hard to come by and mostly rudimentary). These were also useful for the researcher's contextual memory and understanding and as aides in some interactions with respondents.

3.5 CHALLENGES AND LIMITATIONS

Certain practical and logistical limitations along with some gaps in data had to be put up with during the study. There was a concern for personal safety during fieldwork and the managing authorities of the conservation units advised against (or disallowed) accessing the forest alone. On some occasions, the researcher was accompanied by a personal associate and at times by an activist or official associated with the area. The interviews with local users were carried out in the more publically used spaces such as areas near the boundaries of the forest, where there were breaches in the wall. The researcher was unaccompanied during these interviews.

Certain respondents in government positions declined the request to be interviewed given the sensitive nature of the issue. The generally negative portrayal of government agencies in the press with regard to the Ridge issue also may have influenced some government officials to be over cautious in their responses or decline the interview. Some of them obliged after requesting that the researcher obtain the necessary written permission from their seniors be, but left some questions unanswered or inadequately answered. Similarly, local residents often declined to provide information as some of their activities are illegal under forest laws.

A major challenge was in building a consistent narrative around the Ridge in Delhi. The facts and accounts regarding the Ridge were extremely fragmentary temporally and spatially leading to some challenges in filling gaps in details. For example in most cases, central figures involved in the activism to save the Ridge in the 1980s and early 1990s, had trouble remembering relevant names, dates, documents used, sequence of events etc.

The developments around the Ridge are rapidly proceeding at the current moment but for matters of practicality, this study can only account for the data collected until March 2014 leaving some open ends which could provide material for a later study.

3.6 DATA RELIABILITY AND RESEARCHER POSITIONALITY

It is necessary that in a research of an interpretative nature, the researcher also be aware of the politics of research rather than simply the politics of the researched. It has been commented that

> "Discourse analysts tend to be quite humble people who dislike overblown claims and would never argue that their way is the only way of reading a text. In the final analysis, a discourse based study is an interpretation, warranted by detailed argument and attention to the material being studied" (Gill, 2000, p. 188).

In the same spirit, which is indeed the attitude of most interpretative research, it is pointed out that this study is one interpretation of the subject presented here, however it is backed by systematic presentation of data, transparent methods and stated theoretical underpinnings.

Triangulation has been attempted through the use of multiple methods and multiple sources of data as well as an attention to the context of the data (as suggested by Wodak & Meyer, 2008). The question of how to account for the reliability in researching lived experiences is not new (Sandberg, 2005), while some of the data used here can be backed by documentary sources, this is not true for all the data used. Where this is the case, the material is most often presented not as ontological fact but as a discursive text (for example, accounts of memory of lived experience in the Ridge are not taken as factual but as a reading of the how the forest is presented in connection to the author and what are the main storylines discernible).

A researcher also has his or her own subjectivities to account for as it has been noted, "what we call our data are really our own constructions of other people's constructions of what they and their compatriots are up to" (Geertz, 1973, p. 9). An effort has been made to maintain an interpretative awareness (Kvale, 1996) by accounting for the researcher's own subjectivities based on theoretical and methodological perspectives and making these clear through the various sections of this text. An effort was also made to maintain an epistemological, personal and linguistic reflexivity through the study (Willig, 2001).

There are also challenges to be accounted for related to the researcher's gender (Herod, 2005) and social status (McCorkel & Myers, 2003). While in some cases these factors provided obstacles, they were advantages in other ways. As far as gender is concerned, the most limiting factor was safety concerns while making observation visits to the forest. As most of the areas where field work was carried out are restricted areas, there are very few people around. Additionally, these areas are associated with high crime rates and illegitimate activities. Apart from the researcher's own concerns, other people (authorities, local residents for example) expressed concern and often requested or strongly advised the researcher to leave. In other ways, gender was a facilitating factor. Women from the informal settlements were more open to conversation (it is mostly women who collect the firewood) and to inviting the researcher into their homes.

All the information gained from the above mentioned data collection methods were not used explicitly in the writing of this book. However, each method and the various forms of data generated by these methods, have contributed to the understanding and analysis contained in the following pages.

4 A HISTORY OF DEFORESTATION, AFFORESTATION AND PROTECTION

4.1 BRIEF OVERVIEW OF DEFORESTATION AND AFFORESTATION PRE-COLONIAL AND COLONIAL PERIODS

The Ridge is often presented in discourses of the agencies responsible for it (Sinha, 2014) and environmentalists who seek to conserve it (Agarwal, 2010), as an old pristine forest that in modern times is being lost to urbanisation. However, the Ridge has had a chequered history rather than a linear one of deforestation. A historical overview reveals that any reference to an imagined past of a wilderness safe from human intervention is misplaced. Neither was the Ridge a lush wild forest (lying in the semi-arid climactic zone, the short, thorny, open canopied trees of the Aravallis are hardly the kind that the word 'forest' conjures up[36]) and neither was it devoid of human presence as the protective forest laws seek to impose. Moreover, as this study points out, the efforts at afforestation have privileged certain sections of society and institutionalised certain practices and stigmatised others depending on the main storylines that drove them and the main agenda of the dominant discourse. Therefore, the Ridge is better described as a socio-nature hybrid rather than a pristine environment. This way of looking at urban 'nature' would have implications on how problems and solutions are framed.

The history of habitation in the Ridge has been traced back to the stone-age; around 100,000 years ago (Singh, 2006). Tool making sites have been discovered in the Ridge in Delhi and historian Upinder Singh postulates that forest cover provided food (from animals and plants), shelter and tool material to its inhabitants (Singh, 2006). By the fourteenth century much of the forest on the Ridge had been cleared to create orchards and gardens in line with the sensibilities of the ruling classes (Ashraf, 2004). Ghiyasuddin Tughlaq built a fort in the Ridge around 1321, which stands next to the present day wildlife sanctuary on the Ridge. Feroz Shah Tughlaq (1351-1388) is said to have taken some afforestation measures in the Ridge in keeping with his interest in maintaining this as a hunting ground (Ganguli, 1975). The placment of his hunting lodges in the area signals faunal diversity at the time (Singh, 2006). Other historical monuments in the Ridge also suggest a long standing settlement in this area: Lal Kot Fort (eight century), Qutab Minar (twelfth century), Shah Jahan's walled city (seventeenth century), Lutyen's administrative enclave (twentieth century) are all in and around the Ridge. Of the

36 Mandal & Sinha, (2008) suggest the one of the reasons for lack of public engagement with the Ridge may lie with the uninviting and unsoothing vegetation type

seven cities that have existed in Delhi,[37] the first four were located directly on the Ridge (Dhawale, 2010).

Babur, the first Mughal ruler in India (1526-130), was less than appreciative of the already denuded Ridge he encountered, he notes:

> "A range of hills runs from North to South. It begins in the Dihli country at a small rocky hill on which Firuz Shah's residence called the Jahan-nama, and going on from there, appears near Dihli in detached, very low, scattered here and there rocky little hills".

He described the hills as "low, rough, rocky and jungly" (Trans. Beveridge, 1922, p. 485). Thus we can glean that the Ridge was not thickly forested even in the early Mughal times, but continued to serve as an important source of ground water recharge and an as grazing grounds for pastoralists (Sinha, 2014).

There were notable green spaces in Delhi under the Mughals, this was on account of its many parks, gardens and orchards (Gupta, 1986), but it has been assumed that the forests were heavily denuded given the population increase and the attendant cultivation pressures in the late Mughal and colonial periods (Mann & Sehrawat, 2009). In 1833 a British official wrote that the forests in Delhi had been

> "nearly destroyed from indiscriminate cutting since they fell under our authority; and one is allowed to cut what he pleases, and where he pleases, on payment of a merely nominal duty, and the whole country resorts here for supplies. Formerly it was not so" (Major Colvin, 1883, p. 126).

In 1857, the first major uprising against British rule took place and Delhi was an important site of conflict where the rebels rallied around the old Mughal ruler as their titular head. The Ridge was one of the main centres of the confrontation between the British troops and the Indian 'mutineers' since the cantonment was shifted near the Northern Ridge in the 1820s (Dalrymple, 1997; Spear, 2002). The elevated situation of the Northern Ridge provided the British troops the advantage in defending their camp and much of the fighting was centred around it (Dalrymple, 1997). After 1857, the Ridge in the north was cleared of vegetation in the name of security by the British authorities. This was done to offer a clear line of sight as the attackers had used the cover of trees and bushes to snipe at the British on the Ridge (Krishen, 2006). The pictures taken by Felice Beato[38] in 1958, present a striking visual history to the clearing and regeneration of the forest. One can see vast barren lands where the Ridge must have stood (and stands today) especially striking are the panorama from the Jama Masjid which reveals a brown expanse in the place of the Ridge and the Flag staff house, which stands alone, with a single tree for company whereas today it is situated well within the Kamla Nehru Ridge (Reserved Forest of Northern Ridge).

37 Quila Rai Pithora, Mehrauli, Siri, Tughlaqabad, Firzabad, Sehergadh, Shahjahanabad are said to the be seven cities preceding modern Delhi (Frykenberg, 1994; Hearn, 1906)

38 See Beato's Delhi: 1857, 1997 (Masselos & Gupta, 2000)

The other parts of the Ridge were also noted to be rocky and bare, the District Gazetteer of Delhi 1883-84 describes:

> "The hills of Delhi, though not attractive in themselves, give a pleasant view across the Jamna, and in clear weather allow, it is said, even a glimpse of the Himalayas. Their surface is generally bare, supporting little or no vegetation save a stunted Kikar (Acacia Arabica), or Karil (Capparis Aphylla), or the small bush of the Beri (Zizyphus nummularia) which, with its prickly thorn, is so inhospitable to the foot traveller. The surface of the ground is sprinkled with thin laminae of mica, which shine in the sunlight like gold. The stone, which juts up from the ground here and there, is hard and often sharp-edged" (Gazetteers Organisation, 1999, p. 2).

It is suggested that since the hills themselves were not seen as of any productive value, they were encouraged as a site of habitation to spare the land of the plains, which was rich in alluvium from the river, for cultivation (Dhawale, 2010).

Mann and Sehrawat (2009)[39] provide a detailed account of the history of afforestation efforts in Ridge between 1883 and 1913. They describe the afforestation of the Ridge in these times as one of the first major projects of nineteenth century colonial arboriculture. It is noted that there were several spasmodic efforts to afforest the Northern Ridge most notably in 1883-84, 1887 and 1909[40]. These were not a part of a larger scheme of state executed afforestation but the efforts of individual colonial officers. The first of these efforts involved a district officer, J.R. Maconachie and an individual by the name of Dr Ross, who attempted to foster a corporation with the village communities and developin them a conservation ethic to achieve the goal of raising forest cover. Although groves were planted with the villagers, it was not an equal corporation as the grazing and ownership rights of the villagers were restricted or revoked wherever it was felt necessary. The letters between colonial administrators referring to the project, use similar language as is used today for local communities, framing them as a problem to be dealt with and their exclusion from the forest as being in public interest. These early efforts did not seem to have been very successful and in 1987 a second effort was undertaken by a member of the Delhi Municipal Committee, but with different goals. This time the explicit goal was to serve the purposes of the European settlement. The afforestation project was based in a part of the Northern Ridge closer to the Civil Station of the English settlers and the forest was seen as desirable for its aesthetic and cooling effects. The villagers were seen as an obstacle even more than before as all pretence of the conservationist ethic was dropped. There were plans for fences, fines on defaulting cattle found grazing in forest land and other means to protect from the villagers, the forest meant for the British. A later survey showed an increase in tree cover from these schemes. Villagers however continued to use the forest clandestinely as it contained "scarce resources available to them in an unfertile region" (Mann & Sehrawat, 2009, p. 554). After another slump, the matter was picked up in 1909 by the then Deputy Commissioner when the authorities in the old cantonment expressed interest in the affor-

39 The following paragraphs in this section are based on this paper unless otherwise stated

40 Northern Ridge gained attention as is where the British settlement was based at the time

estation for its "sanitary and ornamental advantages" (ibid., p. 555). As a result, exclusion and prosecution of surrounding villagers who broke the ban on grazing was strengthened. Fences were erected, guards employed and notices sent out banning the villagers from the forest. This must have placed some strain on local communities who depended on these resources, Mann and Sehrawat (2009) point out that the fact that the ban was easier to implement in years of abundant rainfall, is proof that villagers broke the ban due to the scarcity of resources elsewhere.

A second phase of afforestation under the colonial administration focused on the Central Ridge and drew impetus from the transfer of the British capital in India from Calcutta to Delhi in 1911. In 1913, the land of 25 villages was acquired and the landscape reworked. The colonial aspiration for an English landscape aesthetic was imprinted on the Ridge. Aesthetic concerns and lowering of temperatures remained the chief rationale for the afforestation. This is evident in the choice of trees selected as well; they were to be evergreen and of a certain height (Krishen, 2010; Mann & Sehrawat, 2009) not the stunted deciduous trees common in a scrub forest like the Ridge. The very site for the new capital as chosen keeping in mind that the soil would be suitable to grow trees and the elevation would command a view over the city (Krishen, 2010; Mann & Sehrawat, 2009). The A more abstract motivation could be that elevated location of the new seat of power would also emphasise the difference between the rulers and the ruled (Roy, 2010). The 'Scheme for Afforestation of the Ridge 1913' states

> "(F)rom an aesthetic point of view, the Ridge is not a pleasant sight […] as it forms the most conspicuous object on the horizon when seen from the site proposed for New Delhi, it is desirable that it should be made more attractive in appearance by covering its slopes with a green mantle of vegetation, and at the same time it is desirable that there should be a wild park in the neighbourhood of the new city for the recreation of its inhabitants" (Mann & Sehrawat, 2009, p. 558).

Not much indigenous vegetation had survived apart from the few patches from earlier afforestation efforts and so a capital intensive scheme that involved artificial watering and additional soil along with strict banning of grazing was implemented. Villages were relocated and guards put in place and the Central Ridge was declared Reserved Forest and fenced off (Mann & Sehrawat, 2009). This exclusion and policing was well established as is clear from later documents, in 1938, it is reported:

> "The Ridge is protected by a staff of chowkidars (guards) working under a Forest Ranger. Their main work is concerned with illicit grazing and the prevention of forest fires. To the novice and the uninitiated, it is perhaps difficult to realise that if this work was abandoned, the Ridge would at once become a grazing ground and eventually a scree of bare stones" (Supdtg. Engineer PWD Delhi Province, 1938, n.p.).

Prosopis Juliflora was planted along with other species and an official involved in the afforestation notes that the tree survived well as compared to others (Parker, 1919). This phase of capital intensive, state managed afforestation showed results quickly as the official notes

> "at the time it was fenced it was just as barren as the portion now outside the fence. Except for a stray kikar or karil, there was probably nothing that had not been cut or browsed down to a few inches. The recovery by a few years of protection, has been remarkable" (Parker, 1919).

The first large scale successful planting on the Ridge was therefore exclusionary and done chiefly in keeping with British aesthetic desires (Krishen, 2010; Mann & Sehrawat, 2009) to provide a dramatic backdrop for their new city that was to be a symbol of their power, a victory of grand British design over the chaotic Indian city (Sharan, 2006).

This brief history of the Ridge provides some insights into the (re-)creation of hybrid spaces. Firstly, it is clear that the problem definition (the discourse closure according to Hajer, 1995) defines the outcome to a large extent. Depending on whether the objective was military security, increasing green cover for recreation, aesthetic concerns or biodiversity conservation, very different natures were imagined and created. Not only were certain uses legitimised and others delegitimised, even certain species of plants and animals gained favour or disfavour depending on the dominant discourses of what the Ridge forest was for. Even the location of the afforestation areas and their form was dependent on the way the need for afforestation was framed and on who did the framing. The forests of Delhi were to give way in some areas to Mughal gardens and orchards and in others to village pasture grounds. These would be replaced in part by state owned forests planted with English gardening aesthetic (Krishen, 2010) and later state controlled conservation spaces depending on the dominant discourse of the time on what urban nature and specifically urban forests represent. Secondly, we can see the key role of individuals, especially those in powerful positions having access to discursive and material resources. The importance of their role suggests that looking at institutional actors as a block may not be sufficient pointing to the inadequacy of overly structural frameworks of analysis for this case. Thirdly, there is evidence of certain sections of society being denied access and use of the forest. Their voices are not to be found in the archives but their presence and continued illegal use is constantly noted. There is therefore a need to try and uncover what the alternative perspective is and how is the Ridge used and experienced by various non-state actors. Finally, the attention given to protection and afforestation has been spasmodic with periods of lull. There is a need to trace why the issue comes onto the agenda every now and then.

4.2 LEGAL STATUS OF THE RIDGE: A READING OF NOTIFICATIONS (1913–2006)

Although the Ridge was given some protection in the later Mughal period (Sinha, 2014) and in through colonial times (Mann & Sehrawat, 2009), notifications are available from 1913 onwards when a state led project to create a forested backdrop to the new administrative capital in Delhi was underway. The 1913 notification[41] declares around 1948 acres (788 hectares) as Reserved Forest under Indian Forest Act (IFA) 1878 (see Box 1 for explanation of laws applicable to the Ridge). Removal of nine villages in the Northern Ridge took place under this notification[42]. Access to fuel to and fodder was banned the area was fenced and guarded. A Forest Settlement Officer was to be appointed according to this notification to asses and settle any claims to the land. In 1915[43], around 395 acres (around 159 hectares) in another village[44] in the Northern Ridge were declared reserved under IFA 1878, adding to the area under the previous notification. This was adjusted in 1942[45] to 372 acres (around 150.5 hectares) presumably as a result of some contested claim to land in the adjusted area. Today only about 87 hectares of this remain in the Northern Ridge.

Refugees arriving in Delhi during the partition between India and Pakistan post-independence were given land on the Ridge as a huge amount of land was needed to accommodate the influx of people. The Ridge was blasted to provide space for new settlements, for example around Jhandewalan where Karol Bagh stands today (Agarwal, 2010; Lovraj Committee, 1993).

In 1979 there were effective protests by NGOs and citizen groups against construction in the Ridge and conversion of forest land to horticultural parks. Indira Gandhi, then Prime Minister, intervened after being petitioned by a student group against construction in the Ridge (Rangarajan, 2009). Consequently, the Lieutenant Governor of Delhi (the head of the Delhi Administration) declared 25 sites in Northern, South Central and Central Ridge, Protected Forests under IFA (1927). These included varied existing land uses such as orchards, parks, green belts, residential areas and a university campus (JNU)[46].

A wildlife sanctuary was declared in the Southern Ridge in 1986 covering the villages of Asola, Maidangarhi and Sarupur[47]. Bhatti village was added to this territory in 1991[48]. It is notable that these two notifications used khasra numbers[49] or survey numbers to denote the territory they referred to. The earlier notifica tions denote boundaries by names of roads/villages/landmarks. As this land is in the peri-urban area the khasra numbers may have been the only clear way of denoting boundaries. As detailed in chapter 7, the use of survey numbers has led to a lack of clarity in boundaries resulting from faulty and incomplete record resulting

41 Notification No.8734h-R.& A., 1913

42 Namely the villages of Dasghara, Khanpur, Shadipur, Band Shikhar, Khatun, Alipur, Pilanji, Malcha and Narhaula. Mann and Sehrawat 2009, mention the state had obtained by 1913 the land of 25 villages in total

43 Notifaction no. 5911-R. & A., 1915

44 Village Patti Chandrawal Mahal Delhi

45 Notification No. F.F 14 (122)/41-LSB-In, 1942

46 Notification No. F. SCO. 32 (c), 1980

47 Notification No. F.3(116)/CWL/84/897/to 906, 1986

48 Notification No. F2.(19)/DCF/90–91/1302–91,91AD

49 Khasra numbers are records kept by the Revenue Department to list the land ownership and use pattern. For example a person can own Khasra number 708–712. They are supposed to be denoted by physical pillars on ground, though in Delhi this is seldom the case

in a long (and still incomplete) process of settlement. Around 10,517 acres (around 425 hectares) of additional land was handed to the Forest Department in the Southern Ridge in 1996 through a writ petition issued by the Supreme Court[50] to hand over surplus uncultivated land under the Gram Sabhas (village councils) to Forest Department for the 'creation of a reserved forest'.

In 1993, a joint NGO forum for the protection of the Ridge was formed (Agarwal, 2010; Srishti, 1994). This was in response to the decision to hand over the management of Northern, Central and South Central Ridge to the DDA. DDA's mandate and vision were seen as unsuitable as it was believed that the DDA would use the land for construction at worst and to build landscaped parks at best (Agarwal, 2010; Srishti, 1994). The development of parks on forest land had been and continues to be an issue of tension between environmental activists and the DDA. The Lovraj committee was set up by the order of the Lieutenant Governor to 'suggest a management pattern of the Ridge'[51], in response to the opposition to handing over large parts of the Ridge to the DDA. The committee presented its report in 1993 recommending that all parts of the Ridge be declared Reserved Forest. The Delhi government issued a notification in 1994[52] declaring "all forest lands and wastelands which is (sic!) the property of the government, or over which government has propriety rights" in the Northern, Central, South Central and Southern Ridge, Reserved Forests. These segments as labelled in this notification continue to be used for administrative purposes today and are used in this book. There was no prior assessment of the existing uses of the area before the notification was issued. Moreover, certain segments of the Ridge that had been notified Protected Forest earlier were not included in this notification, especially in the South Central Ridge.

Under the IFA (1927) and the Wildlife (protection) act 1974, a Forest Settlement Officer is to be appointed to settle all claims to the lands that have been declared reserved or protected. Although right from 1913, there have been provisions in the notifications for appointment of settlement officers; however, there is no evidence of such settlements of claims took place. The Lovraj Committee report terms this an 'administrative infirmity' as it notes that "there are no records of settlements of rights or of acquisition of land owned by village panchayats or privately under the Land Acquisition Act 1894" (Lovraj Committee, 1993, p. 17).

50 M.C. Mehta vs. Union of India and ors. Writ Petition (Civil) No. 4677/85. Orders dated 25.01.96 and 13.03.96

51 Notification No. SF.2 (11)/DCF/1990–91

52 Notification No. F.10(42) –1/PA/DCF/93/2012–17(1)

Box 1

FOREST LAWS APPLICABLE TO THE RIDGE

The Indian Forest Act of 1927 is the main forest conservation act in the country. Largely derived from the Indian Forest Act 1878 (extensions are related to duties on timber), it defines three classifications of forests: Reserved Forests are those lands where all non-forest activity (activity not related to afforestation) is prohibited and subject to penalty. The state can allow specific uses in Reserved Forests if deemed necessary. Protected Forests are those within which the state can prohibit certain activities. Certain trees can be declared reserved within Protected Forests and private rights can be suspended for up to 30 years. A further category of Village Forests provides for local village communities to use and manage the forest. The state can regulate the management of Village Forests. Therefore, Reserved Forests are the absolute property of the state, Protected Forests are state property but certain use rights are recognised whereas in Village Forests the state has only a management role (Hazra, 2002). The Act lays down the procedure for notifying areas under government ownership or waste lands as forest. Under section IV, a preliminary notification is to be issued, stating the intention to declare a certain area Reserved or Protected Forest. A Forest Settlement Officer is to be appointed to settle any existing claims to land or produce in that area. After deliberations, the land demarcated is declared Reserved or Protected Forests under section XX of the act. Since no such settlement has taken place in Delhi's Reserved Forests, the notification process is incomplete (this is not an uncommon situation in legally protected forests in India (Gadgil & Guha, 1992)).

The Forest Conservation Act of 1980 is a two-page document which lays down that no diversion of forest land can be allowed by state governments without prior permission from the central government.

The Indian Wildlife (Protection) Act of 1972 provides legal protection to fauna and flora in an area by prohibiting and penalising hunting, picking, uprooting, trade in forest produce etc. The act also gives state government the power to declare national parks and sanctuaries in areas deemed to be of "adequate ecological, faunal, floral, geomorphological, natural or zoological significance, for the purpose of protecting, propagating or developing wildlife or its environment" (Chapter IV of act).

The Delhi Preservation of Trees Act of 1994 prohibits the felling, cutting or removal of any tree in Delhi regardless of property rights over the land on which it stands. Any such activity requires an application to a Tree Officer appointed under this act. If permission to cut or remove the tree is granted, the concerned party must plant a said number and kind of tree to replace the tree cut or may reimburse the Tree Officer to carry out such a plantation. The penalty for cutting trees without permission includes arrest without warrant and seizure of property

Table 1: Summary of the Legal Protection Status of Segments of the Ridge

Year	Segment of Ridge	Level of protection	Additional details
1913	Central Ridge	Reserved Forest	Nine villages removed and forest fenced. The purpose was beautification of the Ridge in view of the building of the new Imperial capital in Delhi
1915	Central Ridge	Reserved Forest	More area added to the earlier notification
1942	Central Ridge	Reserved Forest	Addition to previous notifications
1980	Central and South Central Ridge areas	Protected Forest	25 areas including land use of various kinds declared Protected Forests. This was through the intervention of the then Prime Minister Indira Gandhi responding to student and citizen activism.
1986	Southern Ridge	Wildlife Sanctuary	Village council lands of three villages declared part of the sanctuary
1991	Southern Ridge	Wildlife Sanctuary	Another village added to the sanctuary
1994	Northern, Central, South Central and Southern Ridge	Reserved Forests	All government forest and waste land on which the government has propriety rights, 777 hectares, declared Reserved Forests. This was recommended by the Lovraj committee, in response to opposition to the decision to hand over large swathes of the Ridge to the DDA.
1996	Southern Ridge	Reserved Forest	As an outcome of the M.C. Mehta Case, Supreme Court ordered all uncultivated surplus land of village councils falling in the Ridge to be handed over to the Forest Department. Around 425 hectares were 'to be made available for the creation of Reserved Forest' under this notification.

Conflicting claims and unclear boundaries have made administering these lands difficult for the government agencies. The Northern Ridge owing its smaller size and long history of protection is a more or less stable territory (though horticultural practices of the DDA have converted it largely into a park). The Central Ridge, due to its location in the heart of administrative Delhi and being fenced in, is also relatively protected, though problems of encroachment even by government bodies and debris dumping continue in this area. The South Central and especially the Southern Ridge remain contentious. In 2006, a notification[53] was issued to appoint a Forest Settlement Officer to settle claims pertaining to the 1994 and 1996 notifications. This led to the removal and relocation of two informal settlements in the Southern Ridge[54]. The boundaries of the Southern Ridge are yet to be finalised

53 Notification No.F.10(42)–1/PA/DCF/9/II/181–198

54 See chapter 7

and the National Green Tribunal has been pressuring the Forest Department to complete this task (Jain, 2014).

Apart from the multiple notifications on legal status of various parts of the Ridge, there are also overlapping and overriding notifications regarding the agencies in charge. The Southern Ridge changed hands from the Delhi Administration (which was handed the part in a notification in 1957[55]) to CPWD in 1963 for 'beautification as parks and gardens'. There was no full scale Forest Department in Delhi before 1996 (Dhawale, 2010), which now manages a large part of the Southern Ridge as a wildlife sanctuary. The Northern Ridge was to be managed by to DDA as per the notification in 1968[56] for 'maintenance and beautification' (Sinha, 2014). The landscaping and horticulture based beautification of the Northern Ridge was seen as inappropriate use by most environmental NGOs and activists (Srishti, 1994), an opinion later shared by the Forest Department (Sinha, 2014). This led to a raising of an alarm by NGOs and citizen groups in 1992 when a notification[57] was passed to hand the DDA management of the South Central and Central Ridge as well. The Central Ridge continues to be under the various agencies including the Central Public Works Department and Land Development Office of the central government (which has handed the portion under its control to the Forest Department for management). As an outcome of a case filed in the Supreme Court and demands from environmentalists, the court ordered the formation of a Ridge Management Board in 1995[58]. This was to serve as a platform for bringing together various government agencies and some NGO representatives to make decisions about the Ridge. The effectiveness of this board has been challenged even by its own members given the multiplicity of viewpoints it contains (Agarwal, 2010). The managing agencies currently responsible for various parts of the Ridge are listed in the table below:

55 Letter No. F.11(3)/55–P&D From Secretary (Development) Delhi Administration to soil conservation officer

56 Notification No. D.O. No.Secy/Vac/187/68

57 Notification No.F.2.(11)/DCF/1990–91/5927–5935

58 Notification No. F56(225)/95/Dev./HO/5596

Table 2: Administrative Division of the Ridge (Source: Sinha, 20014)

Segment of Ridge	Area (Total 7777 hectares)	Managing Agency
Northern Ridge	87 hectares	DDA (73 ha), MCD (11 ha), Forest Department (0.3 ha)
Central Ridge	864 hectares	Forest Department (423 ha), DDA (85 ha), Army (202 ha), CPWD (37 ha), NDMC (25 ha), L&DO (70 ha), Railways (11 ha), Wireless station (6 ha), MCD nursery (3 ha).
South Central Ridge	626 hectares	DDA
Southern Ridge	6200 hectares	Forest Department (1938 ha), DDA (120 ha), Sports Authority of India (32 ha), Revenue Department/Village council lands (4207 ha)

4.3 CONCLUSION

This chapter provided an overview of the historical evidence of the deforestation and afforestation efforts in the Ridge to establish the claim that the Ridge must be analysed as hybrid spaced where categories of nature and culture cannot be separated and elements of social and political power are intertwined with the biophysical elements (Latour, 1993). While "urbanization is primarily a particular socio-spatial process of metabolizing nature, of urbanizing the environment" (Swyngedouw, 2004, p. 8) urbanisation is not a taking over of natural environment by built environment, urban political ecology looks at it as a creation of new kinds of environments (which are both natural and social at once) that are a result of and affect social power relations. The conservation and afforestation efforts in the Ridge have entailed discursive construction of the forest, as well as, material construction through capital input and biotic and social manipulation, all of which was based in the geographical (semi-arid) and political (capital city) setting. Through a review of notifications pertaining to this single geographic formation, we find a legal and administrative dynamism. This provides the context for analysing the role of the state in the management and of the Ridge in the next chapter.

5 THE STATE AS MULTIPLE ACTORS

5.1 INTRODUCTION: THE STATE IN ENVIRONMENTAL POLITICS

The state is one of the central actors in environmental negotiations. It has been variously regarded "a theatre in which resources, property rights, and authority are struggled over" (Watts, 1989, p. 4), an "interface between capital and nature as well as human beings and space" (Escobar, 1996, p. 55) and

> "one of the most important loci of struggles over resources: where forests are demarcated, whether local users are excluded, [...] where issues of access to resources are defined in law, where scientific information is produced and selectively wielded in the name of universal objectives such as conservation, justice or economic growth" (Blaikie, 1994, p. 9).

There is particular interest in political ecology studies regarding the place of the state apparatus in directing, legitimizing and exercising power and control (Bailey & Bryant, 2005; Forsyth, 2003; Peet & Watts, 1996; Robbins, 2000, 2012).

The centrality of states in deciding environmental outcomes is not new or surprising, the scale of action required for environmental management as a public good, usually renders uncoordinated individual action inadequate (Madsen, 1999; Walker, 1989). The state is seen as the only institution with the required resources to provide the requisite framework to balance varied and at times divergent interests (Narayanan, 2008; Scott, 1998; Walker, 1989). It occupies a unique position in which it can address ecological problems at various scales, further it has the legitimacy to do so in the name of public good (Bailey & Bryant, 2005; Blaikie, 1995).

The state, apart from being a locus of struggle is also an actor in its own right with complex and dynamic interests that interact with those of other actors. The notion of the Weberian state, which intervenes in a rational manner and is above competing interests has been critiqued and rejected by political ecologists (Bailey & Bryant, 2005; Blaikie, 1995; Bryant, 1992; Robbins, 2000; Walker, 1989). The 'modernisation' role of the state is implicit in this view and depicts an omnipotent state that is charged with making policies to deliver ecological goods (Blaikie, 1999). A form of this self-positioning of the state was prevalent in the colonial style of environmental governance which brushed over the interests embedded in environmental policies by presenting them as technical, scientific and rational (Agarwal, 1997), a stance that is seen to continue in varying degrees, in most post-colonial countries (Adams & Mulligan, 2002; Anderson & Grove, 1987; Lemos & Agrawal, 2006). This 'techno-managerial' approach describes the formulation and implementation of solutions without conflict (Bryant, 1991, p. 164). It is assumed that states will support suitable policies when the facts are made available to them, yet it is clear that despite a definite growth in adaptation of environmental vocabulary and sensitivity, the basic political and economic practices of the state in relation to environmental conservation, do not shift as easily and automatically (Walker, 1989). Moreover, this view misses the contestation the state may face both from within and without and further ignores to ask who formulates policies and to whose benefit (Blaikie, 1995; Bryant, 1991).

In critiques of the portrayal of the state as non-partisan and rational, political ecologists and other social scientists have tried to capture the biases and contradictions embedded in the state (Blaikie & Brookefield, 1987; Bryant, 1999; Chatterjee, 2004; Escobar, 1995, 1998; Robbins, 2012). The state's role has often been presented pessimistically in studies related to environmental management in general and forest management in particular. The liberally inclined authors look upon the state as a monopolistic, inefficient player that results in overuse of resources under its control (Hyde, Newman, & Sedjo, 1991). On the other hand the more populist authors insist that the state plays an exclusionary role and is detrimental to common interest (Gadgil & Guha, 1992; Guha, 1989b; Peluso, 1992; Poffenberger, 1990). In the second strand, common to political ecology studies, the state is often represented, as the instrument of capitalism (Escobar, 1998; Peluso, 1993). In this representation the state is seen as extractive and destructive. It holds that states consider nature a resource for exploitation in the service of capitalist expansion or for the economic and political gain of the elite classes (Blaikie, 1985; Gadgil & Guha, 1992; Peluso, 1992). Bailey and Bryant (2005) suggest that conservation agendas promoted by states in the Global South, are rarely an end in themselves. Instead, conservation is a means to various political and economic ends, for example, commercial resource extraction, eco-tourism, structural adjustment for funds, and personal gain of officials through corrupt practices etc. In short, the state is only interested in its own political and economic gain (and that of the elite that can control it). Myrdal generalises the post-colonial states in the Global South as 'soft' emphasising

> "a general lack of social discipline ... signified by deficiencies in their legislation and, in particular, in law observance and enforcement, lack of obedience to rules and directives handed down to public officials on various levels, often collusion of these officials with powerful persons or groups of persons whose conduct they should regulate, and, at bottom, a general disinclination of people in all strata to resist public controls and their implementation. Within the concept of the soft state also belongs corruption, a phenomenon which seems to be generally on the increase in underdeveloped countries..." (Myrdal, 1970).

A wider characterisation of states was provided by Evans as predatory, intermediate and development states. Countries lie somewhere between these three points on a scale from those states that are characterised by inefficient, rent seeking bureaucracy and fragmented legislations, captured by elite interests, and the on the other end, coherent Weberian states that function through a well-developed bureaucracy and promote social development through strong public-private ties (Evans, 1989).

This study shows that the interests and actions of the states as well as their relationship with other actors are not as simplistic or easily encapsulated. While the state does indeed play an extractive and exploitative role in some cases, this is not an essential characteristic of states, or even of the same state seen across spatial or temporal scales. As we shall see, the Delhi government has played the role of allowing and managing mining in the Ridge at one time and banning such activities including real estate development in the same space at another. Even within the same temporal frame, different control and access regimes exist in different segments of the Ridge. This is not particular to the Ridge. Forests in India have historically been subjected to different forms of control exercised by different state agencies. For example, the Reserved Forests are exclusive to the state and allow no extraction and in most cases no access to non-state-actors, while Protected Forests are open to some level of controlled access and village forests are open to local people for fuelwood and grazing. Therefore there exists a mosaic of forests with a view of catering to different needs through different patterns of access and use under the same state (Marcot, 1992; Rangan, 1997).

It has been noted that states face a conflict in their mandate, in that they are both responsible for economic growth as well as the guardians and stewards of natural resources which are needed to perpetuate this growth (Walker, 1989). Therefore the state often plays a schizophrenic role and subscribes to both developmentalist (nature as raw material) as well as sustainable development discourse at the same time in the same geographical and cultural space (Escobar, 1996).

Secondly, as argued in this chapter, states are not monolithic actors. Conflict and negotiation within state agencies has an influence on what questions are brought up and what solutions are seen as possible. Internal differences of view and unsuitable bureaucratic constraints have been shown to prevent the state from making effective policies. (Bunker, 1980, 1982; Saberwal, 2003). These internal divisions and multiplicity of mandates and roles in modern post-colonial states, have been traced back to the nineteenth and early twentieth centuries when functionally specific bureaucracies were created for different aspects of environmental management in colonies (Bailey & Bryant, 2005). The main aim of such functional division was 'rational' ordering of the state to enhance resource extraction while protecting the resource base for long term extraction[59] (Adas, 1989; Peluso, 1992; Rangan, 1997; Scott, 1998). What this meant for environmental governance, however, was a conflict-prone situation where the complex and overlapping real world environment did not confirm to the neat categories of the government divisions (Bryant & Parnwell, 1996). The same space is often viewed differently by different agencies, for example, an agency responsible for conserving forests may view a wooded expanse as a forest, another responsible for animal husband-

59 A relevant example is provided by the genesis of the Forest Department in India. The need for forest conservation and therefore the formation of a Forest Department in the nineteenth century is linked to the political and fiscal pressures faced by the East India Company in India and Europe. On one hand, the losses faced due to the impact of the free trade lobby in Britain had driven a need for economic compensation from their colonies. On the other hand, they were wary of running out of resources, in this case timber (important for large scale projects as well as ship building for the navy) as had happened even in England where the forests had dwindled and been replaced by fields and meadows (Rangan, 1997)

ry, may view it as pastureland and a third may desire to facilitate mining in the area, leading to an inbuilt propensity for confusion of mandate. To add to this, the colonial system also initiated certain modes of professional training in view of efficient management of resources through specialist knowledge. This led to most departments having a tunnel vision related solely to their own area of specialisation and normalised an ignorance of a broader picture of how their own work was a part of the whole (Furnivall, 1956). This historical evolution has therefore resulted in a complex governance structure characterised by a sectoral understanding of natural resources and consequently resulted in a fragmented institutional structure in terms of policies, laws and organization (Narayanan & Chourey, 2007).

There are "a variety of logics, intentionality and assemblages" (Pieterse, 2012, p. 20) that factions within local states navigate. Further, the various agencies of the state also interact with non-state actors especially in a democratic set up (Williams & Mawdsley, 2006b). The literature on environmental governance captures some of this expansiveness and complexity by looking at the state and its actions not in isolation, but broadly includes all institutional solutions for resolving conflicts over environmental resources (Paavlova, 2007). Environmental governance models explicitly include non-state players; the market, NGOs, communities etc. and the role of state is seen relative to various actors (Armitage, de Loë, & Plummer, 2012). However, even within the so called conventional governance models (where no multi-actor form of governance is institutionalised), other actors continue to impact the states discourses and actions including policies (Hajer, 1995; Robbins, 2012). As Willams and Mawdsley (2006) point out, the state in India is a complicated institution given not only hierarchical and authoritative colonial roots but also because it is a democratic state, open in part, to social and media pressure. Further, the process of formation of governance models is also of interest as it is the site of negotiation and contestation between actors. Models of governance are after all a reflection of the view of society they embody, i.e., ideas of who should address which problems and how, are implicit in them (Glasbergen, 1998). How have these governance models evolved, and what pressures do they face? Who is included within them formally, and who is left out? These are questions this research engages in.

India has some of the toughest legislation aimed at environmental protection in the world coupled with a poor record of implementation, raising questions about the state's agenda and capability (Saberwal, 2003). Earlier studies in Indian conservation pit a state with a singular will and interest against the poor communities that depend on natural resources; since states monopolise resources like forests in favour of a protectionist agenda or for the use of powerful groups (Gadgil & Guha, 1995, 1992). This standard environmental narrative often forwarded by studies based in the Global South (Madsen, 1999) does not provide the required nuances to explain the fractures in motivation within the state. The division between state and society is also not as clear. Historical and ethnographic studies in forest administration in India have pointed out that the extent and nature of the influence of state policy on the ground is often limited and is enmeshed in local specificities (Sivaramakrishnan, 1996; Someshwar, 1995). Groups of citizens can influence the state at the level of formation of policy and means including corruption and vote-bank politics can effect implementation (Benjamin, 2008; Chatterjee, 2004; Madsen, 1999; Saberwal, 2003).

The point that the state is not a seamless, unitary body, has been brought up by several studies (Baviskar, 2003b; Fuller & Bénéï, 2000; Rangan, 1997; Saberwal, 2003 for example). In epistemology, intent, instruments of action and modes of functioning, different agencies of the state bring different things to the negotiating table. This is legible in their discursive practises and in the material outcomes that they result in or contest. In Migdal's formulation, states are ineffective because they coexist with different centres of power (industrialist, landlords, traditional leaders) that undermine their ability to implement policies (Migdal, 1998). However, as the case of the Delhi Ridge demonstrates, the state itself is fraught with contestation and this has consequence for its policy outcomes. Further, policies and outlooks of various bodies of the state are affected by informal and institutionalised interactions between agencies of the state as well as with other actors. At times, this forces the state to change policy, at other times, it may push implementation of existing policy. This makes the state a dynamic player and policies and actions of state agencies are an outcome of these interactions.

Some advocates of multi-actor governance may argue that states are no longer viewed as the only actor capable of addressing environmental problems (Lemos & Agrawal, 2006). Neither are they always successful in achieving their goals and implementing policies. Yet it is largely acknowledged that states are, and shall continue, at leat in the near future, to be central actors in environmental issues (Bryant, 1991; Rangarajan, 2001; Robbins, 2000). Even with the growing interest in cooperative models of governance, it is not expected that the regulatory approaches will be replaced as multi-actor governance will still need an enabling system of state regulations to be effective (Meadowcroft, 1998). State control is seen as necessary even by its critics for setting parameters and regulating economic and social action as well as adjudicating disputes related to forests and other resources (Marcot, 1992; Rangan, 1997).

The following paragraphs, describe and analyse the three main state actors in the case of the Ridge; the planning bureaucracy and the Forest Department which manage large parts of the Ridge between them and the Supreme Court which mediates disputes and also at times participates actively in policy formulation and enforcing implementation (Rosencranz & Lele, 2008). This chapter is not a comprehensive discussion of their roles with regard to the Ridge, this will be developed through later chapters as their role in local politics of conservation is detailed.

5.2 THE DELHI DEVELOPMENT AUTHORITY

The planning bureaucracy of Delhi, the Delhi Development Authority (DDA) has drawn much criticism by academic studies (Baviskar, 2003a; Roy, 2004; Verma, 2003), social movements[60], the media (Diwedi, 2006; Seth, 2006) and the general citizenry of Delhi for its vision and implementation (or lack of the same) of a planned Delhi. The DDA is a central actor in the politics of space in the city by virtue of being the single largest landowning body in Delhi as well as holding the responsibility for formulating and implementing the master plans for the distribution and utilisation of space (Chaturvedi, 2010). Attempt to discipline urban space in Delhi by subjecting it to a clear zoning pattern and the problems in the realisation of this goal have a long history. With the shift of the British capital in India to Delhi in 1911, began an attempt to plan a rational and modern city as distinct from the crowds and squalor of Old Delhi (Sharan, 2006). The new city was to symbolise a government that was in complete control of its environment (Sengupta, 2007). The New Delhi Municipal Committee invented a series of bye-laws that "strangled organic development in favour of a petrified landscape" (Legg, 2006, p. 192). Nevertheless, by 1939 these neatly planned functional zones were found to have turned to mixed use and 'haphazard developments' (New Delhi Development Committee, 1939). The problem was identified as one of lack of regulation rather than a lack of supply of urban amenities (shortage of public housing that led to the growth of slums for example) (Legg, 2006).

The grand narrative embodied in the plans, linked to larger changes in the national and international political climate, shifted from that of Delhi as the Imperial capital to post independence socialist modernisation (Batra, 2010; Sharan, 2006) and then to the aspiration of a 'world class city' (DDA 2007, p. 2) after the neoliberal reforms in early 1990s (Dupont, 2011). In 1957, the DDA was instituted by an act of parliament to 'check the haphazard and unplanned growth of Delhi'[61] by implementing a master plan prepared with an outlook for twenty years by the Town and Country Planning Organisation (TCPO), with the help of the Ford Foundation. It is interesting to note that the immediate trigger for this phase of modernist planning was in fact an environmental crisis. The contamination of the Yamuna waters with untreated sewage led to a jaundice epidemic in 1955 that resulted in more than 700 fatalities in Delhi (Dennis & Wolman, 1959) driving home the point of the need for a planned system to control the city's development trajectory (Batra, 2010; Roy, 2004).

Under the DDA, the earlier colonial project of zonal planning was continued with the nationalist modernising zeal of the post-independence era (Baviskar, 2003a). The project was to organise space in the principles of classic modernist urban design through enumeration, zoning and slum control (Sundaram, 2004). Delhi was thus emulating the characteristic of all modern states to divide complex, overlapping territories under their control in zones to control people and resources

60 Such as the Sajha Manch and Delhi Janwadi Adhikar Manch

61 Delhi Development Act 1957 dated 27.12.1957

within these zones, through regulations and agencies that have territorial and functional jurisdiction (Scott, 1998; Vandergeest & Peluso, 1995).

Though there are several areas of continuity with the colonial phase, the exclusionary city-planning process is considered "far more pervasive and powerful now" (Batra, 2010, p. 18). The first master plan was presented for public comments in 1960, but the later plans were not translated into local languages or distributed publically prior to becoming statutory (Roy, 2010). Although the DDA is not an elected, representative or publically accountable body, it carries vast powers to transform Delhi's landscape. The authority was given complete monopoly over the acquisition, classification and transfer of all land not privately owned and lying within the city limits by the Delhi Development Act, 1957, making it the central agency concerning land use in most of the city (Baviskar, 2003a; Verma, 2003). This compulsory acquisition of land was justified on the grounds that it makes "planning and implementation of the plan easier" and stipulates that "all this land will remain under public ownership and […] be leased out […] on an equitable basis" (DDA, 1962, p. 7). The land of several villages was vested with the DDA and it could decree the zones of appropriate use of these lands including turning it into green space or earmarking it for residential or commercial purposes (Batra, 2010). However, what was envisaged as an empty canvas by the master plans was on ground, a city full of lived spaces of various sections of the populations and replete with places and practices of mixed use (Baviskar, 2003a). The plans had considered abstract space to create homogenous units that can be measured and compared to each other[62] (Tuan, 1997). But people do not experience space as abstract and homogenous. Such planning ignored the specific interactions, histories, and lived experiences of the 'planning unit' (Baviskar, 2003a; Vandergeest & Peluso, 1995). It is these contradictions that are seen as contributing to the failure of such plans (Vandergeest & Peluso, 1995).

With the liberal economic reforms in the 1990s, the post-independence welfare-oriented urban development phase was replaced by one oriented towards garnering investment and creating spaces of consumption and leisure for elite citizens (Baviskar, 2003a; Ghertner, 2011a). This led to a radical transformation in the urban landscape where large scale infrastructure like flyovers, malls, high rises and business centres took precedence over creation of housing and provision of facilities to the urban poor and at times even over preservation of green spaces (Dupont, 2011; Ghertner, 2011a). The aspirations of Delhi's well-to-do citizens, connected to global consumption patterns, has also helped create a supportive atmosphere for the DDAs treatment of land as a resource to be exploited for symbolic and material capital (Batra, 2010; Baviskar, 2003a; Ghertner, 2011a). These classes of citizens have also actively used the courts to implement their idea of an aesthetic, sanitised city in which slums and the urban poor have increasingly been posed in terms of illegal squatters on public land rather than as bearers of rights

62 For instance, middle income residential plot A should be comparable to middle income residential plot B and can be planned in the same manner

who must be provided for by the government (Baviskar, 2003a; Bhan, 2009; Fernandes, 2004). Environmental concerns, including air and water pollution, have been espoused by such middle class activists effecting the dislocation and dispossession of the urban poor in many cases (Baviskar, 2003a; Bhan, 2009; Ghertner, 2011a). On the other hand, the DDA has pushed construction projects in the urban forests and the river bed of the Yamuna in contravention to its own stated policy of protection of green spaces in the interest of promoting commercial uses and construction of large scale infrastructure to generate investment (Baviskar, 2011; Ghertner, 2011b). Thus both the city's poor and the environment have been thrust into growing insecurity since the 1990s.

It has been pointed out that one justification for the existence of the DDA and its powerful presence, is that Delhi is the capital city and the desire to showcase this as a capital both in "symbolic and material terms" (Baviskar, 2003a, p. 92) has remained high historically (Sundaram, 2004). A practical complication that arises from Delhi being the capital city is that both the central and state governments have stakes in its development. Planning is a state subject[63] and under the 74th Amendment to the Constitution, a very clear process for physical planning is provided which includes provisions for ward committees at the basic level and other agencies to provide recommendations to the elected municipality. Delhi does have two elected municipalities, the New Delhi Municipal Corporation (NDMC) and the Municipal Corporation of Delhi (MCD)[64]. Though these are responsible, along with the DDA, for implementing the plan, they play no part in formulation. The DDA answers not to the state government but directly to the central government which is not as sensitive to local political pressures. Therefore, especially in Delhi, the planning process is particularly exclusionary (Ravindran, 2000).

The disjunction of the plan from reality can be read from the fact that rapid expansion of the city has continued with scant regards for the neat zones. This has been done through a series of informal and illegal means including bribes and political pressure by a host of citizens from the urban poor trying to secure basic service, the middle classes expanding their houses, contractors and businessmen of various scales etc. (Baviskar, 2003a; Benjamin, 2008; Bhan, 2009; Ghertner, 2010; Roy, 2010; Tawa Lama-Rewal, 2007). Baviskar (2003a) makes the point that the building of a planned, orderly city envisaged in the master plan necessitated the labour of poor workers for whom no basic provisions had been planned. The growth of 'illegal' unplanned housing thus mirrored the construction of planned Delhi, not in spite of the master plan but because of it. Most of Delhi's working class therefore exists in this shadow area of illegality and insecurity in contravention to the plans (Bhan, 2009; Ghertner, 2011a; Roy, 2010).

63 A subject on which state governments can make rules and not the central government

64 The NDMC was instituted in 1913 when the new capital was being built and is only responsible for 'Lutyen's Delhi' in the administrative center of the city. The MCD is responsible for 8 of the 9 districts in Delhi. Apart from these the Delhi Cantonment Board is responsible in the cantonment area

The master plan is a legally bound document as it has been notified in a gazette. However, the Master Plan for Delhi (MPD) 2001 was amended more than two hundred and twenty times and openly flouted on several occasions (Ravindran, 2000). The agencies of the government, including the DDA, have also often violated the plans or ignored the violation of plans in several cases. Many large, emblematic projects have been carried out by the government, against the provisions of the plan including the building of flyovers, stadiums, housing complexes etc. for the Asian Games in 1982 and again for the Commonwealth Games in 2010, the large scale constructions on the Yamuna riverbed including the Akshardham Temple, two large metro depots and the Commonwealth Games Village (Baviskar, 2003a, 2011; Follmann, 2014; Srivastava, 2009). A particular case of blatant disregard of the plan by the DDA is discussed in detail in chapter 7, in the case of the Mahipalpur Ridge where the construction of a hotel complex and later a series of high end malls were proposed and supported by the authority. The violation of the plans is also evident in the presence of informal settlements of various classes of society on the Ridge, despite its designation as protected urban green space. The DDA has also been criticised for developing recreational parks and landscaped gardens in the Ridge, which is not permissible under forest laws (Kothari & Rao, 1997; Srishti, 1994).

The DDA is seen as a "hydra headed monster, involved in every aspect of Delhi's development and (is) considered, justly or otherwise, responsible for most, if not all, of Delhi's ills" (Sengupta, 2007, p. 52). A monopoly on vast land resources and it's custodianship of the master plan, makes it a key player in land based politics in Delhi. So much so, that despite many complaints and differences with the DDA over the way it was handling the Ridge, the committee set up to propose an alternate management pattern for the Ridge in 1993, suggested that the DDA should be central to its management given its resources (both legal and administrative as well as financial) and given the fact that no landscape can be managed as an integral part of the city without the participation of this central land managing body (Lovraj Committee, 1993).

All three master plans have included considerations of protecting and expanding of urban green space. The first master plan, apart from stressing the need for local parks in each district, laid down a green belt[65] around the city to control urban sprawl. However, by 1994 already, 34 per cent of the district parks had been converted by the DDA into commercial real estate, and the 1.6 km wide green belt circling Delhi was only on paper (Roy, 1994). The logic of the liberal economic climate and its demands for land as real estate on one hand, and the need for land for shelter and services to the poor not provided for by the planning body on the other, would have suggested a grim picture for the remaining forest cover. However, after a series of conflicts and negotiations described in the following chap-

65 The green belt consists 11,000 hectares of peripheral village land. Certain uses have been specified for these areas which are classified as 'low intensity development' by the MPD 2021. Most of these uses are to do with agriculture, animal husbandry, civic amenities like police stations and fire stations and forest lands including sanctuaries and biodiversity parks

ters, the conservation discourse was institutionalised in the Ridge under the DDA and has taken the form of a heavily guarded, scientifically managed biodiversity park in one area and a reforestation project undertaken with a volunteer organisation in another. A reading of the master plans shows this shift in stance from the recreational to ecological discourse regarding the Ridge.

5.2.1 The Ridge as a Plan Entity in the Master Plans (MPD 1962, 2001, 2021)

The Interim General Plan was prepared by the town planning organisation in 1956 as a precursor to the master plan. The Ridge has been a distinct planning entity right from this stage. This plan takes the Ridge into account as the dominating physical feature of Delhi (Town Planning Organisation, 1956, p. 7). In the section on recreation and open spaces it declares that the Ridge is to be maintained as a natural area with a few recreational and picnic spots. While listing the 'unused' spaces available for development[66], it is emphasised that the Ridge should not be developed but remain open to the public. It is clear, that the plan sees the forest as an important green space in the city where the only active use by citizens that the planners envisaged, is recreational (the only example of how it is used by citizens is that of a morning stroll (ibid., p. 24). It is in this plan that the term 'Regional Park' (ibid., p. 13) first finds mention, which will be a recurrent phrase in later master plans, though its implications remain unspecified. The plan specifies that its view of the value of the Ridge is that it is the last remaining "breathing space left in the heart of the city" in the sense that it provides recreational relief and it should be maintained as such permanently with no further encroachment of construction (ibid., p. 22). In the various master plans this view has been repeated in part or in whole.

In the first master plan prepared by the DDA, the MPD 1962 (notified on 1 September 1962), a clear stance of Ridge areas as a site for recreational park development has been taken. The plan clearly states that the Ridge and Yamuna riverfront are to be actively developed for recreational use. While it provides the example of Budhdha Jayanti Park[67] as the ideal aim for some parts, it calls for large areas to be left in the 'natural state' albeit with undergrowth removed for hiking. It is recommended that after the Northern Ridge is 'developed' as a park, the Southern Ridge could follow suit and be built into a Regional Park. It goes as far as to envision the area of the Ridge to become a "central public park in Delhi, comparable to Hyde Park in London, Bois-de-Boulogne in Paris or Central Park in New York" (DDA, 1962, p.34). Clearly recreation was seen as the valid use of this space, no sense of conservation through restriction or prohibition of public access seems to be present in this phase.

The second master plan MPD 2001 (notified on 1 August 1990), displays a change in discourse from the 'Ridge as a park for recreation' to the 'Ridge as a

66 Here the term development can be seen as synonymous with creation of built spaces
67 A large, landscaped park in the Central Ridge

site of ecological balance' and 'Ridge as a space for conservation of the ecosystem'. The Plan states in its preamble-

> "Ecological balance to be maintained. Delhi has two distinct natural features- The Ridge which is the rocky outcrop of the Aravalli hills and the Yamuna. Some parts of the Ridge have been erased in the central city area. No further infringement of the Ridge is to be permitted; it should be maintained in the pristine glory" (DDA, 1990, p. 3).

It is further prescribed in the section on environment and natural features:

> "Conservation of major natural features in a settlement is of utmost importance to sustain the natural ecosystem [...]. The Ridge in Delhi is defined as a rocky outcrop of the Aravalli range stretching from the University in the North of the Union territory boundary to the South and beyond. The Ridge thus identified should be conserved with the utmost care and should be afforested with indigenous species with minimum artificial landscape" (DDA, 1990, p. 53).

This is a clear shift from the previous plan in which conservation finds no mention and artificial landscape was encouraged. This change in stance could be a reflection of the interaction of the DDA with environmentalists, courts and the Forest Department through the 1980s and 1990s that brought the conservation discourse into the debate in opposition to the DDA's park building tendencies[68]. The master plan shows the responsiveness of the planning process to the discursive interaction of the DDA with other state and non-state actors.

The third and currently applicable master plan for Delhi, MPD 2021, sets out a vision for a "global metropolis and a world class city, where all people would be engaged in productive work with a better quality of life living in a sustainable environment" (DDA, 2007, p. 2). It reiterates the importance of the Ridge as one of Delhi's two natural features. The idea of the Regional Park is re-introduced here; the term is in continuity with the MPD 1962 but is clearly different in thrust. In the first plan the focus was on the word 'park'; public use and landscaped spaces, now the focus is on the word 'regional'. It is stated that the importance of conserving and rejuvenating these ecosystems (of the Yamuna and the Ridge) have regional bearing and therefore surrounding states must also contribute. All four sections of the Ridge declared reserved in the notifications of 1994[69] and 1996[70], namely the areas designated as Northern, Central, South-Central and Southern Ridges have been earmarked for the Regional Park. How this area shall be governed as regional and incur inter-state cooperation is unclear as all these sections lie within the territory of Delhi.

The plan continues the strong shift in vocabulary towards conservation seen in the MPD 2001. In the context of recreational green spaces in Delhi, it is stated that "some area shall be developed in the form of formal parks for the community and the rest shall be developed as woodlands and incidental greens to balance the environment" (DDA, 2007, p. 95). The Regional Park has been kept outside the category of recreational greens as it had been legally notified as Protected or Re-

68 These interactions are detailed in the following two chapters
69 Notification No. F.10 (42)–1/PA/DCF/93/2012–17(1)
70 Notification No. F.1(29)/PA/DC/95

served Forests by the time this plan was notified. However, though the term 'conservation' has been used liberally in context of the Ridge and the Yamuna, it is clear that the idea of conservation in the plan is different from the one called for by activists and NGOs and recommended the Lovraj Committee. The permitted uses for the Regional Park include orchards, shooting ranges, picnic huts etc. Though the language of the plans has shifted to a conservational stance, there are clear continuities with the in the previous plans regarding the landscaping and recreational aspirations the DDA holds for the Ridge. Within the DDA, it is the horticulture and landscape departments that are responsible for the Ridge. Their experience, expertise and inclination lies in, as the nomenclature suggests, creating landscaped horticultural parks.

In practice however, the DDA now forwards the model of the biodiversity park and the re-forested Sanjay Van as its showcase models for conservation, both of which oppose horticultural activity in the Ridge and forward scientific conservation and restoration for intrinsic rather than instrumental value of the forest, as their aims. This divergence between stated aims and practice has seen a long negotiation, detailed in chapters 6 and 7 and has come about through the incorporation of a variety of actors into the management of the urban forests like scientific experts from the University of Delhi, the Forest Department, NGOs, the intervention of the Lt. Governor and the even the Prime Minister.

5.3 THE FOREST DEPARTMENT

The Forest Department in India was raised for the purpose of enhancing and rationalising commercial timber production from Indian forests in the nineteenth century (Gadgil & Guha, 1992; Rangan, 1997; Rangarajan, 2001; Saberwal, 2003). A German botanist, Dietrich Brandis, was appointed the first Inspector General of Forests in India in 1856 to introduce principles of scientific forestry practices developed in Europe (Sivaramakrishnan, 1999). Following the beginning of demarcation and characterisation of forests under him, the department was raised in 1864 and in the following year the first forest laws were promulgated aimed at establishing state control over all forest lands that were not clearly privately owned (Rangan, 1997; Sivaramakrishnan, 1999). The formation of a legal and administrative apparatus applicable to vast swathes of forests was a watershed moment in environmental governance in the country, making the state the largest proprietor of forests in India (Agarwal, 1997; Gadgil & Guha, 1992). The aim was to enable long term, systematic exploitation of timber by the colonial government and to control access to prevent a shortage of timber due to local use (Gadgil & Guha, 1992; Rangan, 1997).

After independence from colonial rule in 1947, forests continued to be seen as resources for exploitation; strategic needs of the colonial government were replaced by commercial, industrial and agricultural development demands of the newly independent country and the Forest Department continued to function as an arm of the state that facilitated the use of forests in the name of national interest

(Gadgil & Guha, 1992). In the 1970s and 1980s, a shift occurred in the forest laws in favour of wildlife conservation under Prime Minister Indira Gandhi. This was linked to the rising global concern for environmental issues on one hand and the evidently growing levels of deforestation in India on the other (Rangarajan, 2009). In this phase, the legal control of the central government over forest lands was increased as state governments were seen as abducted by vested interests (Rangarajan, 2001). Any permission for de-notification of Reserved Forests or utilisation of forest land for non-forest uses like building of dams, roads etc. had to be obtained from the central government. This reduced further, the chance of local communities to bargain for access and use possibilities, but led to an increase in forest area and wildlife populations (Rangarajan, 2003; Saberwal, 2003). In 1998, a more inclusive national policy on forests was attempted that aimed at incorporating forest dwellers as stakeholders. While the commercial rationale of timber production is no longer applicable, the command and control style of conservation that is based on the policing function of the Forest Department remains strong (Rangan, 1997; Rangarajan, 2001; Saberwal, 2003).

The role of the Forest Department has historically clashed with other state agencies with opposing interests. The pre-independence Forest Department had been in conflict with the Revenue Department all over India, since the mandate of the former was to preserve the resources that the latter aimed at utilising (Guha, 1999; Rangan, 1997; Rangarajan, 1996; Saberwal, 2003). Different departments were faced with varied legal frameworks, environmental concerns and ecological visions. While agencies like the Public Works Department supported the aim of the Forest Department in sustained scientific extraction of timber for infrastructure projects, the Revenue Department saw value in levying taxes and fees on extraction of forest produce and conversion of forests to agricultural land for a steady stream of revenue (Rangan, 1997). The establishment of different kinds of forests in India (ranging from reserved and protected to village forests) is a reflection of these competing interests (Rangan, 1997). Though studies documenting this particular conflict are mostly historical studies from the colonial era, this divergence of mandate and the rationale for it, is applicable even today (Saberwal, 2003). In Delhi, the modern planning mandate and vision of the DDA that sees land a resource for generating revenue through development of real estate or as a space for horticultural practices of creating urban green spaces, clashes directly with the mandate of the Forest Department, with its colonial management roots, to create and protect forested land, preferably with little human interference. This is especially true in areas where the Wildlife Protection Act applies, as the act follows the American ideal of nature reserves (Baviskar, 2003b) where the landscape can be preserved from the effects of human impact by "legislatively creating a bounded space for nature controlled by a centralised bureaucratic authority" (Neumann, 1998).

When such clashes between agencies occur, the position of the Ministry of Environment and Forests and the Forest Department has remained weak within the governance structure as compared to other agencies focussing on developmental or commercial activity, since it seeks to monopolise control over a resource that various sections of the electoral base may want to utilise for different purposes (Kothari et al., 1995; Saberwal, 2003; Williams & Mawdsley, 2006b). Until 1985, the Forest Department was under the Ministry of Food and Agriculture, revealing the thrust on revenue generation rather than conservation. In 1985, a separate Ministry of Environment and Forests was established at the central level, but its actual powers shrunk as forests were now a concurrent subject[71] (on which both state and central government can make laws) and the weak coalition governments in the centre could do little to prevent states from de-notifying and converting forest lands (Rangarajan, 2001; Saberwal, 2003). It is documented that the Indian Board for Wildlife did not meet even once between 1989-1997 showing the recession of political backing and concern for conservation which had been strong under Indira Gandhi's government (Rangarajan, 2001). During the 1990s, in the new liberalised economic climate, commercial and industrial interests worked through the stated to gain access to forest resources making significant forays into conservation territories against the law (Saberwal, 2003) resulting in the opening of mines and changed land use in many areas (Rangarajan, 2003; Sahgal & Thapar, 1996). The weakening of the conservation agenda on the national level, however, was taking place at the same time as the push for conservation in the Delhi Ridge by local activists. This may have to do with the fact that the most consistent constituency for the conservation at the national level, has come from the urban middle classes (Guha & Martinez-Alier, 1997; Rangarajan, 2003) and the general loss of forests and attendant local ecological conflicts in the rural areas in the period, drew students and middle class environmentalists to take notice[72].

Forest laws often leave great discretionary room for varying management strategies and practices leaving space for foresters to negotiate the regulatory regime in any protected area (Baviskar, 2003b; Mawdsley, Mehra, & Beazley, 2009). Thus the Forest Department and other local actors must negotiate the areas that can be accessed, who can access and for what and the department must decide where to focus its efforts for conservation (Robbins, 2003a). Historically as in the present day, the fixing of conservation territories has been dependent on a variety of factors including the political clout of local communities and the larger conflicts surrounding the land under question (Gadgil, 1991). The working of the Forest Department, therefore, needs to be studied in context. In Delhi, this context is provided by the interaction with other state agencies like the DDA, enforcement and directions by the court, pressures from environmentalist citizens (often routed through the court) and local users (routed through their political representatives and other state agencies).

The Department of Forests and Wildlife in Delhi[73] is a peculiarity in an urban setting. The greening of urban environments has traditionally been the job of arboriculture and horticulture departments since colonial times and not of foresters (Krishen, 2010; Mann & Sehrawat, 2009). There was no Forest Department in

71 By the 42nd amendment to the constitution in 1976

72 See chapter 6 and 7

73 Referred to simply as the Forest Department

Delhi initially as there was no commercial forestry in the city, the reforestation of the Ridge in colonial times was undertaken for sanitary, aesthetic and health concerns of the colonial settlers by District Officers (Mann & Sehrawat, 2009). A forestry wing formed under the Central Public Works Department was shifted in 1985 to the Delhi Administration under the Development Commissioner (Sinha, 2014). A full-fledged department with senior forestry officials was only instated in Delhi in 1996 (Dhawale, 2010), symbolising the gaining of ground of the conservation discourse and the institutionalisation of the separation of forests from the broader category of urban recreational green space.

The responsibilities of the Forest Department in Delhi include development of forests (afforestation and protection of forest lands), protection of wildlife and their habitat and prevention of loss of tree cover through implementation of the Delhi Preservation of Trees Act (DPTA) of 1994, the implementation of forest laws in notified forests and the implementation of directives of the Supreme Court to the department[74]. Certain imperatives from the national scale also have to be implemented by the department in Delhi. The National Forest Policy 1988[75] directs all states to achieve a minimum of 33 per cent of the total land area under forests. This is the responsibility of the Forest Department. At the city level, the Delhi Preservation of Trees Act charges the department for planting ten trees to replace every tree that it gives permission to cut in the city (where the applicant cannot carry out the plantation himself and can pay the cost of plantation to the Forest Department) . This has had material implications for Delhi's landscape, the need to increase green cover provided impetus for large scale plantation of Prosopis Juliflora (see Box 2), since it has high survival rates and low demand for maintenance (Sinha, 2014). Secondly, the Forest Department has developed forty two 'city forests' (the term has no legal definition) on 1,443 hectares[76] spread across the city, in order to enhance green cover and meet part of the DPTA obligations (Sinha, 2014).

74 Website of the Department of Forests and Wildlife, NCT, Delhi: (accessed 23 December 2014) http://www.delhi.gov.in/wps/wcm/connect/DOIT_Forest/forest/home/rti/particulars+of+organization

75Policy No. 3–1/86–FP, Ministry of Environment and Forests

76 This is vacant village council land and land under other city authorities; city forests are not covered by forest laws and are open to public recreational use

Box 2
PROSOPIS JULIFLORA

Delhi has about 252 species of trees, a majority of which are naturalised (Krishen, 2006). Only six species native to the Ridge continue to occur, four are extinct and one (Dhau) is extremely rare (Krishen, 2006). Parts of the Aravalli in the adjoining state of Rajasthan and a preserved sacred grove near Delhi in the state of Haryana (Mangar Bani) provide examples of what the original vegetation of the Ridge would have been (Krishen, 2006; Singh, 2006).

Many urban ecosystems are characterised by exotic species where native species cannot cope with dramatic environmental changes (Dearborn & Kark, 2010). In the Delhi Ridge, Prosopis juliflora, locally known as vilayati (foreign) kikar or vilayati babul, has become the dominant tree species (Ekta, 2014). The spread of this tree has to do both with its ability to survive in harsh conditions and propagate rapidly, as well as with institutional practices leading to the plantation of this species in vast areas of semi-arid regions. Originally from the Americas, plantation of juliflora was encouraged by international institutions like the World Bank as well as the Indian government to counter firewood shortages, increase green cover and avert desertification in dryer states (Gold, 1999; Robbins, 2001). The Forest Department has actively propagated this species to increase green cover in semi-arid states such as Delhi, Gujarat, Madhya Pradesh and Rajasthan since it grows quickly, requires little input and the wide canopy of the tree can be seen in GIS images used to prepare green cover statistics (Robbins, 2001, 2003a). Local populations associate its spread with government activity to such an extent that its it known as 'sarkari babul' (government babul) in some parts of Rajasthan (Robbins, 2001).

In Delhi, there are accounts of juliflora being planted along with other native species during the greening activity in preparation for the shift of the colonial capital (Parker, 1919). Large scale plantation of this species to increase green cover took place in the 1970s (Dhawale, 2010). The tree has become invasive and is found to reduce biodiversity as it competes with other species for sparse moisture (Hocking, 1993). Moreover, its leaf litter contains water soluble chemicals that prohibit the germination of local species including grasses (Noor, Salam, & Khan, 1995). It also negatively effects bird (Chandrasekaran et al.,2104) and herbivore populations (Abebe, 1994). While juliflora does provide a high yield of fuelwood, the leaves and shoots of the tree are not browsed by wild or domestic herbivores; this is one of the reasons it grows so well (Abebe, 1994; Gold, 1999). The pods of the tree, however, are eaten by grazing animals that then spread the seed through droppings (Abebe, 1994; Mwangi & Swallow, 2005). Once established, juliflora is extremely hard to remove due to its deep roots which enable the tree to regrow even when cut down to the stump (Robbins, 2001).

Due to its effects on biodiversity, the tree has lost favour with key government agencies responsible for afforestation activities in Delhi. With permission from the Delhi High Court in 2009, attempts to clear juliflora began in the Aravalli Biodiversity Park (Mukherjee & Deenadayalan, 2011; Sriram, 2011). While the tree is not being uprooted in Sanjay Van or the sanctuary, it is no longer planted by any state agency in the Ridge and all afforestation activity is focussed on attempts to regenerate native tree cover.

Administratively, this has made it necessary for the Forest Department to work with other land owning agencies in the city including the DDA, municipal bodies, the Public Works Department, the Airport Authority and the Cantonment Board, to encourage and facilitate plantations on land controlled by them[77].

Despite a rather hefty responsibility of achieving a high green cover in a dense urban area, the department suffers severe staff shortages. It has been pointed out that part of the local rootedness of the Forest Department comes from the fact that lower level functionaries are often local residents, implicated in the socio-political structures of the area (Baviskar, 2003b; Robbins, 2000; Saberwal, 2003), this can provide hindrances in implementing policies that face opposition locally, but affords an interaction of the Forest Department with local demands. In Delhi, there are many higher level forest officers but very few ground level functionaries. Of the three hundred and nineteen field level staff recommended for the area under the department, only seventeen were granted by the Delhi Government (Jain, 2014; Lalchandani, 2013a). Moreover, basic equipment is scarce, no uniforms have been issued since 1990, wireless equipment is out of order and few official vehicles are issued. As a result there is almost no monitoring on the ground (Jain, 2014; Sinha, 2014). As one senior forest official puts it, the department is an army of generals with no soldiers to implement policy[78]. While the Forest Department can be a fairly influential state agency on the ground in rural conservation areas (Gold, 1999; Robbins, 2000), in Delhi the pressures of demands for land for urban development and the presence of powerful local agencies like the DDA, have led the head of the department to comment in writing upon the "cumbersome and at times distressing and annoying nature of duty under the Delhi Government compared to my previous assignments" (Sinha, 2014, p. ix).

The tensions between the Forest Department and other state agencies, especially the DDA, were openly discussed in a recent publication released by the department[79], which places the blame for conversion of large parts of the Ridge into built-up area, upon the lack of forest conservation knowledge and lack of will for conservation in other government bodies in the city (Sinha, 2014). The treatment of the Ridge as a Regional Park has also been criticised as it implies "an area of land preserved on account of its natural beauty, historical interest, recreational use, other reasons and under the local government" while forest laws applicable to the Ridge would imply control by the Forest Department for forest conservation rather than recreation (Sinha, 2014, p. 44). It is therefore suggested in the document that the Ridge be mentioned in a separate section of the master plan and not as a park.

77 The targets for plantations by each government body and some citizen groups (such as students, market associations, Resident welfare Associations etc.) are listed in the Greening Delhi Action Plans released by the Forest Department (See for example Greening Delhi Action Plan 2007–8, 2007)

78 Interview FD:1

79 Introduction to the Delhi Ridge 2014, G.N. Sinha (ed.)

Though they both control different parts of the Ridge, the jurisdiction of the DDA and Forest Department in the Ridge is overlapping in two ways; Firstly since parts of the Ridge are notified as Reserved Forest or Protected Forests, they would be under the jurisdiction of the Forest Department as per provisions of the forest laws. However, given the lack of resources available to the Forest Department in Delhi and the ownership of large parts of land in the Ridge by the DDA, the Forest Department has so far only actively managed the Southern and parts of the Central Ridge[80] while the Northern and South Central Ridge are managed by the DDA. Secondly, the DPTA gives jurisdiction over the trees in Delhi, including those standing on the parts of the Ridge managed by the DDA, to the Forest Department. These overlaps have resulted in some friction. In the deliberation of the committee set up to oversee a management plan for the Ridge, the Vice Chairman of the DDA expressed the view that the proper management of the Ridge would be facilitated by declaring the entire Ridge as 'development area' under the Delhi Development Act, 1957 (Lovraj Committee, 1993). The definition of development under the act entails

> "carrying out of building, engineering, mining or other operations in, on, over or under land or the making of any material change in any building or land including redevelopment" (*Delhi Development Act*, 1957, p. 1).

The demand to declare the Ridge as development area demonstrates the DDA's intention of using Ridge land for uses that would not be allowed under forest laws[81].

The conflict between these agencies on the boundaries of jurisdiction and the permitted land use therein, has been taken to the High Court of Delhi and Supreme Court of India. The DDA restricts the definition of the Ridge to the Regional Park in the MPD 2021 (coinciding with the area under 1994 and 1996 notifications). The Forest Department, on the other hand, takes a landscape view of the Ridge by including all areas notified under forest laws in Delhi as well as areas that the Revenue Department classifies as 'gair mumkin pahad'[82] (hilly areas where revenue generating land use is not possible) and the morphological Ridge area as identified and mapped by the Geological Survey of India in 2006 under a seismic zonation exercise (see Map 3). The morphological Ridge has been identified as "that part of the Ridge not falling under any notification, but having similar features as the Ridge and forming a part of the extension of the Aravallis" (Sinha, 2014, p. 14).

80 The Central Ridge is formally under the ownership to the Land Development Office but has been handed over for management to the Forest Department

81 It was noted by the committee that this would be the anti-thesis of conservation of the Ridge and they in turn suggested the introduction of the term 'Forest Area' as opposed to 'Development Area' where the only power of the DDA would be to remove constructions already existing (Lovraj Committee, 1993). Neither of these classifications applies presently

82 These are largely dry rocky areas and therefore in Delhi, mostly lie in the Ridge area

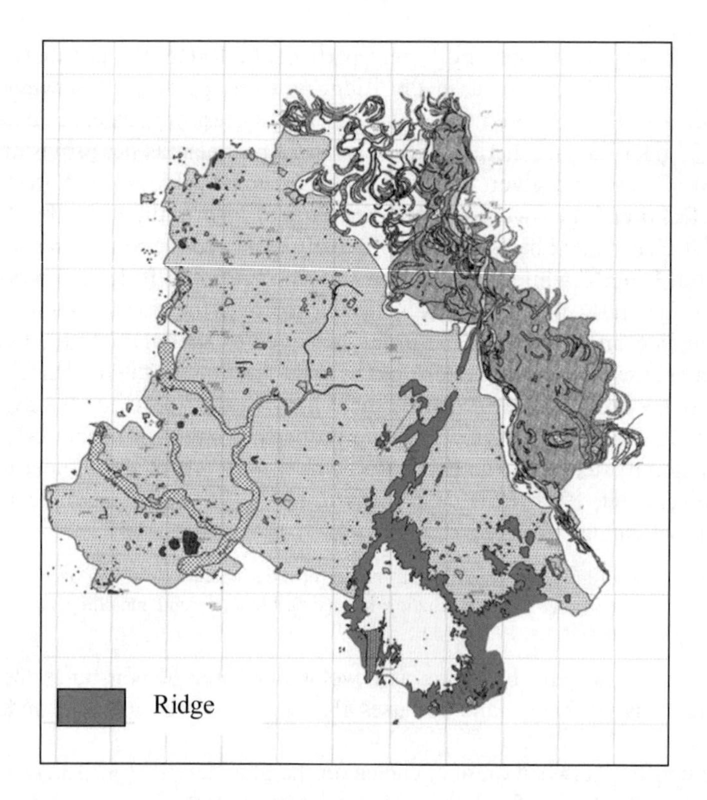

Map 3: Geomorphological Ridge
(Source, Geological Survey of India, 2006, Sheet No. 60)

The stand to extend legal protection to the morphological Ridge was taken by the Forest Department in the case of DDA vs. M/S Kenneth Builders Pvt. Limited[83] in 2010. The case shows a clear friction between different agencies in handling matters of urban development. The DDA gave a tender to Kenneth Builders to construct residential units in Okhla (in South Delhi). The Ministry of Environment & Forests, Delhi (through the Forest Department) and the Delhi Pollution Control Committee refused clearance for the project on the grounds that it was in the Ridge area. In this case, the High court ruled in favour of the real estate company by citing that the said area was not under forest use in the MPD. However, the Forest Department filed an affidavit in the Supreme Court in March 2011, leading to a decision that any construction in the morphological Ridge will need prior clearance from the Supreme Court[84].

83 DDA vs. M/S Kenneth Builders Pvt. Limited, SLP (Civil) no. 35374/2010

84 Another case Ashok Tanwar vs. Union of India (Writ petition (Civil) 3339/2011) reaffirmed that any construction on the morphological Ridge would need permission from the Ridge management board or the Supreme Court

5.4 THE SUPREME COURT OF INDIA

The Supreme Court of India has played an increasingly visible role in forest governance and management in India since 1995 with its extended engagement[85] in the case of T.N. Godavaram Thirumulpad vs. Union of India and others[86] (henceforth, the Godavaram case). While the initial petition was against a case of illegal timber felling in the Nilgiri Mountains, the court combined this issue with several other petitions to deliberate on the scope and implementation of forest laws in India. The court justified this on the basis of the "critical state of forest cover and the non-responsiveness of the governments concerned" (Rosencranz & Lele, 2008, p. 12). In doing so, the court has taken on the task of policy maker as well as administrator of forest related issues rather than restricting itself to its role as interpreter of law (Rosencranz et al., 2007; Rosencranz & Lele, 2008; Sharma, 2008). The court re-interpreted the meaning of the term 'forest' under the Forest Conservation Act of 1980[87] to apply to all forests according to the dictionary meaning of the term, regardless of legal status or ownership[88] (before 1996, this Act applied only to notified Protected or Reserved Forests). This was done keeping in view that several areas that were forested were not notified (Lele, 2007). This widening of the definition helped overcome messy and incomplete records dealing with land classification in most states and sought to extend the implementation of forest laws to all physically existing forests (Rosencranz & Lele, 2008). However, this universal definition has been pointed out as problematic as it does not take into view the existing uses and diversity in forest lands, negating both the ecological as well as the social diversity of forests in India and as such, this definition has been hard to implement (Lele, 2007; Sharma, 2008). Secondly, the need to consult the central government for all areas so defined empowers the central government against the states in matters pertaining to large areas of land (Rosencranz et al., 2007). An immediate downside was seen when encouraged by the stand of the court, on 03 May 2002, the central Ministry of Environment and Forest (MoEF) ordered eviction of all encroachments on forest land within five months. Given the scale and complexity of the issue, this was an impractical deadline and many state governments reacted hurriedly, resulting in atrocities that led to a backlash from tribal communities and civil society (Sharma, 2008; Thakuria, Joshi, & Barik, 2003).

To deal with the spiralling management and administrative work resulting from these cases, the court created a Central Expert Committee (CEC)[89]. The CEC

85 Under 'continuing mandamus', whereby the court, continues to pass orders and directions with a view to monitor the functioning of the executive (Rosencranz et al., 2007; Sharma, 2008)

86 W P (Civil) no. 202 of 95

87 Section 2 of the act decrees that any conversion of forest land must be approved by the central government i.e. from the Ministry of Environment and Forests

88 T.N. Godavaram vs. Thirumulpad vs. Union of India and others Order dated 12 December 1996

89 Under section 3 (3) of Environment Protection Act 1986. Notification, No. 1–1/CEC/2002

is a 'quasi-executive' body (Rosencranz & Lele, 2008, p. 13) with wide ranging powers to investigate cases, demand implementation of laws and direct state agencies in matters of forest governance, with the prospect of attracting penalty for 'contempt of court' for non-compliance. Actions of the state or central government can be challenged before the CEC where they pertain to forests[90]. The committee is responsible only to the Supreme Court and is therefore insulated from pressures from the executive as well as voting populations. The CEC is composed of three MoEF officials (separated from their duties in the ministry during their tenure at CEC) and two civil society members from environmental NGOs. It has been noted that the members of the body are inclined towards wildlife protection and tend to ignore social justice issues linked to forest governance (Rosencranz et al., 2007; Thakuria et al., 2003). Further, through the CEC, the Supreme Court is "micro managing" issues that would have been dealt with by government agencies (Rosencranz et al., 2007, p. 10033).

The CEC was approached by environmentalists against the building of malls in the Mahipalpur Ridge[91] as it was seen as a body sympathetic towards conservation. In response to which the committee, without hearing the residents of informal settlements in the area, recommended turning the parcel of land into a national park (CEC, 2004) resulting in the eviction of these settlements. The pattern noticed in the results of CEC decisions in side-lining social justice demands in the interest of wildlife and forest conservation (Rosencranz & Lele, 2008; Thakuria et al., 2003) held true in this case as well[92].

The frictions caused by the wide-ranging exercise of power by the Supreme Court, infringing on the role of the legislative and executive, resulted in the MoEF tabling a bill in Parliament in 2007 calling for a separate environmental tribunal and abolition of the CEC (Mahapatra, 2007; EPW, 2007; Venkatesan, 2007). The bill was passed in 2010 as the National Green Tribunal (NGT) Act and resulted in the setting up of tribunals in five cities including a principal bench in Delhi. The cause cited for the need of a tribunal for environmental cases was to expedite judgement and rectify the lack of technical expertise required for environmental litigation[93]. The tribunal consists of judicial members and expert members (a minimum of 10 and a maximum of 40 of each)[94]. The Chairman is selected by the MoEF and all members are to be appointed in consultation with the Chief Justice of the Supreme Court, though his recommendations are not binding on the government[95]. The tribunal's jurisdiction extends over "all civil cases where a substantial question related to environment is involved"[96] though the act does not clarify what constitutes such a substantial question. The tribunal can also demand

90 Notification No. 1–1/CEC/2002
91 Application to CEC No. 331
92 See chapter 7, section 7.1.2
93 National Green Tribunal Act, 18 October 2010
94 National Green Tribunal Act, Chapter II, Section 4 (1)
95 National Green Tribunal Act, Section Chapter II Section 4 (2)
96 National Green Tribunal Act, Section 14 (1)

compliance of government bodies including penalising heads of departments for non-compliance[97].

The NGT, appointed by the central government and is seen as a counter to the Supreme Court's CEC which also continues to exist and which is viewed by the government as a body heavily biased towards wildlife conservation while blindsiding social justice issues (EPW, 2007). The NGT has been pressuring the Forest Department to finalise the boundaries of the Southern Ridge, settle existing claims where valid, and demolish defaulting structures including estates of the elite (Jain, 2014; Singh, 2013b)[98].

5.5 CONCLUSION

This chapter argued that the state works within broader socio-political and economic structures and is characterised by multiple visions and aims. The state agencies often have competing agendas and try to push very different policies based in their different mandates and resources. Thus, the state consists of multiple actors rather than being an actor unto itself.

While this internal incoherence leads to contradictory laws and plans and at times conflicts between state agencies, it can also provide space for dynamism. The variety of motivations and legal and institutional frameworks, within state agencies can provide different possibilities of allegiance with groups of non-state actors who may want to forward an agenda which one or the other state agency has affinity to. The tensions within the state therefore provide space for negotiation and, in this case, results in the multiple kinds of conservation spaces existing in the Ridge, depending on the specificities of negotiations between actors.

97 National Green Tribunal Act, Section 28

98 See chapter 7 , section 7.3.1

6 CONSERVATION OF THE RIDGE ENTERS PUBLIC DISCOURSE

The issue of loss of forest cover in the Ridge first drew the attention of non-state actors in the late 1970s, followed by to a more concentrated effort in 1992-93 to prevent conversion of forest land to other uses by state agencies, private and commercial interests. The interaction between state agencies and these environmentalists resulted in the institutionalisation of the discursive framing of the Ridge as a space for conservation including important measures in favour of legal protection of large parts of the Ridge as Reserved Forests. This chapter pieces together the history of this phase of citizen and NGO engagement with the Ridge, through interviews, journal articles (Agarwal, 2010; Kothari & Rao, 1997), newspaper reports and reports produced by the civil society organisations at the core of the movement to pressure government agencies to conserve the Ridge (Kalpavriksh, 1991; Malik, 1998; Srishti, 1994).

6.1 KALPAVRIKSH AND SRISHTI

The engagement of citizens and NGOs in demanding the conservation of the Delhi Ridge began at the end of 1970s and continued in a dispersed manner through the 1980s, culminating in a more targeted movement in the early the 1990s before petering out. In the 1970s, there was a rising concern for environmental issues both at a global scale (Jasanoff & Martello, 2004; Rome, 2003; Rootes, 2003) and at the national level, beginning with the Chipko Movement in Northern India (Agrawal, 2005; Rangan, 2000; Williams & Mawdsley, 2006b). The Indian urban middle class, incorporated into global and national discourses (Gupta, 2000; Lakha, 2000) also responded to such concerns (Mawdsley, 2004; Rangarajan, 2001) as is evident from the case at hand.

The first recorded agitation around the issue of the Ridge was in the October 1979, when students organised a march in Delhi to demand that the government protect the Ridge as a 'natural heritage site' (Kothari & Rao, 1997, p. 123) in response to proposed construction of a school on Ridge land (Kalpavriksh, 1991). This particular protest provided a rallying point for students concerned with conversion and degradation of forests in Delhi (Kalpavriksh, 1991). In order to convince and pressure government authorities to prevent tree felling and construction in the Ridge, lectures, slide shows, bird watching trips and nature walks were used to enrol citizens into the conservation discourse (Kothari & Rao, 1997).

An organisation called 'Kalpavriksh' was formed in 1979 to organise these students interested in nature conservation. As an effect of the active lobbying and demonstrations, the Lt. Governor issued a notification declaring 20 parcels of land in the North and Central Ridge as Protected Forests. This move was later criticised as adding to the list of ineffective 'paper promises' (Kalpavriksh, 1991, p. 27) since it made no effective change in the way these areas were being administered. Instead, it further complicated the messy situation since some of the areas had earlier been Reserved Forests and previous notifications had not been cancelled (Kalpavriksh, 1991; Srishti, 1994). One of the founding members of the organisation recalls:

> "Kalpavriksh got set up with us. We used to go birdwatching; we found some (tree) felling going on. We tried to organise some marches but initially they failed. In (19)79 we got together with a lot of the people in new Rajender Nagar near an area called Pamposh, [name of politician] got permission to build a school, there [...]. There was a huge protest[99]. Then in 1980, April when Mrs. Gandhi came to power, we went and met her[100] and she got it declared. Then nothing much happened after that, couple of places we protested. Kushaknagar by protesting we saved in September 1980, now it's all gone, that's a different matter. Then things died down"[101].

Several of the students who actively engaged with the conservation of the Ridge, were from the University of Delhi which is situated in the Northern Ridge[102]. The involvement of some of the core members of this phase began through recreational activities in the Ridge including birdwatching and nature walks[103] and was initially restricted to opposing tree felling and reduction of habitat for birds and animals (Baviskar, 2010). Members of Kalpavriksh eventually began to engage in environmental issues in various parts of the country including the movement against the Narmada Dam (Kalpavriksh, 1991; Srishti, 1994)[104]. This led to a broadening of the discourse of Kalpavriksh from one that was focussed on conservation and biodiversity issues to including issues of social justice linked to conservation such that, by 1991, the organisation describes itself as one that

> "sees environmental problems as emanating from unequal social structures and believes that a country can develop meaningfully only if ecological sustainability and social equity are guaranteed" (Kalpavriksh, 1991, n.p.)

Another student organisation called Srishti was founded in 1987, and undertook, on one hand, bird counts and nature walks in the Ridge and on the other hand (what the founder refers to as the activist side of the organisation[105]), tried to forward the agenda of preventing the DDA from converting forest areas into built-up area or horticultural parks with ornamental species of trees and plants[106]. Unlike Kalpavriksh, this group focussed its efforts on Delhi and mostly on the Ridge[107].

99 250–300 people according to a later report by Kalpavriksh (Kalpavriksh, 1991)
100 The petition to Indira Gandhi by Kalpavriksh in 1980 to push for conservation of the Ridge is also mentioned in a journal article (Rangarajan, 2009)
101 Interview EV:9
102 Interviews EV:11 and EV:9
103 Interviews EV:11, EV:9, EV:4, EV:8 and EV:2
104 Interviews EV:8 and EV:4
105 Interview EV:11
106 Interview EV:11
107 Interviews EV:11 and EV:5

These environmentalist groups sought to bring attention to the ecosystem services the forest provided the city as motivation for conservation; the importance of the forest was shown in terms of the 'lung function' of the forest; reduction of pollution, absorption of dust that blows into Delhi from the deserts of neighbouring Rajasthan as well as other functions of temperature reduction, water retention and acting as a noise buffer. The intrinsic worth of the forests in harbouring biodiversity was a main concern as well (Agarwal, 2010; Kalpavriksh, 1991; Kothari & Rao, 1997; Srishti, 1994). Apart from this, the forests were seen as an opportunity for environmental education of urban residents as well as, a source of firewood and fodder for the poor residents in the vicinity of the Ridge (Kalpavriksh, 1991; Srishti, 1994).

These organisations met with local officials, lobbied with various authorities, sent out letters and memoranda to officials and well known citizens. Members of both organisations kept the media abreast of the activities of the movement[108] and themselves wrote articles for several major national English medium newspapers (Malik & Agarwal, 1993, 1994a, 1994b; Menon, 1993 for example). Television spots about the Ridge were made for the national television channel (Kalpavriksh, 1991; Srishti, 1994). Direct mobilisation of other citizens was also attempted by approaching eco-clubs in schools and colleges and resident welfare associations, carrying out door-to-door campaigns, conducting nature walks and workshops (Kalpavriksh, 1991; Srishti, 1994). Student groups patrolled the area to gather information about any tree felling, horticultural practices and construction activity in the Ridge and compiled reports to put together fragmentary or missing information about the administrative and legal background (what laws apply to which part of the Ridge, which government agency is responsible for the implementation of these laws) as well as basic listing of flora and fauna found in the forest[109] (Kalpavriksh, 1991; Srishti, 1994).

The basic problems identified by both groups revolved around three main points; firstly, the conversion of forest land, including protected and reserved forest land to built-up area, at times with permission from the government or by government agencies themselves. Other non-forest activities were also noted including dumping of garbage and construction rubble in forest areas as well as collection of fuel and fodder by poor residents. The second main area of concern was the conversion of forests to parks as the DDA had laid out in the master plan. It was pointed out that the removal of undergrowth and planting of ornamental species results in loss of eco-system services which a forest could provide, as well as reduces the faunal and floral diversity. Thirdly, the administrative and legal confusion surrounding the governance of the Ridge was leading to lack of fixing of responsibility and preventing of a landscape level plan, rather each agency was handling the Ridge as it saw fit, resulting in fragmentation and depletion of the forest (Kalpavriksh, 1991; Srishti, 1994) .

108 Interviews EV:11 and EV:5
109 Interviews EV:5, EV:11, EV:2 and EV:6

Kalpavriksh and Srishti (Kalpavriksh, 1991; Srishti, 1994) both demanded that the exact boundaries of the Ridge be marked, all park creation and construction on the Ridge be banned and existing constructions be reviewed to ascertain their impact on the forest and the possibility to shift them elsewhere. Further, it was suggested that a large part of the forest be protected for intrinsic purposes and a buffer area be maintained on the outskirts for recreational and educational purposes as well as for collection of fuelwood and fodder. Kalpavrisksh specified that this zoning would have to be done by government agencies in collaboration with NGOs and local residents. Thirdly, since the Forest Department was the only agency trained in forest conservation, they should be given charge of the Ridge and their staff should be upgraded in terms of strength and training to handle this task (The department was only working in a small area of the Southern Ridge at the time).

6.2 FORMATION OF THE JOINT NGO FORUM FOR THE RIDGE

In December 1992, a decision[110] was made to hand over all areas of the Ridge apart from the Southern Ridge, to the DDA for management, citing administrative confusion resulting from the multiple agencies in-charge of various parts of the Ridge. The Forest Department (which was a small office under the Development Commissioner at the time) was to have a watchdog function to ensure forest laws are followed according to the notifications. The environmental groups saw the DDA as an inappropriate body to handle the Ridge as it had demonstrated an inclination towards horticultural practices and had the mandate to undertake 'development' of the city, therefore clashing with a conservation responsibility (Agarwal, 2010; Kothari & Rao, 1997; Malik, 1998; Srishti, 1994). A former national Minister for Environment and Forest was quoted as saying this move was like handing over "chickens to the lions to look after" (Srishti, 1994, p. 45). The opposition to the Ridge being managed by the DDA catalysed a major campaign against mismanagement and encroachments in the Ridge, bringing together the initially involved student groups and professional environmental NGOs to form a Joint NGO Forum to Save the Delhi Ridge (JNGO) (Agarwal, 2010; Kothari & Rao, 1997; Srishti, 1994). A discussion paper[111] was drawn up by Srishti, and circulated to all environmental NGOs in Delhi on the basis of which a resolution was drawn to form the JNGO forum to coordinate action on conservation of the Ridge. By May 1994, 18 NGOs[112] were part of the forum. The resolution passed

110 By notification by the Lt. Governor (No. F.2 (11)/DCF/1990–91/5927–5935 dated 3.12.91)

111 Menon and Agarwal, 1993, Delhi Ridge Issues and Perspectives

112 These NGOs were: Kalpavriksh, Srishti, WWF (I), INTACH, Conservation Society of Delhi, Conservation Action Group, Development Alternatives, Development Research Action Group, All India Women's Conference., Jan Sewa Ashram, Kindness to Animals and Respect for Environment, People's Commission on Environment and Development, Science aur Kainat Society of India, Society for Conservation of Delhi Ridge and Yamuna (Srishti, 1994, pp. 51–53)

at the first meeting of the JNGO, demanded that the Ridge should not be given over to the DDA, rather the ownership and management of the entire Ridge should be handed to the Forest Department. Secondly a committee comprising of both governmental and non-governmental organisations should be set up to monitor the Ridge and all decisions regarding the area would be reviewed by it (Srishti, 1994).

The single programme of the JNGO; to conserve the Ridge in Delhi, was sought to be fulfilled through a multipronged strategy. On one, hand they lobbied with key executives of the government to influence policy in favour of long term conservation. On the other hand, they ran public campaigns to spread awareness on the Ridge through the media and through activities involving citizens like marches, nature walks, signature campaigns, street plays etc. A third side of the campaign sought to collect secondary data on the Ridge, its legal and institutional history and as well as biodiversity and topographic details. Gaps in available data were to be filled by data collection exercises by the members of the JNGO, these exercises included surveys to list the extent and type of encroachments (built-up area) in each section of the Ridge, interviews with those who use the Ridge for various purposes and documenting the state of the biodiversity as compared with previous species lists (Srishti, 1994).

6.2.1 Lovraj Committee, Recommendations and Results

Following pressure from the JNGO, a committee was formed by notification from the Lt. Governor in April 1993 to "suggest a management pattern for the Ridge"[113]. The committee, known as the Lovraj Committee after its chairman, had representatives from various government agencies (members of the DDA, Development Commissioner of the Government of NCT Delhi, Inspector General, Forests) as well as from four environmental NGOs (namely, Kalpvriksh, Srishti, WWF(I) and INTACH). It was noted that

> "despite clear intent for the last 80 years to conserve the Ridge, there has been no systematic effort to formulate and implement concerted legal, administrative, managerial and scientific measures to achieve this objective" (Lovraj Committee, 1993, p. 41).

The committee's responsibility was thus to identify and recommend this "alternative management pattern to preserve the natural habitat" of the Ridge[114]. Along with administrative structures, the recommendations were to identify means to afforest the Ridge with indigenous species and to provide a plan for monitoring the area (Lovraj Committee, 1993).

The committee held a series of meetings and consultations where detailed discussions took place on which parts of the Ridge had legal protection under forest laws, which encroachments existed, when had they come up and which of them

113 Notification No. F.2. (11)/DCF/1990–91
114 Notification No. F.2. (11)/DCF/1990–91

could be shifted (the NGOs argued that units such as petrol pumps and military camps could be easily removed but other buildings like schools and hospitals could be given some consideration to continue)[115]. Based on the negotiations between members, the committee made the following recommendations[116] (Lovraj Committee, 1993):

1. The four remaining segments of the Ridge should be declared Reserved Forest (at this point only the Northern and parts of Central Ridge were Reserved Forests. Several areas were either Protected Forests or have no legal protection). Pillars should be erected to demarcate the reserved area and settlement officers should be appointed to settle claims under IFA, 1927. If any area of the Ridge should be excluded from the Reserved Forest, a guideline should be formulated to identify it. The settlement officer should determine the legal status of the present occupants and uses of the area.
2. The legal protection of the four areas was to be further strengthened. It was recommended that the Environment (Protection) Act of 1986[117] should be invoked during the notification of Reserved Forest. Along with the power to create an overseeing body, the act empowers the government to regulate land use on private land, which was seen as a useful provision for regulating the Southern Ridge, where much area fell under private or village land. The report also looked to the future by recommending that the Wildlife (Protection) Act 1972[118] to be applied when the conservation of the Ridge had reached a level to make this relevant (when there is a higher level of wildlife). In suggesting this, the committee was drawing an example from the Borivalli National Park in Mumbai and the Van Vihar in Bhopal and hoping to declare at least some parts of the Ridge a National Park in the future. These suggestions were made in the view that the application of the Indian Forest Act had not been effective. Even though the report notes that the problem was in the non-implementation of the IFA, it sought to correct this by application of more laws rather than proper implementation of the existing laws.
3. In one of its foremost recommendations, the report calls for the formation of a committee for the management of the Ridge under the Chief Secretary's chairmanship. The committee was to be given wide ranging responsibilities including decisions on land use, legal issues and laying down means to ensure conservation in the long term. Referred to as the high level committee, this body was to include non-state actors as well as government agencies.

115 Interview EV:5

116 The recommendations are presented here grouped according to the theme

117 Environment Protection Act aims to implement the decisions of the United Nations Conference on the Human Environment of 1972. It is an "umbrella" legislation related to the protection and improvement of the human environment and the prevention of hazards to living beings and property. The act gives power to the central government to co-ordinate various agencies working under previous legislations. It enables the creation of a coordinating authority

118 This act provides varying levels of protection through penalties to species and areas identified under the act

4. The committee recommended, contrary to the demands of the NGOs, that the DDA be handed the management of the Ridge. The only exception being the Asola Sanctuary, as it was believed the management system required for a wildlife sanctuary was different; the Forest Department was to continue to be in-charge here. The report justified this on the basis that no other agency at the time had the resources or man power to take on the task. Further, the DDA had the legal power to remove encroachments and this was seen as a key requirement. Since there was no 'practical alternative' (ibid., p. 29) to the DDA, it was recommended that the DDA set up a separate division to oversee the Ridge, headed by a member who would represent DDA on the Ridge Management Supervisory Committee as member secretary. This member has to be of the minimum rank of Conservator of Forests[119] to ensure valid experience and authority and the department would be led by other members trained in forest conservation. It was also recommended that the DDA consider conservation of the Ridge while planning other uses, for example while planning parks, schools, hospitals etc., the integrity of the Ridge should be kept in mind and alternatives should be used to reduce pressure (recreational or other) on the Ridge. The uses of the Ridge for recreation were also to be restricted by DDA, this could be attempted by publicising other parks nearby. Landscaping and horticultural development of the Ridge was to be curbed.
5. Checks and balances were sought to be installed in a way by having the Ridge management supervisory committee review the plans and progress of the Ridge management division of the DDA.
6. The report also called for an amending of existing overarching plans of the city to include the Reserved Forest as a separate planning category. It demanded that the MPD 2001, be amended to show the Ridge as designated Reserved Forest as well as the Delhi Development Act of 1957 to be amended to include a 'forest area' (as opposed to 'development area', a term used in the plans for the areas where the DDA intends to actively change the landscape according to the needs of the plan). In forest areas, the powers of the DDA to remove unauthorised construction would extend, but not its powers to allow or carry out construction or park creation.
7. The participation of non-governmental organisations in a watchdog role was sought to be formalised by appointing 20 honorary wardens from the NGOs active in environmental conservation. The role of these wardens would be to process information and expertise, monitor, prevent and report any felling of trees or other non-forest activity, create public awareness for the conservation of the Ridge through meetings, discussions etc. These wardens would be appointed for two years and the tenure could be renewed based on performance. Other ways to include NGOs in the management of the Ridge, included recommendations like having the Centre for Environmental Law of WWF-India lend its expertise on the Indian Forest Act 1927.

119 A high ranking official in the Indian forest bureaucracy

8. To encourage citizen participation, it was recommended that Ridge monitoring volunteers be appointed. These would be people from resident associations, schools and colleges around the Ridge. The function of these volunteers would be regular patrolling of the forest to assist the wardens. As boundary walls are capital intensive and are breached often, there is a call for not just physical but 'social fencing' through volunteers. The government was to provide funds for monitoring and environmental awareness campaigns
9. The role of science and expertise was also pushed by the committee in various recommendations. Firstly, it was recommended that the Ridge Management Division of the DDA, once set up should seek the constant and long term engagement of ecologists. A scientific advisory committee was to be set up to guide the Ridge Division on scientific principles of conservation. The committee also picked up specific instances where the role of experts would decide the exact mode of action. Regarding the mining pits of Asola Sanctuary for example, it was suggested that wildlife experts and ecologists would determine the right reclamation methods and in particular, the Wildlife Institute of India in Dehradun should be consulted. Certain organisations like WWF-India and the National Environment Engineering Institute were recommended for mapping through remote sensing.
10. Further concrete recommendations were made, such as, construction and widening of roads to be prohibited around the Ridge, existing roads seen as damaging to the forest should be closed. Where this is not possible, a suggestion was made to convert these roads into toll roads to discourage traffic and to use the income generated for conservation (one could point out that this would restrict use of the road to those who could afford to pay the toll).

The final outcome of the negotiations between state agencies and the JNGO is reflective of compromises by various actors involved. Reflecting the stand of the NGOs, the report chastised government agencies for allowing construction in the Ridge and asserts that it is an ecologically important area for the city that should not be converted to other uses. It also backs demands to declare the Ridge a Reserved Forest and to institute an overseeing body that includes non-state members. It further provided recommendations for the inclusion of citizens at ground level through monitoring and awareness activities and suggests the inclusion of scientists to ensure that the Ridge remains a space for conservation of biodiversity rather, than being managed by those trained in horticulture.

However, citing practical (power to remove and resettle encroachments) and resource (financial and manpower) reasons, the DDA was handed the management of the Ridge rather than upgrading the Forest Department to handle the job, as the NGOs had called for. The demand was repeated by NGOs who did not agree the committee's recommendation (Malik & Agarwal, 1994b; Srishti, 1994). The Centre for Environmental Law of the WWF (I) filed a petition in the Delhi High Court demanding the transfer of the Ridge to the Forest Department since legally protected forests are to be handled by the state governments according to forest laws (Srishti, 1994; The Pioneer, 1994). The Delhi Administration also lobbied in the Delhi legislative assembly and with the Lt. Governor for the Forest Department to be given charge of the Ridge (Mohapatra, 1994; ToI, 1994b). The DDA however continues to manage the parts of the Ridge vested in its control in 1992-93. As a newspaper report noted at the time:

> "DDA may not readily relinquish its own hold on thousands of hectares of prime green land which have future potential for economic exploitation whether for creation of exclusive NRI[120] golf course, Disneyland, restaurants, corporate farms" (Mohapatra, 1994).

The Forest Department has recently again argued that the DDA is unfit to manage Delhi's forests and the task should be entrusted to them (Sinha, 2014).

After the release of the report, no action took place by government agencies to implement the recommendations. The report specified no time limit for the appointments to the multi-party board and scientific advisory council or for the setting up of the department to handle the Ridge in the DDA and the government was dragging its feet. Further, there was no specification of what exactly the powers and responsibilities of the NGO members of the joint committee would be (Srishti, 1994). On 4 February 1994, the JNGO organised a march and a street play in front of the Delhi Legislative Assembly and sent memorandums to the Chief Minister, the Lt. Governor and the Development Minister of Delhi demanding the declaration of the Ridge as a Reserved Forest, removal of encroachments and the setting up of the joint management committee (Kothari & Rao, 1997; Pioneer, 1994; Srishti, 1994; ToI, 1994a).

A week after this event, assurances were sent out from the DDA and Development Minister of Delhi that the Ridge would be declared reserved as soon as a survey had been completed to ascertain the boundaries of encroachments lying within and that all encroachments except slums would be removed (Srishti, 1994; ToI, 1994b). On 24 May 1994, an area of 7,777 hectares across the four segments of the Ridge was declared Reserved Forest[121]. The notification denoted boundaries in terms of landmarks rather than khasra numbers as in the other notifications (unlike in the case of demarcation of Asola Bhatti Sanctuary) this ensured three parcels of forests had clear boundaries for the most part, where forest laws would make construction or park creation illegal. Certain parts of the Ridge were left out of this notification leading to further conflict in these areas[122]. An addition to Reserved Forest area was made in 1996 when surplus land of village councils of 14 villages was added in the Southern Ridge to wildlife sanctuary.

In the meanwhile, the Supreme Court responded to public interest litigation[123] that had been filed in 1985 by a prolific environmental lawyer, M.C. Mehta, reminding the court that areas under the Ridge had been notified as Reserved Forest

120 NRI stands for non-resident Indians, here it refers to a wealthy class of expatriates

121 Notification No. F.10. (42)–1/PA/DCF/93/2012–17 (1)

122 See chapter 7

123 Public Interest litigation (PIL) is a means by which third party individuals or groups (who are not necessarily the affected party in a case) can move the court in public interest. This can be done through informal means like letters or postcards (Dias, 1994). PIL is often used in environmental cases as discussed in chapter 8

in 1913 and 1980, all encroachments in violation of forest laws, therefore, should be deemed illegal, irrespective of government permissions obtained for the same. Special attention was brought to the area of Bhatti Mines in the Southern Ridge, where mines and labourer settlements had been established. The Supreme Court ordered[124], that all post 1980 encroachments be removed from the Ridge as the MPD 2001 mandates that the Ridge be kept free of encroachments to be maintained in its 'pristine glory'. Following court orders, large scale demolitions and evictions took place in various parts of the Ridge and the JNGO had to point out that these demolitions were targeted at poor residents of informal settlements while other encroachments were allowed to continue (Agarwal, 2010; Kothari & Rao, 1997; Malik, 1998; Srishti, 1994). A notice was sent by the JNGO to the President of India that the presidential Polo Grounds also stood on the Ridge and should be demolished (Kothari & Rao, 1997).

The court orders related to the M.C. Mehta case also led to the setting up of the Ridge Management Board (RMB) in 1995 as had been recommended by the Loveraj Committee. The RMB is a seven member non-statutory body chaired by the Chief Secretary of Delhi and consisting of senior representatives of the DDA (Vice-Chairman of DDA), Delhi Government (Secretary Environment and Forests, Secretary Finance), Forest Department (Conservator of Forests, Delhi), and representatives of NGOs[125]. The board was given wide ranging duties of executing the management scheme for the Ridge including maintaining boundaries, controlling use by public, control on encroachment and "upgradation of the Ridge based on sound silvicultural practices"[126]. It was therefore expected to play a developmental, regulatory and advisory role (Sinha, 2014).

However, until June 1997, the board had only met once (Kothari & Rao, 1997). According the NGO members interviewed (both of whom resigned in 2013), meetings continue to be irregular and are largely concerned with granting (or denying) clearance for various requests on forest land[127]. Moreover, there is no clarity on how NGO representatives are chosen, who nominated them and on what basis (neither of the members of the board knew the criteria)[128]. Two out of the four NGOs on the board between 2002 and 2013 (Prakriti and Society of Environmental Management) could not be traced, no member from these organisations has ever attended the meetings of the RMB according to the other two civil socie-

124 M.C. Mehta vs. Union of India and ors. Writ petition (civil) No.4677/1985 orders dated 09.05.1996

125 These representatives have changed since the inception of the board. In the beginning it was Indian Society of Environmental Management and Prakriti. In 2002, Ravi Agarwal from Srishti/Toxics Link and Ajay Mahajan of Kalpavriksh were added. In 2013 they were replaced by Sunita Narain from Center for Science and Environment and Ravi Singh from WWF (I).

126 Constitution of Ridge management board, Office of the Development Commissioner, Government of NCT, Delhi Notification No. F56(225)/95/Dev./HO/5596

127 Interviews EV:2 and EV:8

128 Interviews EV:2 and EV:8

ty members[129]. Often, the board is not notified about diversion of forest land or mismanagement by state agencies, in one case an NGO member of the board had to file a court case against a landfill being created in the wildlife sanctuary by the Municipal Corporation of Delhi, while poor settlers were being removed in the name of conservation[130] (Agarwal, 2010). In other cases, the NGO members have had to carry out inspections on the ground to document horticultural practices and construction by the DDA in parts of the Ridge (Roy, 2005; Samanta, 2004). The RMB is a toothless body in terms of achieving its responsibilities, as no state body has been penalised for violation of forest laws even as they are represented in the RMB which is charged with conserving the forest (Agarwal, 2010).

It is alleged by the civil society members that the RMB has become a body concerned largely with passing land clearances for conversion of forest land (outside the Reserved Forest area) to other uses, as opposed to fulfilling its broad functions[131] (Agarwal, 2010). This is due to its composition, whereby, the DDA is represented on the board and the Chief Secretary of Delhi is the chairman of the board leading to a heavy tilt towards DDA's agenda rather than the conservation related goals set out for the board. Apart from the two civil society members, all other members report to the Chief Secretary in the course of their duties. This conflict of interest has not gone unnoticed by the activists concerned who claim that "[...] the same officer is proposing land diversion for government's projects and also deciding on them" (Singh, 2013d, n.p.). While most requests by private parties have been denied, the state has become the largest requisitioner of forest land for infrastructure purposes (Agarwal, 2010) and the structure of the board makes it difficult to deny permission to state agencies especially when 'public purpose' is invoked which has been done for large public transport projects, security installations, police stations, parking lots, stadiums and other constructions for the Commonwealth Games for example (Agarwal, 2010). The only time the board met regularly according to the NGO members, was when several clearances were needed for diversion of forest land in order to carry out constructions related to the Common Wealth Games held in Delhi in October 2010[132].

6.3 LEGACY OF ENVIRONMENTAL ACTIVISM 1979–1995

After the notification of the four sections as Reserved Forests was passed and the RMB was instituted, the active engagement of environmentalists documented in this chapter, fell away. Larger NGOs like WWF and INTACH did not continue their engagement in keeping a check on the diversion of forest land in Delhi and the smaller student based organisations either disbanded (Srishti) or shifted focus to other localities (Kalpavriksh).

129 Interviews EV:2 and EV:8
130 IA in Almitra Patel vs UOI and Others, Writ Petition (Civil) 888/1996
131 Interviews EV:2 and EV:8
132 Interviews EV:2 and EV:8

Srishti and Kalpavriksh had been at the helm of activism around the Ridge. There were certain tensions within and between these student groups regarding the deep ecological agenda of some core members and the engagement with environmental justice issues on the part of the others. Members of Kalpvriksh were engaging with other environmental movements in India at the time, which had a core element of social justice and livelihood concerns[133], disagreements on the basic philosophy of the organisation, led to major splits in the organisation where the more conservation and wildlife centric members left the group, as it moved to a more social justice based conservation stand[134]. Similarly in Srishti, which started off focussing on biodiversity issues, splits occurred when focus shifted to connections between environment and politics. The founder of Srishti ascribes his decision to leave the organisation to other core members increasingly interested becoming in wider issues of urban environment, while he considers himself a naturalist and not an activist[135]. Furthermore, since these groups largely comprised of students, many of those involved with the issue of the Ridge, moved away from the city as they graduated and could no longer continue their engagement[136].

Once these student groups were no longer active, the movement for the conservation of the Ridge died out since it was never a mass movement. The general apathy on the part of most citizens on the issue of the Ridge had been noted during the attempts to spread awareness on the need for conservation by these groups and the movement remained limited to a small number of environmentalists (Kothari & Rao, 1997; Srishti, 1994). Moreover the contingent elements that made the interaction in this phase possible, changed. One NGO member suggests that the receptiveness of the Lt. Governor at the time made interaction effective as in this phase

> "NGOs had an upper hand and the administration was painted as ignorant and villainous. Then the new Lt. Governor, Vijay Kapoor came in and he found that his officers were demoralised so he put his foot down and said we will not entertain the NGOs to that extent. So the pendulum swung the other way. Today the pendulum has come to the middle; the authorities pay lip service what NGOs are saying"[137].

Nevertheless, there were certain long-term outcomes accruing from these years of pressure from environmentalists on the government. Firstly, the separation of the Ridge from other urban green areas was established as the four segments were notified as Reserved Forests and stricter legal protection was applied to them. Secondly, a space for members of civil society in the negotiations for deciding the use of forest space was institutionalised in the RMB (even if this space is constricted by the structure and composition of the body). As a result, in the four segments of the Ridge, conservation became the dominant discourse since it has achieved both discourse structuration (actors must refer to conservation goals

133 Interviews EV:4, EV:5 and EV:8
134 Interview EV:4
135 Interview EV:11
136 Interviews EV:1, EV:2, EV:4, EV:8 and EV:9
137 Interview EV:10

while speaking about the Ridge and position themselves according to it) and discourse institutionalisation (conservation has solidified into institutional practises as seen in the RMB and official notifications) (Hajer, 1993). Another reason explaining the receding of environmental activism around the Ridge in general, is that some resolution of the issue was seen after a few of the main recommendations of the Lovraj committee were implemented[138]

Within the conservation discourse, the main storylines included that of the Ridge as natural heritage, Ridge as a source of essential ecological services (the metaphor of the 'lung of the city' being often used) and the Ridge as a space for biodiversity. Social justice concerns were not ignored, both student organisations argued for integration of social justice and conservation demands (Kalpavriksh, 1991; Srishti, 1994). Srishti argued that conservation would not be feasible in the Ridge unless local residents were part of decision making and livelihood demands were catered to (Srishti, 1994). Both organisations suggested a resolution by allowing a buffer area on the outskirts to allow public access while retaining a core area for biodiversity conservation (Kalpavriksh, 1991; Srishti, 1994). However, the social justice discourse did not attain dominance and was not well integrated with the conservation discourse once it became dominant. In part, this was due to strategic choices made by environmentalists during the bargaining process where it was felt that initial steps were needed to be made with regards to conservation at this stage[139] and a more complex social message may have blocked resolution[140]. This lack of integration of conservation discourse from social justice discourse had implications for many people who live and work in the Ridge as elaborated in the next chapter.

138 Interviews EV:2 and EV:10
139 Interview EV:5
140 Interview EV:2

7 THE THREE CONSERVATION AND RESTORATION PROJECTS IN THE RIDGE

The environmental movement involving student organisations and the NGO Forum discussed in the previous chapter, demanded conservation of the Ridge in general at the level of the city. This chapter turns the focus to a more localised scale by looking at negotiations and transformations that have occurred at the level of three bounded spaces within the Ridge. Although the discourses and socio-political dynamics identified here often refer to and are intimately connected with those of the city and the Ridge as a whole, these contestations are essentially concerned with the particular parcel of forest they bear upon. Three segments of the Ridge in particular have become sites for different modes of conservation practices resulting from particular strands of conflict and deliberation and the different actors involved. The discursive and material construction of conservation landscapes in these bounded spaces within the Ridge, namely; Sanjay Van and the Aravalli Biodiversity Park in the South Central Ridge and the Asola Bhatti Wildlife Sanctuary in the Southern Ridge are documented and analysed in the following sections. Each of these has been subject to different trajectories of historical development, transformation of land use in the recent past and conflicts regarding use and access to the forest. The analysis engages with one of the central lines of investigation of political ecology; the "sources, conditions and ramifications of environmental change" (Bryant, 1992, p. 13). In particular, the following lines of questioning were explored: What logics and methods were applied and what challenges and oppositions were put up by actors to realise their demands? Which discourses were effective and are reflected in the current landscape and regimes of access and land use and why? Which discourses and actors were side-lined?

The first three sections are dedicated to detailing the transformation of the three above mentioned parcels of forest land and the 'environmental brokerage' (Blaikie, 1995, p. 213) that this resulted from. The last section discusses the discursive interactions associated with the struggle to shape the landscape.

The multiple claims, factual discrepancies, legal confusion and covert negotiations make the reconstruction of events difficult. The following account is based on official government reports, court documents, NGO campaign material and newspaper and magazine reports. This information is supplemented with personal interviews with various actors. The discourses and in particular the argumentative structures and storylines regarding land use and forest conservation in written and spoken statements were analysed to identify how various actors positioned themselves in the debates around the formation of policy and practices concerning the Ridge (Fischer & Forester, 1993; Hajer, 1995) and trace the change in the physical environment that resulted from these contestations. The effect of this change on various sections of society is varied, this is of consequence for the landscape itself as it is, to an extent, the material manifestation of socio-economic contestations (Blaikie & Brookefield, 1987; Escobar, 1996; Robbins, 2012). This strand is developed further in chapter 7, which explicitly considers the question of power, access and control, a central pillar of political ecology studies (Blaikie, 1999; Forsyth, 2003; Robbins, 2012).

7.1 THE MAHIPALPUR RIDGE: OF MINES, MALLS AND BIODIVERSITY PARKS

The Mahipalpur Ridge, a parcel of land measuring about 640 hectares in South Central Delhi, has had an eventful transformation in the last two decades. A look at its micro-history illustrates the negotiation of interests and conflicts that shaped the local landscape. This segment, is in geological terms part of the Ridge (Geological Survey of India, 2006), but it is not mentioned in the 1994 notification that delineates the Reserved Forest areas in Delhi and therefore bears a lower status in terms of legal protection under forest laws. It is located between the middle class and elite enclaves of Vasant Vihar, Vasant Kunj, the populous Mahipalpur village and the campus of Jawaharlal Nehru University (JNU). Its proximity to the international airport of Delhi has made it prime real estate. In this study, this area of the Ridge will be referred to as Mahipalpur Ridge to avoid confusion with Sanjay Van which is known as the South Central Ridge in administrative parlance.

This expanse lay beyond the city limits at the time initial plans for Delhi were drawn up and was a part of the village commons of Mahipalpur. The Master Plan 1962 mentions this area as part of the green belt that is supposed to border the city and is considered a 'no development zone' due to the undulating terrain and scarcity of water. Between the first and second master plans, the upper middle class and elite residential enclaves of Vasant Kunj and Vasant Vihar, housing mostly salaried employees came up around this plot of land as unauthorised settlements but these were regularised in the second master plan (Soni, 2000; Verma, 2004e). The parcel of land in question was mined extensively in the 1970s and 1980s for mica, sandstone, china clay and gravel (Soni, 2000) by a private company that had obtained a lease from the government. In 1989 mining was banned within the city limits leading to closure of the quarries. In the Master Plan 2021, the plot where the biodiversity park is located is earmarked for residential purposes and a district park[141]. However, the current land use includes a series of three Malls, a Hotel, certain public institutional buildings, an army compound and residential units (both formal and informal), surrounding a biodiversity park. The biodiversity park is dotted with fourteen defunct quarry pits that have been planted with trees or serve as water reservoirs. The former miners and their families, many of them

141 A district park is a recreational open space designated for each district in the master plans. They often have some horticulture and landscaping elements and are different from biodiversity parks that are marked for conservation

from Rajasthan, settled around the mines and built a cluster of informal settlements[142] around 40 years ago, of which one cluster was demolished to form the biodiversity park. Currently, two informal settlements, namely, Bhawar Singh Camp and Kusumpur Pahadi continue to exist, straining at the borders of the park (see Map 4 in Appendix B). The control of the plot is divided between the DDA and the army in almost equal proportions.

7.1.1 Proposal for the International Hotel Complex

The Mahipalpur Ridge has been a site of contestation between citizens, including activists and NGOs, and the administration, primarily the DDA. The peaks of this contestation can be traced to certain proposals for changes in land use by the DDA and the response to these by other actors. The first traceable agitation occurred in 1995 against the construction of a road through this segment, meant to connect the middle class residential enclaves of Vasant Vihar and Vasant Kunj. Professor Vikram Soni, a resident of Vasant Vihar, who would become a central figure in the agitation against construction in this part for more than a decade, filed public interest litigation in the Delhi High Court against the DDA to prevent the construction of the road citing the Delhi Preservation of Trees Act of 1994 which prohibits felling of trees without permission from the Forest Department. A stay order was granted and the building of the road was prohibited on environmental grounds[143].

The road issue proved to be a small challenge compared to a proposal the very next year to build an 'international hotels complex' according to which thirteen luxury hotels were to be constructed on the same segment of the Ridge, covering an area of 315 hectares (EPCA, 1999). It became clear that the area left out by the 1994 notification was seen as open for construction by the DDA (Agarwal, 2010). This led to accusations by environmental activists, the media and even the Supreme Court of the omissions in the 1994 notification being deviously motivated to use the land for commercial purposes[144] (EPCA, 1999, 2000). An NGO titled Citizens for Protection of Quarry and Lakes Wilderness (CPQLW) was formed and with support from some other NGOs[145] agitated against the construction of the hotel. The campaign against the hotel proposal brought together some well-connected, highly educated, elite citizens, most of them residents of Vasant Vihar or Vasant Kunj. A list of the membership of CPLQW primarily contains academics, lawyers, journalists[146]. The proposal for the hotel complex took place simulta-

142 The term 'informal settlement' is used here to refer to squatter settlements on publically owned lands also called slums and Jhuggi Jhopri clusters (Hindi term literally translating to shacks and huts) that are characterised by economic poverty and physical fragility (Bhan, 2009)

143 Interview MEV:1

144 Interviews MEV:1 (b) and MEV:2

145 including Srishti, Kalpavriksh, Tapas, Parivartan, Chetna, Pani Morcha

146 http://naturalheritagefirst.in/the-team/ (accessed 02.02.12)

neously with the notifications regarding the Asola Bhatti Sanctuary[147], the learnings and connections built from the push for stronger legal protection of the Ridge were used by certain members of civil society to argue against this high profile proposal[148].

One of the central grounds for opposing the project was the loss of 'natural heritage' embodied in the forest area[149]. Another key issue was the huge water demand the hotel would place on the area of up to a thousand litres a room per day (Agarwal, 2010). Water scarcity being an everyday problem in the surrounding residential area, this second point of opposition added an element of urgency and helped garner support from local residents. A public interest petition was filed by Kuldip Nayar, an eminent senior journalist, ex-diplomat and resident of Vasant Vihar, in the form of a letter to the Supreme Court. The letter demanded that the DDA be 'chained' by the court citing the arbitrary changes made in the Master Plan to accommodate commercial interests and the need to preserve New Delhi's lung[150]. In response, the Supreme Court ordered a stay on the proceedings and demanded that the DDA obtain the necessary clearances before any further construction takes place (EPCA, 1999).

A major objection was raised by the petitioners in court to the change made by the DDA in the planned land use from 'green area' in the MPD 1962 to 'urban area' in MPD 2001. Further, the MPD 2001 was amended in 1995 (June 15th) to increase the area for commercial purposes in the land in question from 8 to 65 hectares. The DDA argued in court that there was an acute shortage of tourist accommodation in Delhi and four and five star hotels were needed for this purpose. They also pointed out that the construction was not in contravention to the law as the area earmarked for the hotels was not legally or administratively a part of the Ridge[151].

In September 1996, a three judge bench of the court ruled that 'development and environment protection must go hand in hand' and therefore an environmental impact assessment (EIA) must be conducted and the 'whole area must be surveyed from an environmental point of view' before any further developments take place[152]. For this purpose, it was ordered that the Environmental Impact Assessment Authority (EIAA) be created by the central government under the Environmental (Protection) Act of 1988. The EIAA would deliberate upon environmental protection issues arising from this case and also any other case in the National Capital Territory. This was seen as a landmark judgement by the activists[153],

147 See section 3 in this chapter

148 Interview EV:1

149 Interviews MEV:1 and MEV:2

150 Letter from Kuldip Nayar to Justice Kuldip Singh, Judge of Supreme Court of India, New Delhi Dated–6th August 1996

151 M.C. Mehta vs. Union of India and ors. Matter regarding Kuldip Nayar's Letter. Writ Petition (Civil) No. 4677 of 1985

152 M.C. Mehta vs. Union of India and ors. Matter regarding Kuldip Nayar's Letter. Writ Petition (Civil) No. 4677 of 1985, Order Dated 13 September, 1996 (p. 3)

153 Interviews MEV:1 (b), and MEV:2

aimed at laying down a high level agency that would have the final say in matters of environmental clearances institutionalising environmental priorities in projects in the city. The DDA was instructed to carry out an environmental impact assessment (EIA) and submit it to the EIAA for perusal. The EIAA heard several bodies including the Geological Survey of India (GSI), The Central Ground Water Authority (CGWA) and the Forest Department of Delhi. The first two identified this complete 640 hectare area as part of the Ridge and the Forest Department deemed it forest land. In the report of the public hearing, the EIAA declared that the environmental impact assessment (EIA) prepared by DDA was faulty. It was recommended in no uncertain terms to scrap the hotel complex project (EPCA, 2000).

In 1997, the EIAA ordered disconnection of electricity and water supplies to residential units in this area, on the grounds that they were also in contravention of environmental regulations due to their location on the Ridge. The residents of the middle class Hill View Apartments & populous Kusumpur slums protested this move and applied political pressure through their local MLA[154],[155]. To resolve the issue, the DDA coined the term 'constraint area' for the 15 hectares on which these colonies lay and decided that the current land use would be allowed to continue in this area. The constraint area was thus effectively considered outside the Ridge and exempt from the environmental legalities attached to locations considered forest land. The DDA then took the opportunity to extend this category to enable the commercial building that they were attempting to push through and the constraint area was extended to 92 hectares without any explanations. The construction of one hotel continued in this extended area of exclusion (EPCA, 2000). The term 'constraint area' has no legal precedent and has not been defined anywhere. Only the implication of the term, that forest laws would not apply here, comes through in the DDAs stand.

In August 1997, a two judge bench modified the earlier judgment given by a three judge bench (of September 1996) and permitted construction on the 92 hectares (additional to the residential built-up area already in existence)[156]. This was in response to a petition from Unision Hotels Limited challenging the recommendations of the EIAA to disallow any construction[157]. The said company had continued the construction of one hotel ignoring the stay ordered by the previous judgment (EPCA, 2000). Despite allowing construction in the constraint area, the 1997 judgement also re-emphasised the pre-condition that all legally required environmental clearances and pollution control measures were to be obtained. It would seem that in effect, the second judgement would not have been any different form the first as the EIAA had already declared that the area lay on the Ridge and therefore was legally closed to construction, thus, no clearances could be

154 Onkareshwar, K., (2006, 9 August) Press release, Ridge Bachao Andolan

155 MLA stands for Member of Legislative Assembly and refers to the local elected representative to the state legislature

156 M.C. Mehta vs. Union of India and ors. Writ petition (civil) No. 4677 of 1985. Orders dated 19.08.97

157 SLP.No.8960/97 in M.C. Mehta vs. Union of India and ors Writ petition (civil) No. 4677 of 1985

granted. However, the judgement was taken as a go ahead for construction of the one hotel in progress despite notices sent by the EIAA to DDA on the violation of the stay order and continuation of construction in January 1997 (EPCA, 2000).

The institutionalisation of environmental checks with the establishment of the EIAA was a short lived victory. In a surprising move, the EIAA was dissolved by the court as an 'ad hoc body', though a reading of the previous judgement[158] makes it clear that it was intended to have wide ranging authority and application. The court then ordered the formation of another body, the Environment Pollution (Prevention and Control) Authority (EPCA), and transferred the work of the EIAA to it[159]. By the time the transfer of documents to the newly formed authority had been carried out, the construction of the hotel had almost been completed. On the question of allowing construction on the rest of the segment outside the constraint area, EPCA reduced the issue to two basic points which would decide the matter: Firstly, was the given area a part of the Delhi Ridge? Secondly, is there sufficient water availability to sustain the project (EPCA, 1999, p.3)? Regarding the first question it noted that according to the GSI, the area was in geological and geomorphological terms undoubtedly part of the Ridge. Legally, the matter was more confusing. The DDA does not consider this area as Ridge land as it was not included in the 1994 notification of Reserved Forests that it considers synonymous with the Ridge. However, the 1980 notification does refer to parts of this parcel as Protected Forest land. A third legal definition in favour of labelling this forest land was drawn from the Supreme Court's judgement in the Godavaram case in 1996[160] declaring that the IFA of 1980 should apply to all forests in the country (adapting a dictionary meaning of forests) irrespective of ownership or classification. This judgement included directions to state governments to identify areas that were "earlier forests but stand degraded, denuded or cleared"[161] and ensure an end of all non-forest activities in these areas. The EPCA recommended that under this judgement, the Mahipalpur Ridge would be termed forest and provisions of the IFA 1980 would apply (EPCA, 1999). The reports of the EPCA (1999, 2000) also chastised the DDA for falling to include parts of the geological Ridge as Reserved forests and pointed out that this went against the spirit of the Supreme Court's orders to declare the Ridge reserved and the DDA's own stated intention in the master plan of 'maintaining the ecological balance of the city' by 'maintaining the Ridge in its pristine glory' (DDA, 2007). It further noted that there was a need to preserve recreational forests and 'green sinks' as the population pressure on land increases and the planned urban green space is already short of expectations (EPCA, 1999).

158 M.C. Mehta vs. Union of India and ors. Writ Petition (civil) No. 4677 of 1985. Orders dated 13.09.96

159 Notification No. S.O.93 (E)

160 T.N. Godavarman Thirumulkpad vs. Union of India & Ors. Writ Petition (Civil) No. 202 of 1995 2SCC267 No.5

161 T.N. Godavarman Thirumulkpad vs Union of India & Ors. Writ Petition (Civil) No. 202 of 1995 2SCC267 No.5

Regarding the question of water availability, it was clear that there was a scarcity of water in the area, which was affirmed by the Central Ground Water Agency (EPCA, 1999). The DDA sought to tackle the issue by suggesting in the EIA submitted by them, that the project would be implemented in two phases, the first phase would continue as such and the second phase would be contingent with measures like water harvesting, thus moving the burden of environmental checks to a later stage. The EIAA had already investigated and reported that the actual water requirements of the hotel would be much higher than the DDA had stated, while the water availability in the area was much lower than assumed by the DDA's EIA. Given the already prevailing water scarcity in this part of the city and the danger of further reducing the water quality of this region[162], it was suggested that the project was not feasible. In any case, the DDA had not been able to provide adequately for the existing residential and commercial units in this water scarce locality and has no record of water harvesting and recycling projects to demonstrate its capability and will. The EPCA demanded that if water could indeed be made available through such measures, then it should be first made available to the existing users rather than to the upcoming hotel (EPCA, 1999).

Based on legal status of the land and water availability issues, the EPCA recommended that the Ridge should be maintained in its 'pristine glory' and the "environmental factors are not in favour of any urban use of this 223 hectares of land and the entire parcel of land should be developed as green" (EPCA, 1999, p. 14). Notably, the deliberations of the court seemed to hinge not on the question of ecological balance, which receded in the background to more immediate matters of legal land classification and water availability.

The question of continuing construction in the 92 hectare 'constraint area' remains. Here too, the EPCA declared that the hotel (now in the final stages of completion) was in contravention to law on the grounds that it did not have sufficient clearances (EPCA, 2000). The Supreme Court bench of 1997[163] and DDA had colluded to enable the construction of this hotel in spite of clear indications from the EIAA and later the EPCA that such construction should not carry through. The removal of the hotel according to the EPCA would also raise questions about whether the residential structures would be allowed to remain on this land (EPCA, 2000). However, citing the investment already made by the hotel and the fault of the DDA (and not the hotel) in flouting its own master plans, the court allowed the hotel to function as long as it complied with the post-construction environmental clearances. Cosmetic fixes (like water harvesting, partial sewage treatment and using recycled water for certain purposes) were ordered to be put in place and the matter was closed[164].

162 The CGWA had mentioned that the water here is already brackish and water table low due to over exploitation (EPCA, 1999)

163 Comprising of Chief Justice J. S. Verma and Justice B. N. Kripa

164 T.N. Godavarman Thirumulpad vs Union of India and Ors. Writ Petition (civil) 202 of 1995. Order dated 17.10.2006

The DDA had approached the issue as a question of legal classification and the question of environment was discussed in court in terms of brown issues like water capacity, sewage, and pollution, even though the petitioners had brought up the issue of natural heritage and ecological balance along with planning violations. The ultimate tilting of the debate towards of brown issues enabled, on the one hand, a cosmetic fix to the hotel issue that included limited technological solutions rather than the much broader responsibility of ecological balance that the MPD 2001 had aspired to. On the other hand it was the pressing and tangible issue of water scarcity that effectively countered the construction of the twelve other hotels proposed.

7.1.2 Malls on Land Cleared for Conservation

The argument regarding, whether this land was part of the Ridge or not, did not end with the hotel issue. Another major agitation took place in the very same area starting in December 2003 after DDA auctioned away land for the building of shopping malls and offices in the 92 hectare constraint area. The stakes for the DDA were high as the plots were reported to have been sold for a massive sum of 1,100 crore[165] rupees (Jyoti, 2006; Sharma, 2006). At the same time, the army also started construction on a part of the land under its control. This time the CPQLW filed a petition with the Centrally Empowered Committee (CEC) [166] of the Supreme Court stating that this area was part of the Ridge and must therefore be preserved as forest and the IFA 1980 would apply to it[167]. It was expected by the petitioners that since the CEC is responsible for checking violations of forest laws in the country and had shown an inclination to environmental protection, it would be more sympathetic towards considering the area as forest land[168]. A list of respondents was called upon by the CEC in response to the application including the central Ministry of Environment & Forests (MoEF), the Forest Department of the Government of Delhi, the DDA, Central Pollution Control Board and the Ministry of Defence (as representatives of the army). The chairman of the investigation, Shekhar Singh, visited the Mahipalpur Ridge a single time and was guided by a member of CPQLW (the key petitioner in the case) during this visit[169] (Verma, 2004e) demonstrating that the activists had the ear of the CEC from the beginning.

The CEC directed the DDA and the army to stop construction in the entire 640 hectares of the Mahipalpur Ridge and proceeded to hear the case. The DDA, the MoEF and the army argued that the area was not part of the Ridge. While the

165 A crore is equal to ten million

166 As mentioned earlier, CEC was instituted to promote forest conservation by investigating violations of forest laws "in the background of the tremendous pressure on forests and the unsustainable removals and the real threat of massive destruction of wildlife habitat" (http://cecindia.org/aboutcec.php, accessed 20.03.13)

167 Application to CEC No. 331

168 Interview MEV:1 (b)

169 Interview MEV:1 (b), MEV:2

Forest Department, the petitioners and ultimately the CEC argued otherwise. It is interesting to see the basis on which these claims were made by each side. The Government of Delhi, represented by the Forest Department confirmed that the area was part of the Ridge based on its geomorphological characteristics as certified by the Geological Survey of India. They further claimed that since the area fulfils the vegetation characteristic of a dry tropical thorn forest it should be declared a Reserved Forest under the IFA 1927[170]. Contrary to this position, the DDA argued the area had no ecological value and declared that there was no basis to call this a forest as it is a "degraded habitat with no soil cover and exposed rock that does not support even a single clump of grass" and that "there is no logic to declare this area a Reserved Forest since it does not support any ecosystem, nor supports any unique species and does not perform any ecological functions"[171]. On the other hand, the DDA admitted that the area is "technically similar to the Ridge" but should be excluded from the attendant protective measures, since there had been "discontinuity due to historical and biological factors", namely, human settlement, mining and the dominance of an invasive tree species in the ecosystem[172]. While in the case of the hotels the DDA had focused on legal classification, this time it shifted the argument to biophysical and anthropological characteristics of the land to convince the conservation-inclined CEC. Ironically, the DDA sought to underline its responsibility by submitting that what it had earlier declared a 'dead ecosystem'[173] with no important ecological functions, would be turned into a "biodiversity park with 3000 species of plants and microbes etcetera" and that the entire area, except the constraint area, would be declared as a bio-heritage site[174]. The DDA's stand was evidently full of contradictions; they agreed that the area was ecologically important enough to deserve the highest protection as a biodiversity park, but at the same time argued that the patch was far too degraded to be of any use at present and suggested that on this basis construction should be allowed. Being under pressure to adapt the conservation mantle, perhaps that the DDA was counting on a trade-off in which the construction would be allowed in exchange for the protection for the rest of the area.

The MoEF added another dimension by basing its objections on alleging doubt over the scientific and professional capability of the CEC special invitee, Shekhar Singh, who was chairing the investigation.

> "The question of proposing an area to be declared as Reserved Forest or National Park has to be dealt with utmost scientific rigour, sensitivity, and care, taking into account the stability of the area in terms of its extent, location, geomorphology, and biological resources to provide environmental services that are expected of a Reserved Forest or a National Park. In addition, the present value or future potential value of the societal benefits provided by the area in comparison to the societal value of the environmental services that would accrue after declaring of the area as Reserved Forest or a National Park has to be taken into account. The value

170 Secretary, Environment & Forest in letter to CEC dated 24.03.04 (CEC, 2004, Annexure B)
171 Letter by Mr. R.K. Jain, Director, DDA to CEC dated 5.7.04 (CEC, 2004)
172 DDA written submission, final hearing by CEC 26.07.04 (CEC, 2004)
173 Letter by Mr. R.K. Jain, Director, DDA to CEC dated 5.7.04 (CEC, 2004)
174 Letter by Mr. R.K. Jain, Director, DDA to CEC dated 5.7.04 (CEC, 2004)

> of these benefits have to be computed following formal scientific and economic methodologies to facilitate a comparison of the societal benefits available before and potentially in the future from alternatives, and from changing the legal and ecological status of the area"[175].

The MoEF in turn proposed to constitute a Technical Committee of Experts to examine the proposal to convert this area into Reserved Forest[176]. The CEC pointed out that neither had such a procedure been deemed necessary in the past, nor was there an assurance that the construction and tree felling would stop in the interim while this new committee was conducting its study. This exchange is to be read in the light of the tensions that existed at the time between the MoEF and the CEC, the former was pushing for the setting up an alternative green tribunal on accusation that the CEC was passing unscientific, sweeping directions to the government regarding conservation, resulting in social unrest in forest areas (as described chapter 5, section 5.3).

The army on its part argued that it was not aware of this being considered forest land and that it only intended to build on 156 of the 825 acres under its control (increasing total built area from 4% to 19%) and afforest the rest. But it centred its cause on the emotional appeal that it had to build residences for the families of personnel returning from fighting in difficult areas like Siachen[177] (a reference to the high altitude Kargil conflict of 1999 between India and Pakistan). The CEC recommended that the army be allowed to use the said area for "improving their strategic requirements and objectives" (CEC, 2004, n.p.) as long as they forested the rest of the area, though it placed no measures to check the fulfilment of this condition.

In the absence of clear legal protection as a Reserved Forest, the question regarding the definition of the Ridge had been provided with different answers by various actors. It was presented as a legal category, as land available for construction in a land starved city, as a geological formation, as land with a certain kind and amount of vegetation, as the last stand of biodiversity in Delhi, as an important source of ecological services in a resource stressed city (mainly that of a water reservoir and that of being the 'lungs of the city'), and as 'natural heritage' for Delhi's citizens. Most of these points are in effect 'storylines' (Hajer, 1995) and each actor picked the ones that suited their argument in this conflict. A discourse coalition can be seen between the DDA, army and the MoEF even though they do not have similar interests and do not necessarily communicate their interests to each other. The CEC, itself an actor, sided with the opposing discourse coalition, including the NGOs, activists and Forest Department, which sees the Ridge as forest.

175 Note titled 'Objection of the Ministry of Environment & Forest (MoEF) in the matter of report prepared by Shri Shekhar Singh, Special Invitee', Central Empowered Committee (CEC) in IA No. 331 Regarding the preservation of environment and bio diversity in the Delhi Ridge Area, Dated 26.7.04

176 Note from MoEF to CEC, 26.7.04 (see full description in previous footnote)

177 Affidavit submitted to CEC dated 28.7.2004 (CEC, 2004)

The final decision of the committee included the considerations of the hotel complex issue and ordered that apart from the 'constraint area' already allowed, the rest of the parcel should be declared Reserve Forest and be converted to a National Park as it is

> "legally as well as ecologically forest area and is critical to the ecological health of the citizens of Delhi. Moreover it is an important water catchment area and is part of the very sparse remaining forest cover in Delhi, it unquestionably needs to be conserved" (CEC, 2004, n.p.)

In 2004, shortly after the CEC gave its recommendations, the DDA evicted one of the clusters of informal settlements from the area in preparation for the creation of the biodiversity park. The CEC had not heard the residents as part of its investigation and the price of conservation was paid by those who had no say in the decision[178] (Verma, 2004e).

Meanwhile the construction of the malls in the constraint area carried on. Environmentalists, Resident Welfare Associations[179] and individual citizens, including students came together to form the Ridge Bachao Andolan (Save the Ridge Movement) and approached the Supreme Court challenging the constructions and demanding legal protection for the entire Mahipalpur Ridge, including the constraint area. This was in continuation with the petition to CEC filed by CPQLW, the members of which were working in close association with the Andolan. This time the court ordered an expert committee of the MoEF to submit recommendations. In the public hearing, residents opposed the mall on grounds of traffic congestion, air pollution and water and power shortages that the malls would impose on the area. Tough these were additional to the RBA's core demand of forest conservation; they formed a discourse coalition opposing the malls. The DDA cited the high architectural standards, construction quality and the fact that these would be green buildings based on energy and water conservation norms as reason enough for the mall to remain (Ghertner, 2011b; The Hindu, 2006a). The situation at the hearing was tense, the parties clashed and threat of physical violence led to the police being asked to intervene (Dainik Jagran, 2006; HT, 2006b). Those opposing the mall complained that the hearing had been one sided with the ministry representatives hearing the developers in private despite requests for public hearings and not giving the opposing party the same chance[180] (Dainik Jagran, 2006; The Hindu, 2006b).

The final report affirmed that

> "the project site has topographical features similar to that of the Ridge. Various studies, including EIA documents submitted now for obtaining environmental clearance, establish the environmental value of this area, particularly as a zone of groundwater recharge. Therefore,

178 Interview SA:1

179 AResident Welfare Association (RWA) is a neighborhood association that represents the interests of the residents of a specific urban or suburban locality (Jumani, 2006). Tough they could also exist in informal settlements, they are generally seen to represent "an activism of the 'middle class', for the 'middle class'"(Tawa Lama-Rewal, 2007, p. 53). In this case as well, RWAs included were form planned neighborhoods of Vasant Kunj and Vasant Vihar

180 Interview MEV:2

> DDA should have exercised adequate environmental precaution based on sustainable environmental management approach. There is no evidence that the environmental impact of the construction of malls was assessed beforehand and that the development of this area for commercial activities is in accordance with the Master Plan"[181].

However, the report ironically went on to recommend that clearances should nonetheless be given on the basis that much construction had already taken place:

> "In hindsight it is evident that the location of large commercial complexes in this area was environmentally unsound. Now many proponents have constructed very substantially and really speaking awarding clearances even with conditions is largely a compromise with de-facto situation. The Expert Committee is of the opinion that at this stage only damage control is possible by strict implementation of effective environment protection measures and resource conservation measures in the project construction and operational stages"[182].

Despite the fact that the construction took place in contravention to the stay orders, and that the builders did not obtain the environmental clearances beforehand as required[183], the court accepted this view and judged that it was too late to remove the said construction[184].

It must be pointed out here that in a very similar case, the Supreme Court had decided otherwise. This case pertained to the construction of a shopping mall in the area of a park in Lucknow and the court decided in favour of the environmental and historical importance of the park stating that:

> "This court in numerous decisions has held that no consideration should be shown to the builder or any other person where the construction is unauthorized. This dicta is now almost bordering the rule of law [...]. Unauthorized construction, if it is illegal and cannot be compounded, has to be demolished. There is no way out [...] In the present case we find that the builder got an interim order from this court and on strength of that order got sanction of the plan from the Mahapailka (Municipal Corporation) and no objection from LDA (Lucknow Development Association). It has no doubt invested considerable amount on the construction *which is 80% complete and by any standard is a first class construction [...] The primary concern of the Court is to eliminate the negative impact the underground shopping complex will have on the environmental conditions in the area* and the congestion that will aggravate on account of increased traffic and people visiting the complex. *There is no alternative to this except to dismantle the whole structure and restore the park to its original condition*"[185] (Emphasis added).

181 Report of the MoEF cited in T.N. Godavarman Thirumulpad vs Union of India And Ors. Writ Petition (civil) 202 of 1995, orders dated on 17 October, 2006

182 T.N. Godavarman Thirumulpad vs Union of India and Ors. Writ Petition (civil) 202 of 1995. on 17 October, 2006

183 The previous judgement in the matter of the hotels made clearances even in the constraint area mandatory (Order dated 13.9.1996 on I.A.No.18 in Writ Petition (Civil) No.4677/85). Moreover, Environmental Impact Notification S.O.1533 (E), dated 14 September 2009, issued under Environment (Protection) Act 1986, has made it mandatory to obtain prior environmental clearance for all major projects

184 T.N. Godavarman Thirumulpad vs Union of India and Ors on 17 October, 2006. Writ Petition (civil) 202 of 1995

185 MI builders Pvt. Ltd. Vs Radhey Shyam Sahu and others (1999) 6 SCC 464

Several objections were raised to the court's final judgement in a review petition filed by the RBA. Nowhere had a definition or grounds for validity of the all-powerful 'constraint area' been provided. Even if the buildings were not on the Ridge or forest land, no clearances had been given to the buildings before construction; the developers were allowed to seek post environment clearance for the sake of damage control. The decision had been based purely in favour of costs incurred by the private parties in a commercial profit seeking venture and suggested remedial rather than punitive measures, in effect rewarding the developers for their own fault. The question of the construction by the army had been ignored as it was seen as necessary construction for strategic reasons. Moreover, the Ridge Bachao Andolan argued that the court had ignored to deliberate on the environmental aspect which was central to the initial petition[186]. Once again, while the DDA and the developers argued in terms of technological fixes for pollution and ground water shortages, the environmentalists spoke of the importance of forests and the ecological services they provide.

7.1.3 Environmentalists against 'Devious Planners and Corrupt Builders'[187]

The conflict around the Mahipalpur Ridge involved for the most part, local residents opposing the developments in their area. However, the long drawn campaign against the mall with the CPQLW and the RBA at its helm, garnered support from some prominent figures as well as the media. Member of Parliament, Raghunath Jha, questioned the MoEF in an open letter, on the continued construction and granting of clearances to the mall (Utpal & Bhan, 2006). Another MP, Brinda Karat, brought up the issue for discussion in the parliament demanding answers regarding the contravention of laws (HT, 2006a). Medha Patkar, a well-known socio-environmental activist in India and founder of the Narmada Bachao Andolan (Save the Narmada Movement), also joined the protests against the mall, bringing much media attention to the issue (Roy, 2006; ToI, 2006). Physicist and Noble laureate, Anthony Leggett, wrote to the President in support of protecting the Ridge as a bird sanctuary and source of fresh water (Dastisdar, 2007). He was introduced to the area by fellow physicist, Vikram Soni of CPQLW on a previous visit to India[188].

The main modes of protest included public interest litigations, demonstrations, public opinion building through the media and personal communication by leaders of the agitation trough letters and petitions addressed to persons of influence. In January 2006, a human chain protest on the construction site brought to-

186 Review petition filed by the Ridge Bachao Andolan - Review Petition INI.A. NO. 1463 of 2006 In Writ Petition (Civil) NO. 202 OF 2005 and Onkareshwar, K., (2006. 9 August) Press release, Ridge Bachao Andolan

187 The phrase is taken from a video on the subject of the Mahipalpur Ridge titled 'On the Brink' produced by CPLQW

188 Interview MEV:1 (b)

gether students of schools and colleges, RWA members and individual citizens; a protest the RBA coordinator claims was the largest environmental demonstration in Delhi until then[189]. A broad alliance was sought to be built in support of the issue. Letters and memorandums were sent to the President, the Prime Minister and the leader of the ruling coalition among others to enlist their support[190] (Indian Express, 2005; The Hindu, 2006c). Apart from persons of influence, environmentalists and students, the surrounding village councils (those of Mahipalpir, Masudpur, Vasant Gaon) and displaced residents of the Lal Khet slums were also part of the Movement in the later stages[191] (Asian Age, 2006).

The single point agenda of the RBA was to retain the Ridge in its 'natural form'[192]. The portrayal of the Ridge as Delhi's 'ancient natural heritage' that once lost can never be replaced[193] (Indian Express, 2005) creates an narrative that fits nicely with the demand to protect it in its 'pristine glory', an objective repeated by citizens, the DDA, the court and the various committees. The idea that the Ridge is two and a half billion years old, older than the Ganges and the Himalayas[194](HT Estates, 2006) was repeated by environmentalists and the media to draw a connection with the emblematic natural features of the country. The 'natural heritage' storyline was so central to the environmentalists that CPQLW eventually morphed into a group called Natural Heritage First. The immediate crisis of water scarcity in the region (Nayar, 2004) proved an effective means to argue against the imminent developments. These storylines, invoking antiquity and crisis, were used as a means of recruiting people into the discourse coalition that opposed the hotel complex and the mall. However, the movement remained limited in its mobilisation of a broader base, including within the non-poor residents of the area, given the conservation focussed agenda of the founders (CPQLW and RBA) and limited inclusion of concerns beyond conservation of forest land and water. Diwan Singh, the coordinator of the RBA wrote to a Vasant Vihar RWA member,

> "I have become accustomed to slow sensitization on issue of Ridge. Firstly, as you said people do not know about it. Secondly, destruction of Ridge doesn't have an immediate tangible effect. The effect is slow and long term. But unfortunately, it is permanent. We can immediately see losing Rs. 200 per month because of increased power tariffs, so we protest. But that loss is something reversible[195]".

Soni also echoes a similar sentiment when he says "we have a tough job because no one benefits directly from conservation"[196].

The movement pitted itself against the DDA and the builders as the main villains[197] (Dash, 2006; Diwedi, 2006; Seth, 2006; Sharma, 2006). The MoEF and

189 Interview MEV:2
190 Interview MEV:1 (b)
191 Interviews MEV:2 and SA:1
192 Interviews MEV: 1 (b) and MEV: 2
193 Interviews MEV:1 (b) and MEV:2
194 Interview MEV:2
195 Letter to RWA Front from Diwan Singh, RBA convener, 20th June 2005
196 Interview MEV:1 (b)

the Supreme Court also received criticism for deciding that the malls should be given clearance since they had already constructed a substantial part (Bajpai, 2007; Sehgal, 2007). This collusion between "big money and corrupt government" (Seth, 2006, n.p.) was the target of criticism by expert committees (EPCA 1999, 2000; CEC, 2004), by citizens in public demonstrations[198], in newspaper articles (Dash, 2006; Seth, 2006) and press releases by the environmentalists[199]. The DDA was labelled the 'Delhi Destruction Authority' (Seth, 2006) and the 'Delhi Deforestation Authority' (seen on a protest placard pictured in a news article, Indian Express, 2005).

CPQLW and the Ridge Bachao Andolan, the two main (and overlapping) organisations involved in mobilisation and litigation in this case were created for and concerned themselves exclusively with the localised agenda of protecting Mahipalpur Ridge. The main members of these organisations, including Vikram Soni and Diwan Singh, would later shift their attention to the construction on the Yamuna riverbed, having cut their teeth with the Mahipalpur Ridge movement, a leading RBA member states: "We learnt from the Ridge Andolan that we must do a lot of research work and work with scientists as well as include affected farmers"[200]. Although some small NGOs (like Tapas for example) also participated, major environmental NGOs of Delhi, WWF India, Centre for Science and Environment (CSE) and Toxics Link, for example, did not involve themselves in the issue. There were, however, individual members from the said organisation that contributed to the agitation, as a member of a large environmental NGO in Delhi puts it:

> "As an institution, there are constraints; you can't go against the government in such a confrontational manner. There is a political economy of institutions (referring to funding relations). As individuals we took things up because we were also citizens"[201].

7.1.4 The Eviction of Lal Khet Slums for Forest Conservation

In response to the petition against the building of the malls, the CEC report had recommended that the Mahipalpur Ridge be declared a national park (CEC, 2004). The DDA, having committed to the idea in court, declared the remaining area of 223 hectares as the Aravalli Biodiversity Park, earmarked for conservation and restoration in partnership with the Centre for Environmental Management of Degraded Ecosystems (CEMDE) of the University of Delhi. Of the informal settlements that had come up around the now defunct mines, a cluster of six camps, collectively known as Lal Khet (literally meaning 'red field', perhaps after the

197 RBA (2006, 16 September), Invitation for Demonstration: You are invited to a last chance to save the urban environment in India, Ridge Bachao Andolan, New Delhi
198 Interview MEV:1, MEV:1 (b) and MEV:2
199 Onkareshwar, K., (2006. 9 August) Press release, Ridge Bachao Andolan
200 Interview MEV:2
201 Interview EV:1

iron rich dust that must have been the most visible feature in its less vegetated past) fell within the demarcated area for the park and was cleared for removal by the DDA.

Lal Khet was one of the oldest settlements in the area preceding even the establishment of Vasant Vihar and was estimated to have housed around 4000 families[202]. Yet, in contrast to the high profile protests against the malls, the demolition of Lal Khet was a rather quick and quiet affair. On 27 July 2004, the residents of Lal Khet were given a verbal notice of eviction by DDA officials and police officers[203]. There seems to have been some intention to protest the decision initially, letters protesting the relocation plan offered were sent to the DDA and the Ministry of Urban Development asking for a public hearing (Verma, 2004c). Apart from requesting a stalling of the demolition, the matter of the ironic legitimacy given to the malls was also brought up in these letters[204]. Demonstrations were carried out along with attempts to enlist support of political leaders including the leader of the ruling coalition of the central government, Sonia Gandhi[205].

Only about a fifth of the residents of the settlement were found eligible for relocation[206] (Verma, 2004c). Of the few that were offered relocation, even fewer could afford to pay the 7,200 rupees down payment required to secure a place in the resettlement colony. Moreover, the relocation area offered to the residents was in Bawana, a distant location in the north-western periphery of Delhi where the recently re-located families from the Yamuna Pushta slum had been sent[207]. Bawana had already achieved infamy after the Pushta relocation, there had been recent reports of deaths including those of 12 children due to water borne diseases fostered by the unsanitary conditions and lack of facilities including drinking water and healthcare (Bhan, 2006; Perappadan, 2006b). Moreover, the residents of Lal Khet had found employment in their current area of habitation after the closing of the mines. The men worked as contract labourers and rickshaw pullers whereas the women mostly worked in the nearby residential areas as domestic helpers. The move would sever these income opportunities without providing alternatives since in Bawana there were few prospects for employment (Bhan, 2006).

This resolve to stand up to the authorities and protest dislocation soon dissipated in the light of a connected event. Interviews with evicted residents reveal that a day before the scheduled demolition of Lal Khet a school in a nearby infor-

202 Interview SA:1, MEV:1

203 A previous verbal intimation had been given on 26.06.04 but the demolition had been postponed for unexplained administrative reasons

204 Letter to DDA vice-chairman, 26th July 2004 from Gita Diwan Verma

205 Interview SA:1

206 Eligibility for relocation is based on availability of formal documents providing proof of residence in the slum since (or before) a cutoff date

207 The Yamuna Pushta case is a parallel case of dislocation carried out on claims of environmental protection in Delhi. This was a large old settlement that was removed from the banks of river Yamuna on orders from the Supreme Court in March 2004, a few months before Lal Khet (Baviskar, 2011; Bhan, 2006)

mal settlement (near Sultangarhi) was demolished without notice. Several children from surrounding slums including Lal Khet were enrolled in this school and the subsequent protests and petitions to the concerned authority were met with no response. The apathy of the administration regarding questions and objections around this demolition led to a sense of helplessness that soon percolated through the area. By the evening several residents started to dismantle their homes and the protest fizzled out after four of the six camps in Lal Khet decided to cooperate with the DDA (Verma, 2004a). The next day, the 29th of July 2004, Lal Khet was demolished[208].

The demolition did not garner much public attention, a social activist working in the area mentions that she was the only outsider present in Lal Khet during the demolition and days preceding it (Verma, 2004a). She points out that none of the environmental group's active in the area protested the demolitions. It must be pointed out that the RBA was yet to be formed and the demolition was sudden affording little time for mobilisation of a protest. At the time of auctioning of the plots to the malls, the Master Plan Implementation Support Group[209] (MPISG), filed a PIL in the Delhi High Court[210] demanding, that the illegal malls as well as the biodiversity park project be stopped as they were in contravention to the master plan and represented a misuse of land acquired by the DDA for planned development. Issues regarding use of bore wells for ground water extraction and excavation of Ridge land for construction and park development were also brought up by the PIL. The High Court, however, declared that the matter was sub-judice in the Supreme Court and therefore no decision could be provided by them.

MPISG blames the environmentalists and the CEC for the demolition of Lal Khet, by pointing out that by calling for conservation in an area which had been lawfully mined for many years, they were implicitly validating the demolition of the informal settlement and in this, they were supported by the media that failed to highlight the flipside of such conservation measures (Verma, 2004e). They point out that their call to save the Ridge does not reflect on the issue of for whom the Ridge is being saved (Verma, 2004e). While the petition was being heard by the CEC, the MPISG was demanding the details of the petition and a public hearing including the residents of the old informal settlements[211] to which they received no reply (Verma, 2004c).

From a planning perspective, it was argued that the area earmarked for the biodiversity park contains residential area of 35 hectares in the master plan. This is enough for at least 5,000 housing units for the economically weaker section according to current social housing schemes. These housing rights were laid down

208 Interview SA:1

209 A citizen platform pursuing the implementation of the Delhi Master Plan: http://architexturez.net/doc/az–cf–22055

210 Shiv Narayan v/s DDA & Ors.Writ Petition no. 8523/2003

211 Letter to CEC : With reference to news reports about NGOs petitioning it for declaring Lal Khet area ridge, to draw attention to available statutory protections, PIL being heard by High Court, etc. by MPISG, 14th July 2004

by due deliberation and process of law in the master plan unlike the biodiversity park which is not mentioned in the plan and was pushed through by the environmentalists and courts. Therefore the biodiversity park, just as the mall, has usurped the area marked for housing and displaced those who had a right to such housing (Verma, ; 2004, 2004a, 2004b, 2004c). The DDA had acquired huge tracts of land under the Land Acquisition Act of 1984 with the express purpose of implementing the master plan of which providing housing is a key objective. The existence of slums is in fact a manifestation of the failure of the DDA to implement the plans and meet its housing objectives and its tendency to lease land for commercial purposes instead (Verma, 2003). The MPISG further asserted that those living and working in the area had been minding local natural resources by reporting illegal mining, tree felling, and ground water extraction by calling the police, lodging complaints, making applications etc. and that they were being removed on the pretext of better protection for the environment even though the existing environmental legislation could not be implemented as seen in the case of the malls (Verma, 2004b).

Although CPQLW and their associates did not protest during the quick demolition of Lal Khet, the irony of the slum being cleared on environmental grounds while the malls were steadily rising just a few feet away was not lost on them or the affected residents. Residents from Kusumpur Pahadi (a slum adjoining the Mahipalpur Ridge), some of whom had been recently evicted from Lal Khet also joined the agitation against the malls and added a social justice discourse to the agitation. It was reported that in the public hearing by the MoEF the displaced residents argued that

> "if our houses could be demolished on grounds that we were encroaching onto Ridge land, how could the malls be allowed to come up on the same land? Are there different sets of rules for the poor and the rich?" (The Hindu, 2006a, n.p.).

The lawyer for the RBA, argued in court that if the slums were illegal, how could the builders' construction be justified by the MoEF report (HT, 2006c). The invitation by the RBA for demonstration against the mall included the removal of the informal settlement as one of the issues for agitation

> "Developers have not permitted government authorities to impose humanitarian and city planning norms or at least providing temporary housing to incoming labour. The labour has to live in jhuggies/shanties which are set in the green areas. The labour is paid less than the daily wage. This is development […] and the jhuggies (Lal Khet and others) are evicted and replaced by malls. Is this justice?"[212].

That said, the conservation demand remained central to the RBA and the slum residents could only add support to this main agenda by pointing out that a mall should not be built in a space cleared for conservation. A social activist working with the evicted slum dwellers who had joined the RBA protests recalls

212 RBA (2006, 16 September), Invitation for Demonstration: You are invited to a last chance to save the urban environment in India, Ridge Bachao Andolan, New Delhi

"there was no decision making power for the people of the settlement. The movement was good but along with the environment, they should have looked at livelihood issues. The environment is also important but it should not be simplified"[213].

Residents of surrounding villages joined the protests on certain platforms as well, though their main interest was different from that of the environmentalists (See Box 3 for an explanation of how village common lands across the Ridge were vested with the state). Lal Khet residents were concerned with highlighting the injustice of their displacement in the face of the building of the malls and the villagers were concerned with water shortages that would be aggravated by it. On the involvement of the village councils, Soni of CPQLW mentioned in an interview to a magazine

"It is not my movement, it is people's movement. People have joined the movement by themselves. They are usually more concerned over water than on general environmental issues. We have not planned our movement as such." (Sarkar De, 2006, p. 7).

Box 3
FROM VILLAGE COMMONS TO STATE MANAGED FORESTS

The expansion of Delhi engulfed several villages into the city. As the villages were urbanised, their residential areas remained exempt from the municipal building bye-laws but the agricultural and common lands have been turned to other purposes. Village commons were usually of a lower quality in term of fertility and were used for grazing, fuelwood collection and other non-agricultural uses (Kaul, 1990). Since almost the entire Ridge fell under the category of 'gair mumkin pahad' (non-cultivable hills) much of this was held as commons. In the northern parts of Delhi, the villages were dominated by Jat agriculturalists and much of the land in these villages was brought under cultivation, but in the drier southern parts, Gujar herdsmen dominated and preferred to maintain consolidated common lands for cattle rearing (Kaul, 1990; Soni, 2000).

The Delhi Land Reforms Act of 1957 cancelled all individual rights in waste lands, grazing lands, forests and lands of common utility in villages of Delhi and all such lands were vested in the village councils (section 7 of the act). The commons and fields of some villages were acquired by the DDA from the village councils (in the case of commons) and individual farmers (in the case of fields) in exchange for monetary compensation, for the purposes of planned urbanisation (Kaul, 1990) as in Sanjay Van and Mahipalpur Ridge. In the Asola Bhatti Sanctuary area, vast tracts of the Southern Ridge were held by the village councils as commons and wasteland. However, village councils have ceased to be active in formal politics in Delhi, the last election for village councils was held in 1983 (Mahapatra, 2014). All uncultivated village council lands in the Southern Ridge were handed over to the Forest Department in the Asola Bhatti area under orders from the Supreme Court in 1996 (Writ Petition (Civil) No. 4677/1985).

Local leaders from the villages also adapted the environmental agenda and fused it with their claim on land. In February 2013, the village council of Mahipalpur

213 Interview SA:1

Village joined the environmentalists of the RBA in decrying the construction activity pushed by government agencies in the Mahipalpur area and asserting that only environment friendly activities should be allowed on the Ridge land, which was earlier part of the village commons but had been taken over by the DDA[214] (Singh, 2013a). To this end a plantation drive was planned to 'reclaim village common land' (HT, 2013). The local politician and co-convener of the farmers collective (Kissan Mahasangh) spearheading the drive, explained the idea behind this plantation as being a protest against misuse of land by the government and a demonstration of what they hold as the solution to the question of the Ridge: leave village common lands as open green areas [215]

Thus a variety of socio-environmental agendas came together in a broad discourse coalition against the DDA and the construction activity it enabled in the Mahipalpur Ridge. The eviction of Lal Khet was the most conspicuous case of social injustice in the process of the formation of the current form of socio-nature hybrid in the Mahipalpur Ridge, namely the Aravalli Biodiversity Park. However, as we shall see in the following section, the conservation discourse put in practise through the park model has led to re-configuration of not just the landscape but also of access and control mechanisms at play. The events that leading to the establishment of the park, have been summarised in the timeline below (Figure 2).

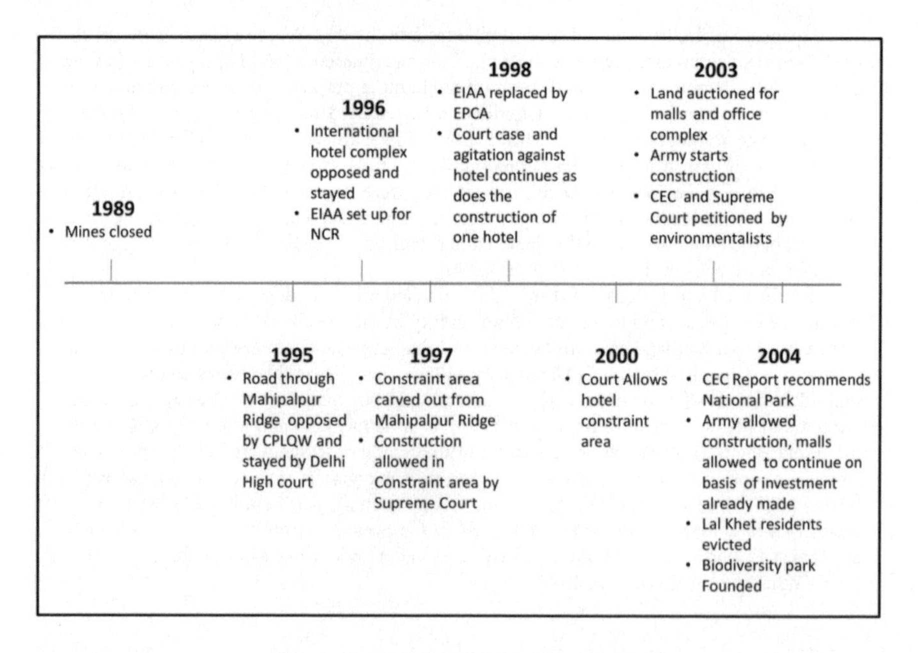

Figure 2: Timeline, Mahipalpur Ridge

214 Interview SA:2
215 Interview SA:2

7.1.5 Current Status: Aravalli Biodiversity Park and its Neighbours

The foundations of the Aravalli Biodiversity Park were laid on 22 February 2004 and the ceremony was attended by the Deputy Prime Minister, the Lt. Governor of Delhi and four Members of Parliament, among others. The park was presented as the new model of conservation that would restore the native ecology of the Aravalli Range to the Ridge (The Hindu, 2004). The park is managed by the Centre for Environmental Management of Degraded Ecosystems (CEMDE) of the University of Delhi and the DDA provides civil engineering and manpower support. The idea of a conservation park in the area had already been mooted in during the deliberations in court as described above and DDA had committed to it. The entry of the CEMDE in such a decisive role had to do with the intervention of the Lt. Governor. According to one of the former members of the CEMDE team

> "Vijay Kumar, the ex Lt. Governor knows Shiela Dixit (then Chief Minister) and Deepak Nayyar (then Vice Chancellor of the University of Delhi) well. He was very keen that these pockets be saved from commercial use. He approached the university through the Vice-Chancellor to do something. The funding was large so a specialised sub-group like the CEMDE had to be brought in. We were asked to present a proposal to the Lt. Governor and he pushed the whole thing through[216]".

In the span of a decade, the landscape has changed significantly through specific efforts to create the ecological imaginary (Peet & Watts, 1996) envisioned by the DDA and CEMDE. The open scrub vegetation has been replaced by a carefully curated mosaic of what is considered natural by the planners and managers of the scheme. The area consists of a patch of acacia dominated forest, a section of moist deciduous forest, a conservatory of medicinal plants, around 150 acres of rangeland including various kinds of grassland and woodland. The open pits from the abandoned quarries have been developed as water bodies in some cases (though some ponds had formed naturally before the park was developed). In other cases the lower temperatures and moist conditions provided by the series of larger pits (up to 700 meters in length and 30 meters deep) are being used to plant a 'tropical rainforest', an orchid conservatory with around 50 orchid species and a fernery. The park also contains a butterfly conservatory where plants and weeds were specially planted for the 90 odd species that exist here today. Sprinklers have been installed to maintain the moist conditions necessary and a butterfly nursery has been built where the caterpillars are cared for to be released later. Water scarcity is indeed one of the problems that the park authorities have to deal with, bore wells have been sunk to provide water to the saplings and trees. There are plans to add a rock garden, a rare cactus collection and a bulbous plant conservatory among other things (Roychoudhury, 2013). According to the scientist in-charge, the park holds around 1,000 different species of plants, 190 species of birds, 24 species of reptiles and over a lakh[217] insect species[218]. The park is presented as a

216 Interview EX:4 also corroborated by DDA:1
217 A lakh = 100,000
218 Interview EX:2 (b)

'home for vanishing flora and fauna' and its goals are spelled out as conservation, restoration of the original Aravalli ecosystem and nature education (of students, teachers, administrative officers and scientists who can receive permission to access the park)[219]. 'Bringing back of lost biodiversity' of the Ridge and 'saving nature from urbanisation' are the main storylines of the CEMDE. As the scientist-in-charge of the park puts it:

> "The Ridge and the Yamuna used to support life when the population was scanty. Today they are fragmented and degraded. We cannot bring back larger animals like the leopard but we are reviving the lost genetic diversity of Delhi"[220].

One of the goals for these parks cited on the official website is providing 'heritage sites' for ecological communities. Even though the park authorities seem share the heritage storyline from the environmental activists, they differ in that they see themselves as creating and 'bringing back' this heritage rather than protecting existing heritage. This active creation of 'nature' and the authority of attesting to what is the 'correct' biophysical nature for the area, provide CEMDE its role and legitimacy.

The whole area is divided into the nature reserve area and the visitors' area. The visitors' area is in fact a trail, about one and a half kilometres long, cutting across the park from one gate to another, fenced on both sides with barbed wire and wooden posts (made with the remains of the infamous vilayati kikar trees that have been chopped down). It is clear that the only public access and use in this area has been conceived as walking across the park on the trail, "guided movement of a few people" as the park authorities put it[221]. The scientists of the CEMDE control the area tightly. As natural scientists with an eco-centric view, they consider most human activity 'interference'. A field biologist of the centre complains of people cycling, jogging, exercising, doing yoga or bringing their dogs to walk on the trail.

> "Birds and animals feed in the morning and evening that is when most people come here too and that does not suit a biodiversity park. People have no understanding of core and buffer areas"[222].

In the list of activities, these recreational activities are not the only ones that the park's management disapproves of. The main conflict of interest has arisen with the two informal settlements at the borders of the park. Located in the 'constraint area', they were allowed to continue to exist. According to a scientist in the park:

> "When we came here, this area was severely degraded, there was only Prosopis juliflora, cows were wandering about and it was a free for all. How to get rid of these people? Slowly, slowly we encroached this area and fenced it everywhere"[223].

219 The website of the Aravalli Biodiversity park is available at- https://dda.org.in/greens/biodiv/aravalli-biodiversity-park.html (accessed 02.12.14)
220 Interview EX:2
221 Interview EX:2
222 Interview EX:3

The residents of these areas used the open space that is now the park for many purposes. Apart from recreational uses, the space fulfils the needs of fuel wood, pasture for cattle and pigs and is used for sanitation purposes as an open toilet. Though these uses continue covertly, the building of walls and strict security enforcement is rapidly making continued access difficult[224]. The park management is often petitioned by groups of residents from the nearby settlements for return of impounded cattle, fuel wood and through fares[225]. One group of women approached the CEMDE scientist in-charge, requesting that a particular path connecting their settlement to Vasant Kunj through the park be re-opened as many of them work as domestic workers in that area. The scientist told them it is not possible to do so. "How can I allow that, it will fragment the whole forest"[226] he explained. The park design demands a complete subordination of all other uses and claims for strict protection of the ecosystem from human activity outside that of the experts.

Not everyone agrees with the park's design even among environmentalists and scientists. An environmentalist from CPQLW mentions that the removal of 60-70 year old Prosopis juliflora trees on account of being invasive serves no purpose. The park does not confirm with the demand of CPQLW of letting the forest regenerate by itself, he states: "I don't know why they are fiddling with it. I suppose they have to show that they spent the funds"[227]. The NGO Tapas filed a public interest litigation petition in the High Court against the park to secure this area as Ridge forest and demanded application of forest laws here. The petition opposed the digging and construction activities that have been carried out by DDA for the purpose of this park, the non-native species grown here and bore wells sunk to extract water for the plantations. All this, according to Tapas, is invasive on the ecology (The Hindu, 2007). The 'let nature take care of itself' demand of these environmentalists is not acceptable to CEMDE since their role as an institution is to provide active management and not solely protection (which is the role of administrative bodies like DDA). A conservation ecologist who was part of the Kalpavriksh movement to save the Ridge in the 1990s and carried out some research work with students in the park in 2011-12, also disapproves of the scheme

> "I don't agree with the design of the project. They have no historical understanding of the flora; they have tried to recreate the larger Aravalli region. For example, they brought species from Gujrat and Rajasthan and not looked at local variations [...]. There is no basis for choosing this vegetation. The area is very rocky and the ground water is very deep since there is a water scarcity. So a thorn scrub forest is natural here, it doesn't have species from all over Aravalli like bamboo. This is more like a museum not a restoration project. But at least they are replacing the Prosopis juliflora"[228].

223 Interview EX:3, interestingly the scientist uses the word 'encroached' to describe the development of the park
224 Field notes FN-ABDP, FN-KP, FN-BS
225 Interview EX:2
226 Interview EX:2
227 Interview MEV:1 (b)
228 Interview EV:6

There exist, different normative conceptions of what the 'natural' in the creation of a space for nature conservation in the Mahipalpur Ridge should entail.

Despite these oppositions, the biodiversity park model seems to have caught on within the DDA and has been institutionalised and extended. The Delhi Biodiversity Foundation was established to overlook the implementation and extension of the biodiversity park model. The foundation includes the highest administrative authorities of Delhi including the Lt. Governor, the Vice Chairman of the DDA and the Chief Secretary of GNCTD[229]. Presently two parks following the biodiversity park model exist, the other being the Yamuna Biodiversity Park set on 457 hectares of the riverbank (AMDA Bulletin, 2011). A substantial part of DDA's budget for Delhi's green spaces has been allotted to the two parks. The Aravalli Biodiversity Park alone was granted funds to the tune of fifty two crore rupees between 2004 and 2011[230]. Four additional biodiversity parks are planned in the near future: the Northern Ridge is to be converted into such a park as is Tilpath Valley adjacent to the Asola Bhatti Sanctuary, Neela Hauz Lake near the South Central Ridge and Sanjay Lake, an artificial lake in East Delhi[231].

That CEMDE shares the idea of the environmentalists that the Ridge is a space for conservation and not a horticultural park. The DDA, as is evident in the master plan, had earlier envisioned the Ridge as a recreational green space beautified through its horticultural expertise. This was modified into discourse of nature conservation through its exchange with civil society in the 1980s and 1990s and has found a concrete model in the biodiversity parks through the introduction of the CEMDE's storyline of deep-ecological conservation and removal of human interference.

As for the contestation, there exists a 'tender peace' (Keil & Boudreau, 2006) between the environmentalists and the park authorities (DDA and CEMDE) regarding the biodiversity park. As a member of RBA puts it "we have managed to save a part of the Ridge that is some solace. I would say battle lost but war won"[232]. The surrounding informal settlements though face their share of struggles for access to the space which by early 2014, seemed close to being walled-in entirely[233].

7.2 SANJAY VAN: CITIZEN RESTORATION PROJECT

The once contiguous South Central Ridge now exists in fragments, of which the 784 acre (around 317 hectares) Sanjay Van is the only one that received Reserved

229 List of members from the biodiversity park website- https://dda.org.in/greens/biodiv/delhi-biodiversity-foundation.html (accessed 02.12.14)

230 Delhi University, CEMDE- Biodiversity Parks Programme of DDA http://www.du.ac.in/du/uploads/rti/Biodiversity_Park_2652011.pdf (06.04.14)

231 Interview DDA:1 , Also listed at https://dda.org.in/greens/biodiv/aravalli-biodiversity-park.html (accessed 02.12.14)

232 Interview MEV:2

233 Field Notes FN-ABDP, FN- KP, FN-BS

Forest status in the 1994 notification. This parcel of forest contains and is surrounded by some of Delhi's prime historical heritage sites; the Qutab Minar complex and Mehrauli Archaeological Park are located on its eastern boundary. The vantage point of the walls of the 12th century fort Quila Rai Pithora within Sanjay Van provides a view of the forest's green canopy punctuated with the domes and minars of the monuments adjacent to it. In the more recent past, the area consisted of forest and village commons (including fields and pasture land) for the inhabitants of Kishangarh, Masudpur, Malikpur Kohi, Sultan Garhi, Rangpuri, Mahipalpur and other villages in the vicinity. The segment also formed an important link in the drainage network of the Ridge and the water bodies here are connected to the Neela Hauz Lake that eventually connects with the Yamuna River (DDA, 2013). In 1989 the Mehrauli area was declared protected heritage zone and Sanjay Van was envisioned as a green space surrounding the monuments (DDA, 2013). After the development of the Vasant Kunj residential area, this segment was cut off from the rest of the South Central Ridge and despite its Reserved Forest status fell into neglect. Today it is surrounded (in clockwise order) by the Qutab Minar, the Qutab institutional area, the large urban villages of Kishangarh and Mehrauli and the middle class housing area of Vasant Kunj that lies across Aruna Asaf Ali Road (see Map 5 in Appendix B).

The declaration of Sanjay Van as Reserved Forest did not immediately or automatically change its material character into that of a forest ecosystem. DDAs horticultural department continued to manage the area as a landscaped recreational park. This is evident in the central lawn, wide gravelled pathways with brick borders, abundance of ornamental plants like Bougainvillea and constructed cement benches, all of which continue to exist today as markers of the horticultural practices of the management before 2010. The growth of Prosopis juliflora was tolerated and even encouraged in this period by the DDA as it increased visible green cover[234]. The surrounding villages continued to use the space as a common land for recreation, fuel wood gathering and grazing. The forest also has several permanent structures that continue to exist in the area. The largest of these are a temple and a dargah (a Sufi shrine). Residents of the Mehrauli (Ward Number 2) and Kishangarh villages recount that there was no hindrance to their use of the forest apart from the general fear of the groups of men who used the open space to abuse alcohol or narcotics. The DDA staff's presence was scarce and the boundary wall was almost non-existent[235].

In 2005, following complaints from citizens regarding the planting of ornamental plants by DDA, garbage dumping by neighbouring residents and encroachments of various kinds, the civil society members of the Ridge Management Board made an inspection visit to the area and chastised the DDA on its lack of management. The main criticisms of the RMB included horticultural plantations, the presence of a motorable road cutting across the forest, heavy grazing pressures from cattle belonging to the surrounding residents, encroachments including several shrines

234 Interview DDA:2

235 Filed notes FN-MRW2 and FN-KG

and temples and a concrete shed that housed DDAs gardening tools among other things (Perappadan, 2006a; Roy, 2005). The report concluded: "under this circumstance the forest character of the area is not maintainable, unless drastic action is taken" (Perappadan, 2006a, n.p.). Not much changed despite this official acknowledgement of the situation in the following years. In January 2010 the state of Sanjay Van was summarised in a popular newspaper thus:

> "Its main gate from Aruna Asaf Ali Marg is blocked but a sign welcoming you to Sanjay Van stands boldly next to a gap in an almost non-existent boundary wall. Once in, a strong stench of decay makes you wonder whether you were actually welcome or this was just a cruel joke. A two minute walk from the entry point leads one to what used to be a huge water body, now is just full of several litres of Vasant Kunj's untreated sewage. Apart from the fact that most parts of Sanjay Van are green (with Prosopis juliflora), it is difficult to imagine the 626 hectare area is actually part of Delhi's Ridge and a Reserved Forest" (Lalchandani, 2010, n.p.).

This lack of forest character showed in apparent disturbance to the wildlife of the Sanjay Van. A pack of six golden jackals (an animal sighted often in this parcel of the forest) was found in 2007 in the Delhi airport and a single adult in the Connaught Place Market a few months later (Singh, 2007). The birds noted in the areas, including migratory ones, dwindled according to local birdwatchers[236].

A short report prepared by a resident living in the close vicinity of the forest (in a middle class housing scheme) was published by the NGO Srishti, listing the non-forest land use and activities in Sanjay Van in 2011. The purpose of the report was to call for a community conservation project involving the various organisations based in the Qutab Institutional Area and other constituencies, such as student groups, bird-watchers, morning walkers, devotees visiting religious structures, cyclist etc. (Chauhan, 2011). This particular project was not followed through. Meanwhile, another group had begun its efforts to remake the forest. In 2010, a voluntary citizens group under the title Working with Nature (WWN) undertook the restoration of the forest with the DDA. The group is led by its founder member, a retired senior Air Force officer and resident of Vasant Kunj. Using a golf cart to move around Sanjay Van on its wide, levelled trails, he directs the DDA horticultural staff on everything from tree selection, maintenance and water body management[237]. He has enrolled a handful of other members to form the group including ecologists (one each from JNU and Delhi University) and bird watching enthusiasts to assist in technical aspects of restoring native trees and reviving water bodies. However, as an individual, his role in the project remains central. His custodianship of the forest and the form of physical nature within it is far reaching, to the extent that he refers to Sanjay Van as 'my forest' at several points in the interviews undertaken in the scope of this research[238].

The Lt. Governor (Tejender Khanna, in this case) played an important role in making possible the re-shaping of Sanjay Van. Both the DDA and WWN highlight his 'kind patronage and guidance' (DDA, 2013; WWN, 2012) as a means to

236 Interview EV:7
237 Field notes FN-SV, Interview SVEV:1
238 Interview SVEV:1, SEV:1 (b)

legitimise the Sanjay Van restoration project and WWNs role in it. It was the close acquaintance of the WWN founder and the Lt. Governor[239] that led to the semi-formal arrangement that allows WWN to direct DDA's manpower and funds to work in the area. Being a frequent visitor to the area for morning walks, the founder of WWN a self-described 'nature lover' wanted to bring the notice of the government towards the need for the restoration of Sanjay Van. He presented a proposal to the Lt. Governor's office demanding removal of invasive trees, restoration of the water bodies and involvement of stakeholders including younger citizens. The proposal demanded that the area be maintained through a 'scientific approach' (as opposed to DDAs horticultural approach) to restoring original flora and water bodies. After campaigning for these changes through 2008-2009, in the words of the WWN leader; "finally, I brought him here and he agreed that something must be done. Then he said why don't you do something"[240]. The Lt. Governor wrote a letter to the DDA asking them to work in association with WWN[241]. It is through this arrangement that WWN, characterised as an 'empowered citizens group (working) in partnership with DDA' (WWN, 2012), continues to play a major role in reshaping the area. The Lt. Governor's patronage was inscribed in the forest by naming a viewpoint 'Tejender Point 2011' after he inaugurated it to commemorate the International Year of Forests[242] (DDA, 2013).

The main agenda of WWN in the Sanjay Van revolves around planting indigenous trees, reviving water bodies and involving certain groups in activities within the forest. WWN is directing the efforts of the DDA to restore native trees to the area including Dhak, Kejri, Babool, Khair etc. which had all but disappeared from this patch according to them[243]. About 100 acres have been dedicated to growing medicinal plants. The group is also attempting to restore the five water bodies in Sanjay Van by using water from the nearby Neela Hauz Lake and treated water from the sewage treatment plant (STP) in Vasant Kunj. Three of the five water bodies had earlier dried up and two were heavily silted and contaminated with sewage leakages and waste water discharge from surrounding areas (Chauhan, 2011). The 'waterman' Rajinder Singh[244] was involved in an advisory role to build simple, small scale water harvesting structures to replenish the water table and contribute to the revival of the water bodies (DDA, 2013; Krishna, 2014). There is some evidence that bettering the water recharge in the area has been on the DDA's agenda before. This can be concluded from reports of the School of Planning and Architecture (SPA) sanctioned by the DDA (SPA, 1997). The involvement of a famous environmentalist like Rajinder Singh, however, would

239 interview SVEV:1

240 Interview SVEV:1

241 Interviews SVEV:1 and DDA:3

242 This viewpoint has been renamed to 'Point 2011' in 2014

243 Interview SVEV:1, SVEV:1 (b)

244 Rajinder Singh was awarded the Ramon Magsaysay Award for community leadership in 2001 and is known for his work in community-based water harvesting and water management in semi-arid villages in Rajasthan

certainly have served an additional symbolic purpose and helped gain some public attention.

Some of the demands and aims of the WWN are in line with those of other environmentalists concerned with the Ridge, for example, the building of boundary walls, employment of security staff and the reintroduction of native flora. The inclusion of bird watching enthusiasts in the WWN contributed an additional dimension to the aims of the organisation. It is hoped to have the area declared as a bird sanctuary in time as the bird count increases. To this end there is special attention paid to the revival of the water bodies, pisciculture and tree selection to attract more birds[245]. The idea is also being highlighted through newspaper articles to garner public attention and support (Lalchandani, 2011b; The Hindu, 2011). The status of a bird sanctuary can be expected to legitimise and legalise a higher level of physical protection to this segment and further decrease levels of access currently possible for the local population.

WWN does not however, subscribe to the need to separate the forest from citizens to the extent seen in the biodiversity park model. One of the main stated aims of the group is to 'connect people to the forest', a phrase repeated on signboards within the forest, in leaflets, and by the founding member of WWN in lectures and interviews to the press (The Hindu, 2011; WWN, 2012). This is 'connection' is encouraged through involving groups of citizens in organised plantation of saplings. School children, local RWAs and groups like the Rotary Club of Vasant Kunj as well as office goers from adjacent Qutab Institutional area are some of the targets of such community plantation drives. School children (including those of 'non-rich schools and special children' stresses the WWN leader) have been regularly involved in jungle cleaning, plastic removal, tree plantation, nature walks, bird watching etc. by WWN as they have a "stake in the future and must gain ownership of the forest in order to care for it"[246].

There are, however, other groups of users that are not as welcome as the ones mentioned above. The main category of people seen as degrading and harming the forest by WWN and DDA staff, are the residents of the two surrounding urban villages of Kishangarh and Mehrauli Ward No. 2. Garbage dumping on the periphery, grazing of pigs and fuel wood gathering are the common grouses against them[247]. Small scale poaching of wild boar, smaller mammals, such as hares, birds and even occasionally blue bulls has also been known to take place according to local residents and field staff of the DDA[248].

Sanjay Van also has two temples, a dargah and scattered old mausoleums of Sufi saints that are frequented by the local residents. Of the two temples, the larger one holds bi-annual fairs in addition to the steady stream of devotees and vehicles it attracts on a daily basis (Chauhan, 2011). This temple is on the outskirts

245 Interviews ENV:7 and SVEV:1
246 Interview SVEV:1
247 Interviews SVEV:1, DDA:2 and SVFO
248 Field notes FN-SV and Interview DDA:2

and therefore WWN and DDA plan simply to fence the temple out[249]. The second temple is small and hosts two priests in one room, a courtyard and a small garden. One of the priests, who has lived here since 1981, recalls the pre-WWN Sanjay Van as a place of degenerate activities focused on alcohol and drug consumption. At the same time the temple dwellers do not share a cordial relationship with the authorities who called the police when they tried to pave the pathway to the temple[250]. The identity of this space to them is not forest but village commons. The priest recounts the myth behind the establishment of the two temples in Sanjay Van as evidence that the space has long been an area of habitation and pasture: A hermit called Gorakhnath (after whom both temples are named) was meditating here when he saw a worried villager looking for cattle he had set out to graze in the area. He told the man to go home and his cattle would be there. In return for this favour the man was to build a temple in his name for the surrounding villages[251].

The Aashiq Allah Nazaria Peer Dargah shares a far more tense relationship with the authorities. The dargah is claimed to be a historical site (more than 700 years according to its website) [252] and enjoys wide-spread patronage including that of certain powerful ministers who are known to frequent it[253]. The dargah board has petitioned the DDA and the Lt. Governor to be allowed to build a paved access road, install bore wells for water and erect streetlights citing the large number of visitors they receive[254]. Problems of access for the devotees, hampered by Sanjay Vans' security staff in recent years, is also on the list of complaints[255]. Access has been denied to them on grounds of being located on forest land[256]. The WWN founder contends that

> "these are not adivasis (native forest dwellers), they have no claim to forest land, the land mafia has sold the land to them and they are not being removed to cater to the vote bank"[257].

According to the president of the board of the dargah, the DDA records show the wrong survey numbers for the dargah structure and despite many requests and signature campaign petitions, this has not been corrected making the dargah an encroachment on forest land as per records;

> "The DDA did the survey to settle land rights when this was declared a forest, they gave the wrong survey number. We cannot simply move the dargah two kilometres away. How can a 750 year old dargah be an encroachment?"[258].

249 Interview SVEV:1 (b)
250 Interview SVEV:1 (b)
251 Interview SVLS:1
252 Website of the dargah- www.ashiqalla.com (accessed 24.07.14)
253 Interviews SVEV:1 (b) and SVLS:2
254 Interview SVLS:2
255 Earlier there were several routes to access the dargah from various settlements around. Now due to the walls built around the space, access has become difficult
256 Interview DDA:3
257 Interview SVEV:1
258 Interviews SVLS:2

The case is yet to be settled by the court meanwhile both sides continue to approach politicians and members of powerful elites to argue their case. According to WWN and the DDA, the dargah only continues to exist in the forest due the political connections it enjoys and the sensitive religious nature of the matter that makes the government wary of forcing the issue[259]. The patronage it enjoys has seen the dargah's main structure expand in the last ten to fifteen years according to local residents and DDA staff[260] (Chauhan, 2011).

The accusation of destruction of heritage has been turned on the WWN by the dargah board. They allege that in order to impose a new landscape the WWN and DDA destroying religious heritage. Repairs are not allowed to be made on any structures within the forest without permission from DDA and the dargah authorities point out that unless constant care is taken the Prosopis juliflora roots will damage both the main dargah structure and the mausoleums around it. There is also active destruction of property in the creation of the forest. The police have been called by the dargah board when they found certain mausoleums had been destroyed during plantations by the DDA[261].

The space is being turned from a village commons to a managed forest, though this change was neither uncontested nor is it complete (as in the other two cases detailed here). The use of the space for various purposes including recreation[262], fuel wood collection, grazing, small scale hunting of birds etc. continues in a clandestine manner. The 2010 report on the condition of Sanjay Van mentions people washing their motor bikes in the forest, tying their cattle in the area and drying dung cakes for fuel to the extent that in areas adjoining Kishangarh and Mehrauli, the report states: "it is hard to distinguish if you are in someone's backyard or the backyard is in the forest" (Chauhan, 2011, pp. 10–11). Access for such activities is increasingly restricted through the construction of concrete walls and an increase in security measures. The DDA staff and WWN at times resort to calling the police when they find fuel wood collectors or grazing cattle[263]. Building of the wall at the end of 2013 raised an alarm among those who accessed the space from the two villages, many admitting that tough they knew that their activities were illegal; they did not expect such drastic changes in their access possibilities[264]. WWN sees this change of access and use patterns as an essential step in realising the forest character they have in mind, as the founder puts it:

> "It was a free for all, not a forest, there was mass opposition to the wall, we got many petitions but I want a clean area so that people can come and form a connection with nature"[265].

259 Interviews SVEV:1 (b) and DDA:3
260 Interviews DDA:2 and SVFO
261 Interview SVLS:2
262 Recreational activities include playing, visiting sites of religious importance, walking and jogging, etc., but also alcohol and drug abuse
263 Interview SVEV:1
264 Field notes FN-KG (b) and FN-MRW2 (b)
265 Interview SVEV:1 (b)

In this case, unlike in the case of the Mahipalpur Ridge, the DDA is seen as an ally rather than an opponent by the environmentalists involved. The problem is seen in their lack of scientific knowledge and experience regarding forestry issues, a fault that can be rectified by capable management of their activities. Thus the positioning of actors (self-positioning as well as perceived positioning of other actors) made alternative storylines (that of stakeholder partnership) and alternative solutions possible. In such a non-confrontational situation, in contrast to the lack of engagement from larger NGOs regarding the Mahipalpur Ridge, WWF India entered a partnership with WWN on nature education for school children in Sanjay Van[266].

In both cases the partnership of external organisations with the DDA was demanded and justified on the grounds of a more scientific approach to forest management than the DDA had demonstrated. The scientific approach and the role of scientists is, however, very different in the two models. Some members of WWN who subscribe to a harder scientific conservation storyline disagree with the citizen tree plantation drives on grounds that it leads to unscientific stewardship[267]. The scientists of the biodiversity park are not convinced about this model of restoration either; the head of the CEMDE dismisses it by declaring it will not succeed in the long run "what can [name of WWN founder] do with the school children? They cannot replace the Juliflora like this"[268]. The WWN leader on his part, points out that their organisation is not paid by the government and still shows results despite not having a free hand like the scientists in the biodiversity park[269]. The DDA mentions Sanjay Van as a model for restoration of the Ridge (DDA, 2013) just as it does for the biodiversity park model though the former has not achieved the same level of institutionalisation as the later.

Interestingly, despite such emphasis on scientific management, in neither case was a study conducted to find out previous uses of the land or their effect on the forest. Uses of the surrounding slums were seen as 'criminal' and illegitimate and therefore a major reason for degradation of the forest. Nonetheless, there was no attempt to actually ascertain their level of impact. An ecologist from Kalpavriksh (former student member) contends: "I don't think the extraction of fuel wood and grazing is on a very high scale. The Ridge can take far higher extraction in my estimation, especially the juliflora"[270] but no empirical data exists in how much biomass is actually extracted. The role of science as embodied in both of the above models is to define and manage native flora and improve the quality of water bodies and the level of the water table. All other concerns are subordinated to this legitimised and institutionalised scientific discourse (within the governance structure of the park and Sanjay Van) that cannot be debated by other actors on the same terms.

266 Under the Young Climate Savers Programme, WWF (I)
267 Interview EV:7
268 Interview EX:1
269 Interview SVEV:1
270 Interview EV:6

7.3 THE ASOLA BHATTI WILDLIFE SANCTUARY

The rural-urban fringe has been presented as the "ultimate battlefield of environmental and socio-political change brought about by urbanisation" (Bentinck, 2000, p. 13) and is an interesting space to study environmental transformations. These peri-urban[271] spaces have been noted to be often ignored in terms of planning, regulation and service provision as the 'dark zone' between the city and the country (Schenk, 2005). The fringes (as well as many parts of forest land in Delhi) remained a relatively less regulated social and legal space until recently (Soni, 2000). The ambiguous legal status of much land in such areas and a relative lack of regulation have resulted in a variety of unplanned land uses serving various sections of society (Arabindoo, 2005; Schenk, 2005). Urban peripheries or fringes in Delhi are thus characterised by heterogeneity and fragmentation (Arabindoo, 2005) as they contain a range of settlements including elite 'farmhouses', overcrowded village clusters (subsumed by the city and housing both rural population and more recent migrants to the city), unauthorised colonies for both low and high income sections of society and slum resettlement colonies, set up by the city authorities for those evicted from urban slums (Dupont, 2005; Schenk, 2005; Soni, 2000). The upper and middle classes are attracted to the periphery for reasons of more space and better environmental conditions while the poorer settlers find relatively less vigilantly policed land to build their settlements in the absence of available affordable housing in the city. Though tenures remain insecure and dependent on political patronage and bribes (Schenk, 2005)

At this southern edge of Delhi also lies the largest contiguous section of the administratively defined Ridge, the Southern Ridge, which contains the Asola Bhatti Wildlife Sanctuary. This part of the Ridge covers almost 80 per cent of the Reserved Forest area in Delhi and is spread over approximately 6,200 hectares[272]. According to the master plans, this section of the forest lies in the designated green belt but as the latest master plan admits, this greenbelt has been heavily violated by both planned and unplanned developments (DDA, 2007). The expanse of the area and the peri-urban location coupled with its status as a wildlife sanctuary, make this a unique space for conservation compared to the more urbanised segments of the Ridge. This is the only section directly managed by the Forest Department[273], though forest laws apply to all sections of the notified Ridge, giving

271 The rural-urban or peri-urban fringe can be defined as an area in close proximity to the city, as they are a zone of spatial contact between city and countryside (Mortimore, 1975). These peri-urban fringes have recently undergone or are undergoing rapid transformation (Rohilla, 2005) from 'self-regarding' localities to localities which exist in a "continuous but subordinate relation" to the city (Jargowsky, 2005, p. 25). This transformation is linked to the dynamics of processes of the city (Arabindoo, 2005; Dupont, 2005)

272 The demarcation of the actual boundaries of the sanctuary is yet to be completed. The figure is provided by the Forest Department (Sinha, 2014) and used in the Master plan (DDA, 2007) provisionally

273 Small patches of the Southern Ridge outside the sanctuary are managed by other agencies like the DDA and the Sports Authority of India

the department limited jurisdiction in all notified forests. The sanctuary carries several layers of legal protection. The stringent Wildlife (Protection) Act of 1972 is applicable through the 1986 notification[274] that declared that the community lands of three villages namely Asola, Maidangarhi and Sahurpur were to be consolidated in the Asola Wildlife Sanctuary. The community lands of Bhatti Village were added to this in 1991[275] when mining was banned in the area. IFA, 1927 is applicable to the area under the 1994 notification declaring the Southern Ridge a Reserved Forest. In 1996, under the decision regarding the M.C. Mehta case, the Supreme Court gave orders to hand over the 'uncultivated surplus lands' of the village councils to the Forest Department for the 'creation of a Reserved Forest'[276]. This was done under the Delhi Land Reforms Act of 1954 (section 154), but since the explicit purpose of the court order was to create a Reserved Forest, the IFA, 1927 was applied to these areas an interpretation of the law[277]. These two notifications and the court orders created the present Asola Bhatti Wildlife Sanctuary[278] covering around 2782 hectares. The southern and eastern borders of the sanctuary coincide with the Delhi state boundaries with Haryana. Towards the north of the sanctuary the large informal settlement of Sangam Vihar is located and the villages of Asola, Fethpur Beri lie on the west. Sanjay Colony[279], an informal settlement, is located within the sanctuary boundaries connecting Asola and Bhatti areas (see Map 6 in Appendix B).
When the court ordered that these village common lands be handed over to the Forest Department, it regarded these spaces as unutilised wastelands or, as the wording of the 1996 notification puts it, 'surplus' lands that could be put to the more useful and desirable function of creating forestland. On the other hand one of the main reasons for degradation and obstacles to the ecological restoration of the area, according to both the authorities and environmentalists commenting on the sanctuary, was identified as grazing (Mailk & Roos, 1994; Sinha, 2014) indicating that the area had clear utility and was being used. In fact, the extensive use of this space as pasture has a long history and the absence of fields does not imply that the space was unutilized. The scarcity of water (especially in the Southern Ridge which falls in the rain shadow area) rendered farming a difficult prospect. The dominant Gujar settlers in this area are traditionally cattle rearers (Kaul, 1990). As the gazetteer of 1883-84 notes in regard to the hills around Bhatti[280]:

> "water of course lies very deep, and irrigation by well is almost everywhere impracticable. A moderate pasture is obtained by flocks of sheep and goats herded by Gujar boys. This tribe

274 Notification No. F.3 (116)/CWLW/84/897/to906, 1986
275 Notification No. F.2(19)/DCF/90–91/1382–91, 1991
276 M.C. Mehta vs. Union of India and ors. Writ Petition (Civil) No. 4677/85. Orders dated 25.01.96 and 13.03.96
277 Interview FD:1
278 Named Indira Priyadarshini Wildlife Sanctuary in the 1986 notification, though this title seems to have fallen out of use now
279 Not to be confused with Sanjay Van
280 Spelt as Bhati in the gazetteer

has appropriated almost entirely the hill villages, as they suit their pastoral traditions [...]" (Gazetteers Organisation, 1999, p. 2).

The second factor cited as being responsible for degradation was mining[281] (Sinha, 2014). Starting in the late 1960s, the area was a site of extensive mining for quartzite sand, locally known as 'badarpur bajri' used extensively by the construction industry largely in Delhi (Jain, 1994; Sinha, 2014; Soni, 2000). Mining was mainly manual and used mules to carry the bajri out of the deep pits (Mohammed, 2010; Soni, 2000). The mines were leased out to private contractors who profited by overexploiting both the land and the labour as a result of which, in 1975, the central government ordered the Delhi State Industrial Development Corporation (DSIDC) to take over the mining activity on behalf of the Delhi Administration (Sethi, 2005; Soni, 2006a). It was in this phase that the informal settlements of Indira Nagar, Balbir Nagar and Sanjay Colony came up, specifically to house workers of these mines. In practice, the takeover by the government unit resulted in little more than a collection of octroi charges from the private mining companies, the conditions in the mines failed to improve (Soni, 2000). In 1983, press reports of death of labourers due to a mine collapse, led to a judicial inquiry instituted by the Ministry of Labour. The mines were then vested with a second public sector agency, the Delhi State Mineral Development Corporation (DSMDC) in 1985 (Soni, 2006a). This too did not ensure worker safety or compliance with environmental protection measures (Mohammed, 2010). Government control over these mines remained a "legal fiction" as the fringe of the city where these mines were located remained largely unregulated and neither the Delhi Administration nor the Union Government interfered with the working conditions or environmental controls in these areas (Soni, 2000, p. 86). The mines of Bhatti were ordered to be shut down by the Union Ministry of Labour after a mining accident that killed seven labourers in 1990 (HIC-HLRN, 2006; HT, 1991). The closing of the mines left around 4,000 miners with no compensation or alternative employment. Their demands to re-open the mine remained ineffective (Soni, 2006a).

Long after the official ban on mining, illegal mining is known to have continued in the area. Moreover legal mining activity continued adjacent to the new sanctuary in Gujriwala located just outside Delhi borders and the produce was transported by truck through the sanctuary making restoration and conservation difficult[282] (Salim, 2010). Some reports pointed out that the ban of mining in Delhi and the declaration of the sanctuary had fuelled illegal mining by increasing profit margins from the illegal operations (ToI, 1993). A Supreme Court ordered investigation[283] of similar mines less than 5 kilometres away from the sanctuary in the Aravalli range across the Haryana border (where mining had continued after it had been banned in Delhi) revealed that violation of the environmental and labour

281 Interviews FD:1 and ETF:1

282 Interview ETF:1

283 M.C. Mehta vs. Union of India and others, decision dated 18.03.2004 in Writ Petition (civil) 4677 of 1985

guidelines was a norm rather than an exception leading to extensive damage to the landscape and water table. Mining within the sanctuary has now almost ceased entirely though locals and authorities report occasional attempts at small scale mining and smuggling[284].

With the closing of the mines a new space had opened up for conservation. The declaration and consolidation of the sanctuary was happening at the same time as the citizen movement to protect the Ridge against encroachments and mismanagement by the DDA, and the conservation discourse was well in circulation in the circles of the courts and the Lt. Governor. Of the environmental organisations in Delhi, two were directly involved in the initial stages of pushing the conservation agenda in the Southern Ridge. The first was WWF (I), that lobbied for the declaration of a sanctuary with the Lt. Governor (Mailk & Roos, 1994) and the second was INTACH. According to the Director of the Natural Heritage division of INTACH at the time (1994-1996):

> "There was nothing in writing but the intention was that it should be conserved. The extreme decision was taken by the decision maker (the Lt. Governor) that this area should be converted into a sanctuary so that if it is notified, no one will dare touch it. De-notifying will send a very wrong message so it is almost never done. (He was a) very forward looking man [...] when we made a presentation he said, I have done hunting in this place, I know the beauty of the area, even if it can be brought back in 50 years, it is worth it"[285].

Water scarcity was a particularly pronounced hindrance to the re-greening of the declared sanctuary and it was by virtue of its expertise in water conservation and water body regeneration that INTACH got involved. INTACH presented to the Lt. Governor[286] a conceptual plan for water harvesting in the area that included creating large water reservoirs by channelling rainwater into the quarry pits and in-situ ground water recharge methods[287]. To this end they rallied some employees of the DSMDC (which had shut down its mining operations in the area), under the general manager, to follow INTACH's direction to channel rainwater into one pit as a demonstration rainwater harvesting model. The ground working expertise of the mining organisation was used to further the conservation agenda in the area. The results were shown to the Lt. Governor; a large pond with a 'waterfall' that provided a fairly dramatic sight in stark contrast from the mine ravaged surroundings[288]. Besides these organisations, the Wildlife Institute of India (WII) in Dehradun, an institute of the union Ministry of Environment, was asked to prepare a management plan for the sanctuary (WII, 1994).

284 A particularly innovative use of unmanned camels to transport mining produce through the sanctuary was reported by the residents of Anangpur village and the ETF officials. Camels are less conspicuous than trucks and do not need to be accompanied by their handlers (Salim, 2010)

285 Interview EV:1

286 The Lt. Governor at this time was P.K. Dave

287 This plan for water harvesting in the sanctuary was presented as a part of a larger project the Delhi Government had commissioned titled 'Blueprint for Water Augmentation in Delhi' (INTACH, 1998)

288 Interview EV:1

The notification of the sanctuary was not met by any opposition, probably since its implications were not clear immediately. There were, however, alternative proposals for the use of the land that continued to be pushed alongside. Firstly, housing clusters of various kinds already existed on this land; these are discussed in detail later in the chapter. The Municipal Corporation of Delhi (MCD) demanded that the pits of Bhatti Mines be sanctioned as landfill sites citing the paucity of land in Delhi to manage its waste and the refusal of neighbouring states to help in the regard by leasing land for the purpose (Agarwal, 2010; Kalpavriksh, 2002; Toxicslink, 2001). The Master Plan 2021 also lists Bhatti Mines as a proposed landfill site to be developed according the plan. This was challenged in the Delhi High Court by environmentalists under the banner of NGO Forum for the National Capital Region[289]. The absence of water bodies, low water table and distance from the city were presented to the court as reasons rendering the site an ideal option. The notification as sanctuary and the consequent coverage of forest laws were the crux of the argument against the landfill bolstered by the Forest Department. The court ruled against the MCD citing that the landfill would be detrimental to wildlife and its habitat including groundwater quality and instructed them to develop better waste management methods rather than open a new landfill in Bhatti Mines[290].

The DSMC also attempted to retain some operations in the area by co-opting the environmental discourse. In 1993, they floated the idea that the area would be greened and the incidental mining that would take place in the process of restoration of the environment would be used to finance the restoration. The Chairman and Managing Director of DSMDC, presented a plan that included further digging of the existing pits, creating a pool of underground water and reducing the slope of the pits by digging away at the sides to plant grasses and trees. By the companies estimation this operation would supply Delhi's need for bajri for three to four years at half the cost of the bajri from Haryana and meet ecological needs as well (HT, 1991). This proposal was criticised as a thinly veiled attempt to recover ore by the WII management plan and never took off (Jain, 1994; WII, 1994).

7.3.1 Delimitation the Boundaries of the Sanctuary

A major issue that continues to be a bone of contention between various actors in the Southern Ridge, is the lack of clarity regarding the boundaries of the sanctuary. As the current master plan notes, there exists a considerable unsanctioned built-up area in the southern part of Delhi (DDA, 2007) and despite the passage of almost 18 years since the last court order adding land to the sanctuary, the demarcation process is yet to be completed. Under the Indian Forest Act, 1927[291], the forest is declared reserved under section four and the settlement of rights and

289 IA in Almitra Patel vs UOI and Ors., Writ Petition (Civil) 5236/210
290 Almitra Patel vs UOI and Ors Writ Petition (Civil) 5236/2010. Orders dated 26.05.11
291 Applied under the 1994 notification of Reserved Forests in Delhi

claims on the said land is to be undertaken after this by a duly appointed Forest Settlement Officer. Following this settlement, the final boundaries are demarcated under section twenty of the act, completing the notification process. The sanctuary is yet to receive the final notification under section twenty. The delay has enabled further changes in land use and the expansion of built area between 2001 and 2013 (Nandi, 2013; Sinha, 2014). Such a delay was made possible by the complex deliberations and oversights by the actors involved, to maintain status quo regarding the boundary decision, according to a senior official of the Forest Department[292]. Land in India is demarcated using numbered survey pillars (khasra numbers) on the ground maintained by the Revenue Department. The notifications for the sanctuary mentioned the boundaries in terms of survey numbers rather than through landmarks as in earlier notifications in case of the Ridge. The Forest Department was faced with a rather messy situation on the ground, as the survey pillars had been removed and there was no clarity on ownership for large portions of the land in question. The Revenue Department also had incomplete, damaged or conflicting records[293] on some areas, leading to many conflicts regarding demarcation[294] (Sinha, 2014). According to the Forest officer in the sanctuary, one of the reasons this situation has been allowed to continue for so long is the scale of built property at stake, he explains:

> "[…] it is very conveniently maintained as a grey area because there are serious stakes involved. The amount of money and the amount of land involved must run into several thousand crore rupees. All the farmhouses, all the rich and famous of Delhi are here. Also all the illegal jhuggies (shanties) have so many people living there, they also have their political options, it is a huge vote bank. It serves them well that there is no clarity on what is what; everyone is clueless so everyone is safe"[295].

This ambiguity in legal status and clear demarcation of land is not particular to Delhi or the Ridge. Roy (2002) in her attempt to ascertain land ownership in the periphery of Calcutta concludes that, "[...] cartography as an instrument of developmentalism, a tool by which modern states supervise and articulate their territories" was found to be absent (Roy, 2002, p. 135). This 'unmapping' of the fringe has enabled the proliferation of multiple overlapping and clashing regulations that foster such ambiguity that land tenure is rendered negotiable (Roy, 2002, p. 139).

The court asked for a compliance report on the handing over of the land to the Forest Department in 2002 for which a ten day time period was given. To meet the deadline of the court, the handing over by the Revenue Department and the taking over by the Forest Department was carried out on paper without clarifying the ground situation[296]. The handing over documents are simple handwritten doc-

292 Interview FD:1

293 In Delhi, the revenue records are still compiled and maintained by hand and some of them are very old. Some older records are in Urdu and no one in the department could read them

294 Interview FD:1

295 Interview FD:1

296 Interview FD:1

uments listing that on this date, this khasra of so many bighas/biswas[297] was handed over to the Forest Department. Thus survey numbers listed were given to the Forest Department without accounting for the constructions that already existed or spelling out where exactly the boundaries of the survey numbers began and ended[298] (Sinha, 2014).

Litigation regarding the classification of land also delayed the matter. The Forest Department filed a few cases on its own accord and some in conjunction with the Revenue Department. Citizens also used litigation to further their claims further stalling the process[299]; according to a senior forest official

> "people start at the Revenue Court, if they lose there, they approach the District Court, this in itself stalls the demarcation for at least five years. After they lose in the District Court, they approach the High Court, where a single judge bench may deny them and they can ask for a review by a double judge bench. After that they normally try political manipulation by approaching political leaders. No case goes to the Supreme Court because if the Supreme Court says this is forest land, then it is final"[300].

Although it is normally the Revenue Department and not the Forest Departments' jurisdiction to demarcate land, the constant pressure from the judiciary and Ridge Management Board, finally led to a renewed effort to demarcate the Southern Ridge in early 2012 (Sinha, 2014). A Special Task Force (STF) was constituted by the Forest Department[301] to identify and demarcate the forest boundaries and the encroachments within them. The STF is a multi-agency body consisting of a survey team, revenue officials (one kanoongo and two patwaris[302] dedicated especially to the project), a Forest Department representative and a representative from the Survey of India. The Conservator of Forests is the nodal officer and the Deputy Commissioner and the Deputy Conservators of Forest (South and West) are the reviewing officers of the STF. A total of 21 villages in the South and South Central Ridge were identified as having problematic or absent boundaries to be rectified by the STF[303]. The survey used the Total Station Method (TSM)[304] to carry out digital mapping to be followed by ground truthing of the maps produced by the Irrigation and Flood Control Department of Delhi which is responsible for building walls and erecting the boundary pillars. Once the digital maps are produced, they are to be signed by revenue and forest officials and the removal of

297 Traditional unit of land measurement still extensively used in India

298 Interview FD:1

299 Interviews FD:1 and MoEFD:1

300 Interview FD:1

301 MoEF, GNCTD, Vide order no. F.8 (118)/PA/CF/RUC/pt.IVrohill/7709–7723 (dated 28.02.2012)

302 Clerk level ranks in local revenue offices

303 These are Nebsarai, Chattarpur, Dera Mandi, Rajokri, Ghitroni, Mahipalpur, Rangpuri, Pulpehladpur, Devli, Aayanagar, Jaunapur, Ladhosari, Rajpurkhurd, Satbari, Asola, Bhatti, Siadulajab, Maidangarhi, Tughlaqabad, Sahurpur and Fatehberi

304 This is a methodology disseminated under the Rapid Training Programme of the Jawaharlal Nehru Urban renewal Mission (Sinha, 2014)

encroachments as ordered by the courts would be expected to commence (Sinha, 2014).

The National Green Tribunal has been pressuring authorities involved to complete the demarcation of the Ridge and settle legal claims of locals. In July 2013, they demanded that the legal settlement be completed in 90 days. The Revenue and Forest Departments were pulled up for causing an "unmanageable situation" by not clarifying the exact boundaries of the forest land and facilitating construction in the area through "corruption, causing violations of notified areas of the Ridge and ignoring the rights of people" (Lalchandani, 2013b, n.p.). This 90 day deadline was not met and in a later hearing in September, the Delhi Government assured a completion of demarcation and settlement by mid-October, 2013 (Singh, 2013c). This deadline was extended to mid-November 2013 but the process was only nearing completion in March 2014 during the last fieldwork for this research, with the boundaries of five villages proving problematic on account of faded, incomplete and illegible records[305] (Sinha, 2014).

7.3.2 Creating a Forest in the Sanctuary: Restoration Efforts

The land handed over to the Forest Department to create the sanctuary was highly degraded in terms of vegetation cover and soil quality due to the scale of mining and grazing[306] (Sinha, 2014). An NGO report noted with regard to Bhatti Mines;

> "It was a surprise decision on the part of the government to declare a barren piece of land without water or vegetation, as a sanctuary. In fact the project was considered to be non-viable" (Srishti, 1994, p. 33).

Initially the Forest Department was solely responsible for afforestation with the Department of Flood Control and Irrigation providing civil works like wall construction (Jain, 1994). Advisory support was provided by the Botany Department of Delhi University (Professor Suresh Babu and his team which later formed CEMDE), the Forest Research Institute in Dehradun and the Wildlife Institute of India[307]. Even though the Delhi Administration had reportedly proclaimed that in afforesting the sanctuary, they would "achieve in 15 years what nature would create in 15,000" (Mailk & Roos, 1994, n.p.), this phase of reforestation was not particularly successful. The main obstacles to attaining the goals of creating a forest in the area were identified by the Forest Department as Illegal mining, heavily leached, eroded soil and water scarcity (including a disrupted water regime due ruptured aquifers from mining), heavy grazing pressures from the surrounding villages and the permanent and semi-permanent structures or 'encroachments' on forest land (Sinha, 2014).

305 Interviews FD:1 and MoEFD:1
306 Interview ETF:1
307 It may be recalled from chapter 6 that this was as per the recommendations of the Lovraj Committee

The expanse of the land and the paucity of staff and resources available to the Forest Department necessitated the recruitment of an external organisation to re-forest the area. For this purpose the department requested the raising of an Eco-Task Force (ETF) in Delhi to work in the Southern Ridge. The ETF is a unique agency which brings together the resources of various government bodies: The Union Ministry of Defence raises a Territorial Army unit consisting of a small core of serving army personnel and a larger body of local ex-servicemen, they are financed by the Union Ministry of Environment and Forests (MoEF) and are deputed to work as per requirements of the state government of their location and are advised and provided material resources (like tree saplings and fencing) by the state Forest Department (MoEF, 2010). The idea was first proposed in the early 1980s to Prime Minister Indira Gandhi by Norman Borlaug, famous as the 'father of the green revolution' (IFDC, 2014; Stokstad, 2009) who saw the rapid depletion of forests as an alarming development that could only be tackled by an organisation like the army (Sharma, 1990). The ETF is deputed in 'difficult terrains' that are either severely degraded or remotely located and have a problematic law and order situation (MoEF, 2007) to meet the double goal of ecological restoration and rehabilitation of retired army personnel (MoEF, 2010). The Forest Department was familiar with the work of the EFT in other areas degraded by mining[308], and requested the Delhi Government to raise such a unit in Delhi[309] (Sinha, 2014). Following the request of the Chief Minister to the Ministry of Defence and the MoEF, the 132 Infantry Battalion (Territorial Army) Eco Rajput was raised in the Red Fort in Delhi in the year 2000[310]. The unit comprises three officers, five junior commissioned officers (JCO) and 139 jawans (junior level soldiers). They were charged with about 850 hectares of Bhatti Mines initially. To this was added about 565.5 hectares of the sanctuary in the Dera Mandi area in 2006 and 607 hectares of Asola in 2011[311]. The main aims of the project were spelled out as prevention of drought and desertification, restoration of ecological balance and developing the biodiversity of the area (Sinha, 2014).

There was another reason the deployment of the ETF was seen as necessary; though mining has been banned in the area, illegal mining continued with support from locals including residents of Sangam Vihar and Anangpur Village, who participated in mining and smuggling of badarpur bajri (Salim, 2010). The contiguous borders with Haryana enabled a porous operation of mining activity since mining was not banned on the Haryana state side (Jha, 2001). The residents of the three mining villages were also seen as a hindrance to the project through their engagement in such activities along with grazing of their livestock and gathering of fuel wood. The Forest Department was unable to deal with this and the pres-

308 For example in limestone mining areas in the mountains near Mussoorie reforested by the ETF of the 127 Infantry Battalion (TA) (Chopra, 2013)

309 Interview ETF:1

310 Interview ETF:1

311 Interview ETF:1

ence of a uniformed force was seen as the solution; "the fear of the uniform helped" according to the commanding officer of the ETF[312].

From 2001-2003, the ETF concentrated on tackling the 'mining mafia'. During this phase they sought to increase the green cover through planting and aerial seeding of Prosopis Juliflora to obtain fast results[313]. The native species preference had yet to be established in Delhi's forestry practices as strongly as it has been done in present restoration projects. The systematic plantation of native species only began around 2008, when a list of 57 such species was handed over under advice of CEMDE, Delhi University[314]. Lessons were learnt along the way as the initial years saw poor results: Between 2000 and 2008 the survival rates of the plantations was only 5-6 per cent (Sinha, 2014). The lack of water sources in the area led to a reported expenditure of 30-40 lakh rupees per year on water tankers (Pandit, 2007). As the commanding officer of the ETF admits, they had no forestry training and thus certain mistakes were made[315]. The environmental obstacles of water scarcity and poor soil quality were tackled by engaging in water harvesting through contour and graded bunding, bench terracing, construction of check dams, building of water reservoirs and soil conservation measures before plantations could succeed (Sinha, 2014). 36 out of approximately 200 mining pits have been converted into lakes through these measures. Open plantations led to high losses to grazing in the initial years according to the ETF, leading to fenced plantation (in pockets from 2006 and in its entirety since 2010). The survival rate of saplings since 2011 is held at 87 per cent by the ETF[316]. According to the Forest Department between 2000 and 2006 there were a total of five days of rainfall in the sanctuary, between 2008 and 2012 this number stood at 33 to 58 days a year, this variation in rainfall is shown as a sign of success in restoring the ecological viability of the area since the Forest Department attributes it to the increase in tree cover (Sinha, 2014). The increase in the fauna and flora species count has led to triumphant headlines in national newspapers for example: 'Nature, Nurture pull Bhatti Mines out of pits' (Lalchandani, 2011a), 'Rare bird spotted after decades' (Chauhan, 2012) marking a stark change from the earlier tone of print reporting that was more pessimistic as exemplified by 'Asola sanctuary grazed into extinction' (Mailk & Roos, 1994) and 'Look for animals in this sanctuary and you will find trucks' (Indian Express, 1999).

7.3.3 Wildlife and other Animals: Monkey Relocation to the ABWLS

In earlier sections, there has been some mention of the role of non-human actors in deciding the form of the space: The invasive Prosopis juliflora that lessens bio-

312 Interview ETF:1
313 Interview ETF:1
314 Interview ETF:1
315 Interview ETF:1
316 Interview ETF:1

diversity legitimising expert intervention, the lack of water that makes an effective argument for afforestation, the aridity and degraded soil that prevented earlier afforestation efforts in the sanctuary and the presence of quartzite sand made alternative use for mining an economically attractive option. The presence of large number of monkeys introduced by judicial orders is one such dramatic example of complex human and non-human factors that construct socio-natures.

In terms of the composition of the biodiversity in the sanctuary, there have been certain curious proposals and executions. The spotting of endemic species like a pack of endangered striped hyena and the occasional visits by leopards from the Haryana side has been some cause for celebration within the Forest Department and ETF as well as environmentalists as the presence of a carnivore shows higher carrying capacity of the eco-system and therefore is a measure of success for the afforestation efforts[317]. There are other creatures whose presence is owed to the broad interpretation of the term 'wildlife'. The sanctuary contains an enclosure for a herd of endangered blackbuck antelope. These have been rescued or recovered from areas near Delhi by the Forest Department and could not be accommodated by the Delhi Zoo (Vij, 1999). There was also a proposal to release recued bears in the sanctuary that had been given the go ahead by the government but was opposed by environmentalists and had to be abandoned as the sponsors for the scheme found the area lacking adequate water sources and green cover (Agarwal, 2010; Pioneer, 1998).

The Delhi High Court ordered the Municipal Corporation of Delhi (MCD) to capture and release rhesus macaques (commonly found monkeys in Delhi) in the Bhatti Mines area in 2007[318]. This case was filed by members of an RWA as monkeys had been a pest in their area and when the Deputy Mayor of Delhi fell to his death while trying to get rid of monkeys on his terrace, the matter of 'simian menace' became well publicised[319]. Other states like Madhya Pradesh were asked to take in Delhi's monkeys but they refused to do so citing lack of capacity and the disturbance to citizens (Sinha, 2014). After a tussle between the municipal body and the Forest Department regarding whose duty it was to control the monkeys, the court ordered MCD to hire monkey trappers and the Forest Department to take care of their feeding at the sanctuary. The monkey trappers, few in number as they are, earn a small sum based on the number of monkeys captured (Bhatnagar, 2013; Jamatia, 2013). There is no way to establish if these monkeys have been captured in Delhi (as was intended) or from neighbouring states like Haryana or Uttar Pradesh.

There was no limit placed on how many monkeys could be released in the area. By February 2014, it was estimated by the Forest Department that the number

317 Interviews FD:2 and ETF:1

318 New Friends Colony Residents vs Union of India (UoI) And Ors. Writ Petition (Civil) No. 2600/2001. Orders dated 14 March, 2007

319 Press headlines in India and internationally use the term 'simian menace' or 'monkey menace' often, for example: 'City at wits' end over simian menace' (Buncombe, 2005), 'VIPs may get protection from monkey menace, but what about rest of Delhi?' (Singh, 2014g), 'Delhi's monkey menace: Simian agonistes' (The Economist, 2007)

stood at more than 17,000 (Sinha, 2014), though the forest officer in the sanctuary pointed out that it is not known how many of those stay in the sanctuary[320]. A fifteen feet high hard plastic wall was ordered by the court, to be constructed along the residential areas for 1,500 m to prevent the monkeys from entering these areas (Sinha, 2014). However, evidently this is not enough to contain the monkeys as neighbouring residents complain about monkeys attacking people and snatching food[321] (Singh, 2013) essentially translocating the problem from Delhi to the villages and settlements on the outskirts.

The release of monkeys in the sanctuary has also had effects on the composition of the biodiversity of the area. For one, the court ordered such trees to be planted that would provide food for the monkeys and grow quickly in order to ensure they had a place to play[322], a period of five years was given for this. Meanwhile about 20 feeding points have been set up for the monkeys. These feeding grounds are raised concrete platforms where they are supplied with fruits brought in by trucks from the Azadpur Mandi (farmers market) at the reported cost of more than 20 lakh rupees per annum (Bhatnagar, 2013; Singh, 2014h). The easy availability of food has led to high rates of breeding according the ETF and Forest Department (Sinha, 2014). Monkeys also tend to destroy young plantations and hamper other species, for example by eating eggs from bird nests[323]. The monkeys are now dominating the sanctuary much to the consternation of the Forest Department which was unable to convince the court these monkeys that have lived among human populations are detrimental to the aims of reforesting the sanctuary[324].

A recent news article cited the new head of the Forest Department hoping for an active re-introduction of leopards and hyenas into the sanctuary by creating corridors to Aravalli forests in Haryana and Rajasthan (Nandi, 2014b). Though this scheme is yet to materialise, it is clear that this idea of wilderness ignores not just the needs but even the presence of those who live and work in the sanctuary's vicinity.

7.3.4 Homes in the Sanctuary: Villages, Slums and Mansions

After mining was banned, continuation of human habitation and the use of the space by residents were seen as a major hindrance to the wilding aspirations that the Delhi Administration had for the area. At the end of 1994, it was estimated that around 100,000 people and at least 5,000 heads of cattle lived around the sanctuary (Jain, 1994). This particular area on the southern fringe of Delhi, is the

320 Interview FD:1

321 Field notes FN-SGV, FN-SC

322 New Friends Colony Residents vs Union of India (UoI) And Ors.Writ Petition (Civil) No. 2600/2001. Orders dated 14 March, 2007

323 Interviews ETF:1 and EV:7

324 Interview FD:1

largest parliamentary constituency in the country (Soni, 2000) and therefore the residents of populous settlements can leverage significant political pressure in their favour owing to their numbers.

For the first two decades after the 1986 notification, there remained within the bounds of the sanctuary three settlements of miners, namely, Indira Nagar, Balbir Nagar and Sanjay Colony, covering about 93 hectares (Sinha, 2014). There were also other residential settlements, including the prominently large unauthorised colony Sangam Vihar and palatial farmhouses on disputed land on the outskirts of the sanctuary. In the initial years, when the afforestation efforts were yet to gain ground in earnest, the intention seemed to be simply to fence out the three miner settlements to reduce disturbance to the plantations (HT, 1993; Mailk & Roos, 1994). There are reports that there were high tensions between the residents of the settlements and the Forest Department staff at the time with the later accusing the former of purposeful disruption of their plantations and violation of rules. The residents on their part were resentful of the restrictions placed on them and the denial of access to a space they had been using for decades (Jain, 1994).

The management plan prepared by WII recommended that

> "the first and foremost priority is to give this area complete protection […] by evolving locally acceptable ways to reduce and then eliminate the dependence of the people, whether it is for a road, forest biomass or for water" (WII, 1994, n.p.).

For this purpose the plan proposed the establishment of a core zone of no disturbance and a peripheral zone for nature education. The report holds that

> "the socio-economic profile of the villagers living in the vicinity of the sanctuaries is not land-based to the extent seen in rural areas. Most […] live a semi-urban life and are not solely dependent on forest biomass." (WII, 1994, n.p.).

These views ignored the fact that some villages are no longer dependent on the land largely because they have been divested of the use of land by the fencing of the former village commons. The exclusion of people from the use of the land had been achieved through first establishing and closing down of the mines and then vesting the village commons with the state. Moreover the continued grazing and fuel wood collection apart from the use of space for recreational and sanitary uses shows that the land is still a part of the use patterns of the locals. In addition to the issue of access to forest land for various uses, there is the more dire consequence of eviction of settlements ascertained to be on forest land, even though they were settled before the declaration of the sanctuary.

7.3.4 (a) Sanjay Colony: Tribal village or illegal slum?

At the time of notification of the sanctuary, neither were the residents of the area heard, nor was any notice given to them regarding plans for their resettlement[325].

325 Under the Wildlife (Protection) Act, 1972, a Forest Settlement Officer is to ascertain claims before the notification is complete

The push for clearing the area of habitation came in 1996 with the Supreme Court demanding the removal of the three mining settlements of from inside the boundaries of the sanctuary on the basis of affidavits provided by the Delhi Government deeming these to be slums[326, 327]. A survey of the area was ordered including the houses in the three mining clusters. The eviction was stalled the next decade due to several reasons. The first reason was the peaceful agitation by the residents of the three colonies which found support in organisations like National Alliance of Peoples' Movements and People's Union for Civil Liberties (PUCL) (Soni, 2006b). Secondly, the government asked court for extensions on the evictions citing the need for more time to complete work in the resettlement project. A site for relocation was identified in Jaunpur (seven kilometres away from Bhatti) for the Bhatti miners. However, work on development of the resettlement project was stalled and finally abandoned. In July 2002, in the fresh eviction noticed served by the MCD slum department the relocation sites were specified as Holambi Kalan, Bawana and Dera Mandi (around 50 km away) eliciting opposition from the affected residents (Soni, 2006b). The justification provided in court by the authorities was the lack of water in the original site to re-settle approximately 20,000 residents. A report by the Hazards Centre (on housing policy in Delhi) suggests that there was also an opposition by the Farmhouse Owners' Welfare Association of Jaunpur as they did not want their ideal green setting disrupted by housing for the poor (Hazards Centre, 2003). Although the argument was made in terms of lack of water, the report points out that these estates were guzzling far more water in the area than the settlements of the poor. Further, it is pertinent that even in other re-settlement sites, such as Bawana, water and sanitation problems were rampant (Bhan, 2006; Perappadan, 2006b). The Meharauli Block (where the Jaunpur site is located) is relatively better off than others in the NCT with regard to water availability (Hazards Centre, 2003). An NGO member on the Ridge management board requested a senior counsel of the Supreme Court to file an Interlocutory Application in August 2002 pleading reversal to the old relocation site. This further delayed the execution of eviction (Soni, 2006b).

These petitions were finally dismissed by the court leaving no further scope for relief. Shortly after the court issued renewed orders for evictions in June 2006[328] since the infrastructure work at the resettlement colonies was deemed complete. A four week deadline was given for the commencement of removal of Sanjay Colony, to be completed within eight weeks. Following these directives given by the Supreme Court, Indira Nagar (around 22 acres) and Balbir Nagar (around 65 acres) were cleared within a day on the 20 April, 2006 (Soni, 2006a). About two kilometres away from these demolitions, Sanjay Colony, the largest settlement of the three (around 145 acres), presented a more complicated case given its size and resolve. It could not be relocated as "the officials were forcibly

326 Specifically, Jhuggi Jhopri Clusters

327 M.C. Mehta vs. UoI and ors. Writ Petition (civil) No. 4677/1985. Orders dated 03.01.1996 and 09.04.1996

328 M.C. Mehta vs. UoI and ors. Writ Petition (civil) No. 4677/1985. Order dated 07.02.2006

prevented by the illegal occupants from undertaking the exercise [...] (which) could not be carried out due to stiff resistance" according to the Forest Department (Sinha, 2014). Sanjay Colony had staged a civil disobedience by refusing to allow a biometric survey for re-location and preventing officials of the MCD and Forest Department from entering the settlement. Residents went on a relay hunger strike against the demolitions and sat on a dharna (sit-in protest). There was also a public hearing on the 23 July 2006 (HIC-HLRN, 2006; Soni, 2006a). The commanding officer of the ETF recalls, "Women came in hundreds, took off their clothes and lay on the road to block us. No outsiders were allowed into the colony for weeks"[329]. The protesters were supported by social activists from the National Alliance of People's Movements and Joint Women's Programme and were given "instant endorsement by leading environmentalists from Delhi" (Soni, 2000, p. 90) Due to this resistance and probably some political connections, Sanjay Colony managed to hang on. Even in 2013 (during the field visit for this study), the slogan "jaan de denge zamin nahi denge" (we will give up our lives but not our land) was painted on a hut in the village centre and every now and then, it was brought up in the interviews[330].

The residents of Sanjay Colony argued in their petition to the Delhi High Court[331] and later the Supreme Court that Sanjay Colony was a village settled in the area since 1976 and was therefore residential land rather than Reserved Forest. The petition pointed out that in 1981 some families were given pattas (lease documents) for 120 square yards plots on village council land under the Twenty Point Programme[332]. In 1983, Sanjay Colony even elected representatives to the Bhatti village council. There is, however, no official record for these two claims with the government (Soni, 2006a), though the residents have old documents to prove their claims. They further argued that the colony had permanent structures including an MCD primary school and a government higher secondary school built in 1978 and 1985 respectively, a veterinary centre, a community hall built in 1989, an Ayurvedic hospital in 1980, a primary health centre built in 2001 among others. These were presented as material evidence that the colony had been developed with support from the government and was thus legitimate[333]. On these counts the residents of Sanjay Colony represented by the Human Rights Law Network (HRLN) argued in court that the colony be regularised as a village. The petitioned gained Sanjay Colony some time but in August 2009, the High Court dismisses the peti-

329 Interview ETF:1

330 Field notes FN-SC

331 Nav Yuwak Gram Vikas Samiti versus Government OF NCT of Delhi and others. Writ petition (Civil) 4362/2007

332 The Twenty Point Programme (TPP) was launched by the Government of India in 1975. The programme was first revised in 1982 and again in 1986 and includes aims of redistribution of wasteland to the landless and providing housing for the economically weaker sections (http://delhiplanning.nic.in/TPP2006.pdf accessed 03.09.13)

333 Photographic evidence of these claims have been presented on the website of Bhatti Mines (Bhagirath Nagar): www.bhattimines.com (accessed 12.02.12)

tion citing the Supreme Courts previous orders in the M.C. Mehta case, to clear the Ridge of encroachments[334].

Sanjay Colony is notified as an illegal slum settlement but its residents resent the classification and insist that their settlement is a village. It is claimed that the ration cards issued to some of the residents before 2001, were converted into to slum cards without an explanation and the classification of the settlement was set at a slum cluster in January 1990. The files for the relevant orders are not to be found. A social activist and anthropologist working with the Bhatti miners claims that these changes were made as the top bureaucrats in the Delhi government knew that mines would be phased out and they would need to clear the land of encroachments to make it available it for other uses (Soni, 2006b). The sprawling settlement is physically structured like a village and has an "unmistakable rural ambience" (Soni, 2000, p. 89) with open spaces, some wide dusty roads as well as snaking lanes, cattle tethered outside some huts and small single storied houses, most of them with a bundle of fuel wood in the yard[335]. The majority of residents now work in construction as contract labourers in Delhi or Haryana or lay cables, some deal in scrap, a few keep livestock. According to the officials of the ETF and Forest Department, the population of the village has shot up from the initial estimate in 1996 of around 4000 to more than 40,000 [336] (Singh, 2014b).

The nomenclature of the settlement is revealing of its history and character. Some of the earlier settlers who came here to work in the mines in Bhatti village recall that they faced opposition from the local Gujar landlords and had no support in forming the settlement from the mine owners[337]. They managed some patronage from the ruling Congress government; the foundation stone of the settlement (still standing today) was laid by Sanjay Gandhi[338] on 12 December 1976, giving the village its name. It was after this that the government also built a school along with a health centre and a police station providing legitimacy to the 'colony'. According to the residents they were promised three thousand pucca (concrete) houses at the time of foundation of the colony demonstrating the intent of those in power at the time to develop the settlement under government patronage[339].

Sanjay Gandhi's name was adopted only after his death in 1980, the other name for the settlement (one that also continues to be used after the renaming) is Bhagirath Nagar. This nomenclature is symbolic of the identity of many of its residents. Some of the residents are newer settlers and the composition is varied in caste and region of origin. However, the majority of the original settlers are of a

334 Nav Yuwak Gram Vikas Samiti versus Government OF NCT of Delhi and others. Writ petition (Civil) 4362/2007. Decision dated 27 August 2009

335 Field notes FN-SC, FN-SC (b)

336 Interview ETF:1

337 Field notes FN-SC

338 Sanjay Gandhi was a politician and the son of Indira Gandhi, then Prime Minister of India. There are conflicting accounts regarding how Sanjay Gandhi came to be associated with the village. Some hold that he was out on a hunting trip and others say he was here for an official visit (Sethi, 2005)

339 www.bhattimines.com (accessed 12.02.12)

particular tribe called Od that formed a large part of the mining labour. The nomadic tribe was spread across Sindh, Gujarat, Maharashtra, Madhya Pradesh, Rajasthan and Haryana (Soni, 2006a). After the partition of India and Pakistan a large number of them came to work in the area as Hindu refugees and found employment in the mines in Delhi and Haryana with their expertise in working the earth. The three settlements in the Bhatti Mines formed the largest Od settlement in the country and Bhatti has since been something of an Od homeland as it is the only long term settlement of the nomadic tribe (Sethi, 2005; Soni, 2006a). They are known to be "hereditary diggers and earth masons by trade, the Ods have been nomads for centuries, and were known in the subcontinent as indigenous 'civil engineers', constructors of ponds, canals and embankments" (Soni, 2006a, n.p.).

Box 4

VILLAGE CONSERVATION: MANGAR BANI SACRED GROVE

Around 10 kilometres south of the Asola Bhatti Wildlife Sanctuary in the state of Haryana, lies a valley where the slow growing Dhau is found in abundance along with other trees like Kala siris and Salai. These trees are native to the Aravallis but while Dhau is extremely rare in Delhi, the other two are extinct in the city (Krishen, 2006). Mangar Bani retains the native ecosystem of the Ridge as it is revered by local village communities as a sacred grove. No wood collection, felling or grazing is carried out in the grove which forms a part of the village commons of Mangar, Bandhwari and Baliawas villages (Baviskar, 2013).
In the mid-1980s, collectively owned lands were partitioned and villagers were given titles to segments of land in the commons. Many locals sold their titles to real estate developers soon after. However, this was done prior to land demarcation and while the amount of land owned by each family was recorded in the title, the exact location of plots was not known and land around the sacred grove was privatised (Baviskar, 2013; Shrivastava, 2011). The villagers formed a Village Development Committee which petitioned the Haryana Forest Department to acquire the Bani to prevent any construction in the area (Shrivastava, 2011). In 2009 the CEC also advised to the Haryana government to acquire the land for conservation (Dash, 2012; Shrivastava, 2011). The Draft Development Plan (TCPD, 2012) for the area however, classified the hills as agricultural rather than forest land and though construction of industries, farmhouses and hotels is prohibited, mega tourism projects and education related construction are allowed in the area. Local villagers and environmental activists are struggling to have the grove declared a legally protected forest, failing which a crucial biodiversity reserve, religious-cultural site, wildlife corridor and water recharge zone will be lost to commercial interests (Dogra, 2013; Shrivastava, 2011).

Bhagirath is a character in Od mythology that pledged never to drink water from the same well twice, and dug a new well every day as a consequence. I was told this origin myth by a group of Od women to justify their claim that theirs is a community used to creating water bodies in dry areas and working the earth[340]. The other major community here is that of the Kumhars (caste of clay potters) who also claim expertise to earthwork and came here to work in the mines (Soni,

340 Filed notes FN-SC

2000). In 1996, 75 per cent of those who faced eviction were Ods and Kumhars (Hazards Centre, 2003). After the closing of Bhatti Mines, 63 per cent of Ods and 23 per cent of Kumhars had continued working in mines in Haryana (Hazards Centre, 2003; Soni, 2002). Currently, residents of Sanjay Colony work as day labourers in the National Capital Region, commuting long distances from the peri-urban hubs to the urban areas like Gurgaon, Ghaziabad and Delhi for work.[341]

There is a solution proposed by the residents to this predicament. Claiming traditional expertise of the main communities in earth work and water management in dry lands, they demand that the villagers be employed to carry out the plantations in the sanctuary[342]. The village headman of Sanjay Colony asks "why have these army people been brought here? We work with the earth, it is what we do. Why can't they employ us to use our traditional ways of water harvesting and planting?"[343]. This demand to be co-opted for afforestation rather than being blamed for destroying the sanctuary finds few takers. For one, the ETF has its own main agenda in providing employment to ex-servicemen. Secondly, the portrayal of these residents as criminals and illegal encroachers, ensures that their demands are not given any weight by the state agencies.

7.3.4 (b) Sangam Vihar

The second settlement identified as a problem area by the sanctuary authorities is Sangam Vihar. This is an 'unauthorized colony', a term which refers to a settlement where private or public land has been transferred or subdivided without permission from the concerned authority; these settlements have come up in contravention to the master plan without clearance for their building and layout plans (*The Gazette of India*, 2008).

Sangam Vihar is one of the largest unauthorised colonies in Delhi (Dasgupta et al., 2010; Zimmer, 2012), housing around 3 lakh registered voters[344]. Sangam Vihar is divided into two voting constituencies and several blocks. The settlement came up around 1979 and developed through the 1980s and 1990s to meet the housing needs of the growing labour force needed in the city. The influx of partition refugees (and availability of land abandoned by those leaving for Pakistan), the demand for labour during the construction boom driven by the ASIAD games

341 Filed notes FN-SC

342 An example of village based conservation can be seen in the Ridge in Haryana, close to the sanctuary. See Box 4.

343 Filed notes FN-SC (b)

344 Population data regarding unauthorised colonies is either non-existent or extremely vague demonstrating the ambiguous nature of such settlements (Zimmer, 2012). It is difficult to ascertain the number of residents but it is clear that the real number is far higher than the 3 lakh people who are registered as voters. Through the application for regularization prepared by various blocks of Sangam Vihar, the population can be estimated to be more than 6 lakhs, other estimates suggest the real number is around a million inhabitants (Sheikh & Banda, 2014)

preparations (1982), the development of the new industrial areas, such as Okhla Industrial Area and the demand for service sector and domestic labourers by the rising middle class in Delhi have been identified as the factors behind the rise of Sangam Vihar over the decades (Vedeld & Siddham, 2002).

The land ownership pattern in Sangam Vihar is ambiguous. The vast area it covers includes both contested government and private lands. The Master Plans (2001, 2021) show the area as part of the designated 'green belt' of the city comprising agricultural and forest land. The older residents remember the area as belonging to Gujar landlords of the Devli and Tughlaqabad villages[345]. The government acquired some of this land in exchange for compensation but this did not prevent the previous landowners from selling the land by transferring 'power of attorney' over haphazardly cut plots, which is a common process of formation of unauthorised colonies in Delhi:

> "In Delhi, land designated as 'agricultural' cannot be converted to any other use without government sanction. The Delhi Development Authority has the first right to acquire such land. The Delhi Lands (Restrictions on Transfer) Act 1972 prohibits the sale of land notified for acquisition. Thus, landowners are legally forbidden from changing land use, and selling to anyone other than the DDA. Decades can pass between DDA's notifying their intent to acquire the land and the actual transfer, leaving landowners in limbo. A common way around this situation is to transfer property on the basis of 'power of attorney', a circumvention frequently used in property transactions in Delhi and elsewhere. Technically, such transfers violate the law and are therefore 'unauthorized'" (Hazards Centre, 2007, p. 13).

Sangam Vihar was one of the fastest growing areas in Delhi in terms of voter count (Vedeld & Siddham, 2002) but continues to suffer from an acute shortage of services. Water, electricity, garbage collection, sewage etc. are all catered for (albeit inadequately) by collective or private arrangements, The insecurity fostered by the ever present spectre of demolitions discourages large scale investment even by long term residents (Gupta & Puri, 2005; Vedeld & Siddham, 2002; Zimmer, 2012). The possibility of regularisation of unauthorised colonies, is presented by the government every now and then to court voters (Mitra, 2003). Between 1961 and 1977, 778 such settlements were regularised (Mathur Committee, 2006). The latest large scale regularisation in Delhi was kicked off in 2007 when ahead of assembly polls in 2008, the Delhi government declared it would accept applications for regularisation of unauthorised colonies (Sheikh & Banda, 2014). The application procedure includes the resident associations submitting layout plans, details of residents, infrastructure etc. Development works (including physical infrastructure development and social infrastructure, such as schools and hospitals are to be carried out in the colonies found eligible before regularisation. However, the guidelines lay down that no colonies (or parts of colonies) that lie on notified or Reserved Forest areas[346], would be eligible for regularisation[347].

345 Field notes FN-SGV

346 Other disqualifications to regularisation are colonies located on land of archaeological importance and covered by the Archaeological Survey of India Act, colonies in the way of infrastructure projects intended in the master plan, such as roads, drainage pipes etc., and colonies housing the affluent sections of society

Zimmer (2012) points out that the state has an ambiguous relationship with unauthorised colonies. One the one hand these colonies represent a failure of the vision of the planned city to provide adequate housing as well as a failure of the government's ability to control the urbanisation process. Thus, they are represented as illegal and criminal. On the other hand regularisation of such colonies provides an opportunity for the government to raise the housing stock and also gain voters sympathy. In line with the argument that the 'state' is not a monolithic body of interests or intent, we see that in the case of colonies located on contested forest land, the Forest Department has vehemently opposed regularisation while those seeking votes are trying to push it through as described below.

In 2008, 1,218 unauthorised colonies were handed provisional regularization certificates ahead of the assembly elections, largely seen as a pre-election move for votes by the ruling party (Lemanski & Tawa Lama-Rewal, 2013). The residents, however, saw this as an assurance of impending regularisation. Of these 1,218 colonies, it was reported that a majority had come up on forest land or in and around archaeological sites and were thus not eligible for regularisation (Indian Express 2010). Of these provisional certificates 40 were cancelled by the government on unspecified grounds in February 2012[348]. Of the 295 applicant colonies in South Delhi only 14 were found eligible for regularisation, 187 were reported to be found to be on forest land and 47 on the Ridge (The Hindu, 2009). In September 2012, 895 unauthorised colonies were found eligible for regularisation by an order clearly stating that the reasons for successful applications lay in the clearance of two conditions, no part of these colonies lay on "forest and Ridge areas and protected area under the provisions of the Ancient Monuments and Archaeological Sites and Remains Act, 1958", and the colony did not pose "any hindrance to the provisions of infrastructural facilities under the Master Plan 2021"[349].

In various petitions[350] by unauthorised colonies in South Delhi to the Delhi High Court, it was argued that the settlements established before the notification of the land as Reserved Forest should be exempted form eviction and be regular-

347 The Gazette of India: Extraordinary [Part II–Sec.3 (ii)], Notification Regulations for of Unauthorised Colonies in Delhi (Under Section 57 of DD Act, 1957), Dated 25th March, 2008, and Revised Guidelines 2007 for Regularisation of Unauthorised Colonies in Delhi, Ministry of Urban Development (Delhi Division), Government of India, Dated 5th October, 2007

348 Government of NCT, Delhi, Department of Urban Development, Unauthorised Colonies Cell: (16 February 2012) List of 40 Unauthorised Colonies whose Provisional Regularization Certificates has been Cancelled. http://www.delhi.gov.in/wps/wcm/connect/DOIT_UDD/urban+development/unauthorised+colonies+under+the+jurisdiction+of+government+of+nct+of+delhi/prc+cancelled (accessed 11 July 2014)

349 Government of NCT, Delhi. Department of Urban Development, Unauthorised Colonies Cell (September 2012) order available at: http://delhi.gov.in/DoIT/DOIT/DOIT_UDD/895uc/895uc1.pdf (accessed 13 July 2014)

350 Shree Hazur Baba Sadhu Singh Ji Maharaj Trust and others vs Union Of India & others. Orders dated 11 November, 2011, Freedom Fighters Social Welfare Association and others vs Union of India & others. Orders dated 15 March, 2011

ised. Only land that was vacant at the time of notification should be used for afforestation. The court, however, decided that regardless of the time of establishment no construction on the Ridge would be cleared for regularisation. It pointed out that according to the notification and the Supreme Court orders preceding it, all uncultivated land belonging to the village council would be vested in the Forest Department and since construction did not equal cultivation the eviction was fair. The point of view of the court is clear from the opening lines of one such decision which states:

> "This batch of seven petitions entails the usual tussle of mankind i.e. of deforestation with a short time perspective to use the land of which no more is being produced for residential and commercial purposes, as against of afforestation with the long term perspective of preserving the environment necessary for the very existence of mankind"[351].

However the sheer quantum of people that face eviction coupled with the fact that the boundaries of the Ridge have yet to be settled make for a complicated legal and logistic situation.

The resident associations of various blocks of Sangam Vihar also applied for regularisation and some blocks have been successful in receiving regularisation certificates in 2012[352]. There are other blocks that have faced stiff opposition from the Forest Department and the Ridge Management Board on grounds that they are located on forest land[353]. There is much confusion amongst the residents and local councillors of these blocks, who have been assured of the regularisation by their politicians, but mostly hear in the news that regularisation is doubtful given their location on forest land[354]. During visits to the area in 2012 and 2013, the residents of these blocks seemed fairly certain that regularisation was imminent, the local councillor and MLA assured them of it (a 'certainty' repeated to in the interviews for this research). At the same time, the senior officials of the Forest Department were adamant that such a regularisation would not be allowed and that the local politicians had no basis for such claims. The forest officer in-charge of the Southern Ridge points out that the notification for regularisation clearly stated that no construction on forest land would be regularised but now the Forest Department is being asked to provide reports on land ownership to enable regularisation, he says:

> "So we have also learnt a little bit of English, so we keep writing that this is a very tentative report and should not be considered for legal purposes that it has no legal value [...]. Now they want me or the other three DCFs[355] in Delhi to say that we have no objections to this

351 Freedom Fighters Social Welfare Association and others vs Union of India & others. Orders dated 15 March, 2011

352 Government of NCT, Delhi. Department of Urban Development, Unauthorised Colonies Cell (September 2012). Order No.F.No.1–33/UC/UD/Policy/07/Part file/1050–1072

353 There are also blocks in Sangam Vihar that are having problems with regularization as the Archeological Survey of India claims they are on or close to heritage land

354 Field notes FN–SGV

355 Deputy Conservator of Forests

land being regularised. I absolutely have objection to this land being regularised. I cannot write I don't have any objection[356]".

It seemed there was a total disconnect between the debate on regularisation at the level of the Forest Department, the Department of Urban Development and various government agencies involved (DDA, MCD) and the information available at the level of the affected residents. It was decided in a meeting involving the Lt. Governor, Minister of Urban Development and officials of the DDA and MCD, that clearance from the central government and the CEC of the Supreme Court will be needed in those cases where the Forest Department has objected[357]. However, in July 2013 the Department of Urban Development of the Delhi government ordered development work (a precursor to regularisation) to be carried out in 90 unauthorised colonies including some located on forest land despite opposition from the Forest Department[358]. It is unclear if regularisation in these cases will be allowed and if so, how long it will take.

The discourse of conservation does not reach the residents of Sangam Vihar. Even the houses located at the edge of the sanctuary, separated from it only by a wall, did not recognise the term 'wildlife sanctuary'. Some residents agreed that their access to the forest for sanitary purposes, grazing and fuel wood had been curbed but not completely stopped. A few households reported collecting fire wood and keep goats that graze in the forest, thought the number of such residents seems to be very small compared to the size of the settlement[359]. At each of the group discussions in Sangam Vihar, the idea of the sanctuary and the role of the Forest Department were all but absent, no resident recognised these terms. Some claimed that their goats are impounded by the 'corporation people' if they were left to graze in the forest and yet it is the Forest Department or ETF and not the Municipal Corporation of Delhi that would have been the agency involved[360].

The older residents were aware that their settlements were located on forest land but their storyline is quite the opposite of the one heard from the sanctuary authorities. In many interviews and group discussions, it was brought up that when the earlier residents arrived the area was indeed forest and field. The residents interviewed (who arrived in Sangam Vihar between 1983 and 1989) described the past of the landscape as containing 'jungle/junglaat' (forest/wilderness), 'patthar' (rocks), 'pahad' (hills), 'kheti' (farms) and 'khadaan'

356 Interview FD:1

357 Minutes of the Meeting Regarding Regularisation of Unauthorised Colonies and Finalisation of Plan for the Special Area Held at Raj Niwas, 5 March 2010

358 Order No.FN 1–33/UC/UD/Policy/07/Part file/1050–1072. Subject: Permission for development work to be carried out in 90+39 more unauthorized colonies including unauthorized colonies affected by forest and ASI, GNCTD. Dated 29th July 2013

359 The housing types in Sangam Vihar are heterogeneous. Though narrow lanes and cramped housing is the norm, the 'Rajasthani quarter' adjoining the sanctuary walls, for example, contains homes with small courtyards where the cattle, goats and fuel wood stoves are located and these residents are more likely to access the forest than those who live in narrower lanes with no courtyard. (Filed notes FN-SGV)

360 Field notes FN-SGV

(quarries). According to them, the workers who came here looking for work and an affordable place to live had 'settled' the area and had built Sangam Vihar out of this rustic, inhospitable past. One resident who came here as a child of six around 1983 recalls:

> "There was nothing where we are now, just hills, fields and forest. Some people used to work in Bhatti Mines, others in construction and garment industries. One government run bus came here once in a while but otherwise we were not connected very well to the city. We walked or took tongas (horse carts). There were poisonous snakes, monitor lizards, wolves[361]. We had to walk to school through this forest and it was very unsafe. Now it is much better, we have some facilities, electricity came five or six years ago. It is more liveable now"[362].

In this narrative, the residents needed protection from the forest rather than the other way round and the current situation is a result of 'progress' made by these residents against the dangers and isolation of the forest as opposed to them having resulted in degradation of forest land as the conservation discourse of the Forest Department and ETF portrays.

The discourse surrounding the forest in Sangam Vihar is thus fairly different from the residents of Sanjay Colony located only a few kilometres away. Firstly, their interaction and direct confrontation with the authorities is far less frequent given that there is no ETF presence in this area as of now. Access to the forest through breaches in the wall is often unchecked. Whereas Sanjay Colony lies clearly on designated forest land rather than on the disputed boundaries of the sanctuary like Sangam Vihar. Secondly, residents of Sangam Vihar aspire to be integrated into the urban citizenry through regularisation of their settlement unlike Sanjay Colony residents who demand rights to be employed in the forest on the basis of being villagers settled with government support and having tribal knowledge in earth working. Thirdly, a small percentage of people from Sangam Vihar access the forest for various purposes in relation to the total population, therefore the problem of forest access is not shared by most of the resident body, making access a side demand of the poorest section of residents, whereas Sanjay Colony faces eviction of the entire settlement. However, the dominance of the conservation discourse, as seen in court decisions, has very real implications for some residents of Sangam Vihar. They not only stand to lose the chance to have their residences regularised and possibly face demolition, but until the decision is reached, they may not receive basic facilities as they are seen as carrying an added layer of illegality as compared to other parts of the same unauthorised colony.

361 Their mentioning of wolves evokes a sense of dangerous creatures lurking in the forest. Other residents mentioned ghosts and snakes to similar effect to contrast the forested past with the safer present (FN-SC, FN-SGV)

362 Field notes FN-SGV (b)

7.3.4 (c) Farmhouses

In protests related to the removal of settlements from the Southern Ridge, both in the courts and outside it, the presence of elite farmhouses on the similarly classified land was presented as a clear case of discrimination favouring the affluent (HIC-HLRN, 2006; Soni, 2006a). In the first master plan, ostensibly to enable the rural residents of Delhi to continue their agricultural pursuits, the rural outskirts were allowed to have farming lots with a small percentage of built area. The house was to be a modest single story outhouse in the field. The total built area was not to exceed 500 square feet for a one to three acre plot (DDA, 1962). The second and third master plans allowed for 'low intensity' constructions, such as farmhouses, in the green belt (excluding the Ridge) and acknowledged that several unauthorised farmhouses had been established (DDA, 1990, 2007). The Delhi 'farmhouse' is in fact a misnomer. Most of these farmhouses have little to do with farming and are private estates containing enormous villas of the affluent classes behind massive, fort-like walls (Soni, 2000). Some of them are used to host marriages and parties or function as spas, resorts, restaurants etc. Several of these structures were found either to be completely unsanctioned or to contain construction in contravention to the norms of the farmhouse policy (Mathur Committee, 2006). Apart from grabbing land designated for agricultural or green cover purposes, these estates also draw huge quantities of ground water affecting the water table in this already water scare area (Hazards Centre, 2003; Mathur Committee, 2006).

Various committees[363] have deliberated on the problem of dealing with existing contraventions of laws and preventing such unauthorised practices by estate owners in the future. They have been consistent in their assertion that no such structures should be allowed to continue where they are located on forest land. One such committee recommended that since

> "affluent persons can afford to have houses and land in regular colonies of Delhi and unlike the case of economically weaker sections of the society, their occupation of unauthorized colonies cannot be said to be a matter of finding basic shelter. Moreover, affluent persons are generally knowledgeable and are fully conscious of the various regulations which they violate when they inhabit unauthorized colonies or undertake unauthorized development in farmhouses. Indeed, in the fitness of things, strong action needs to be taken in regard to unauthorized development by affluent persons whether by way of occupation of unauthorized colonies or in farmhouses" (Mathur Committee, 2006, p. 23).

Despite this, it seemed that it was only the unauthorised constructions of the working classes that were the brunt of the drive to clear the sanctuary of encroachments.

In the recent past, pressured by the NGT to demarcate and establish the sanctuary boundaries, the Forest Department has been pushing for demolition of farm-

363 Prof. Vijay Kumar Malhotra Committee, 1998, Justice Nanavati Commission of Inquiry, 2005, Tejendra Khanna Committee of Experts, 2006, K.K. Mathur Committee, 2006 (Mathur Committee, 2006)

houses on forest land. It was reported in a national newspaper that farmhouses were occupying four hundred hectares of forest land, "think two hundred football fields of forest land" urged the report (Singh, 2014c, n.p.). The Forest Department through its ongoing boundary marking exercise, ascertained that 15 farmhouses were indeed within the protected land of Asola Bhatti Sanctuary and marked them for demolition. This was presented as an early sample, the department estimated that the real number of such encroachments was more than 50 (Singh, 2014c). The walls of some of these compounds were demolished and notices served for the vacation of the main structure (Ashok, 2014; Ghosal, 2014; Manzoor, 2014; Singh, 2014e). This was followed by serving of notices and sealing of some multi-storied residences that were also marked for demolition (Singh, 2014d).

The main trigger for action on the part of the Forest Department was the summoning of its top three officials by the NGT on 22 April, 2014, regarding encroachments in the Neb Sarai area (which is part of the sanctuary). The NGT demanded an explanation on why the boundaries had yet to be finalised and how encroachments had been allowed (Ghosal, 2014; Singh, 2014e). The Ministry of Environment and Forests also demanded the clearing of forest lands at the same time (Nandi, 2014a). According to the Forest Department officials, the delay is explained by the fact that they have only just received clear directions from the state government to go ahead with the demolitions (Nandi, 2014a; Singh, 2014e). Further, with the current exercise of demarcating the boundaries of the Ridge, clear information is now available to enable action. It was also pointed out that the gap in the political corridors in Delhi currently[364] is also an enabling factor (Nandi, 2014a). It is clear that such large structures could not have come up without the authority's knowledge, the Forest Departments admission that they had no clear orders before, points to the political patronage and influence that prevented such demolitions in the past. It had been affirmed by an investigation into the issue that "connivance of field staff and political pressures" were factors in the proliferations of illegal farmhouses (Mathur Committee, 2006, p. 3). The owners of the farmhouses pointed out in their appeal to the NGT for staying the demolitions, that they had been given permission to build on this land by the government and had also been paying property tax (Manzoor, 2014; Nandi, 2014a; Singh, 2014e) further implicating the government.

The demolition of the farmhouses is held as the first phase of demolition as these structures are sparsely populated and cover huge tracts of land. The second phase would aim at Sanjay Colony, a case where the size of the population and political compulsions have so far provided insurmountable complications for the administration's eviction plans (Ghosal, 2014). The Forest Department suggested once again building a boundary wall to fence the settlement out, a suggestion not accepted by the NGT (Singh, 2014a). Orders to begin relocation of Sanjay Colony were passed in mid May 2014 by the Delhi Government (Singh, 2014f). The Chief Wildlife Warden was quoted in a news report saying "We're on the job. We

364 The last elected government in Delhi resigned in February 2014, leaving the reigns directly in the hands of the Lt. Governor

will not rest till the entire forest land currently under illegal occupation is retrieved" (Singh, 2014e, n.p.). It remains to be seen what the fate of Sanjay Colony will be in the face of this renewed drive to clear the sanctuary of habitation.

7.4 DISCUSSION AND CONCLUSION

The purpose of this chapter has been to trace dynamics and contestations that led to the formation of these three conservation and restoration units in the Ridge. A look at the discourses and practices around these spaces enables us to understand whose vision has materialised in the biodiversity park, wildlife sanctuary and the citizen restoration models and conversely what aspirations, representations and practises have been dismissed or excluded. The argument is not that loss of forestland or biodiversity is not taking place but to examine the meaning given to this by various actors and their answers to the question of "Who is responsible? What can be done? What should be done?" (Hajer, 1993, p. 45). The two themes of how landscape shaping polices are formulated and influenced and what effect this has on those who directly use this space have been central to this investigation. These will be developed further in the next chapter.

The three conservation units discussed above have had different trajectories and embody three different models of conservation and restoration supported by varied legal and administrative frameworks enabling different access possibilities (see Table 3 for summarisation), despite the fact that they all lie on the Ridge. Contestation has made possible a biodiversity park in a patch not seen as the Ridge by the managing body whereas parts of the reserved Ridge have been encroached or simply fenced off (as in the Central Ridge). Re-configuration of space has ensued depending on discursive contestations and not just planning and legal status (which are also formed through negotiation).

Table 3: Summarisation of Legal, Historical, Administrative and Access Variation of the Conservation Units

Conservation unit	Legal status	Historical Land use Trajectory	Main authorities in-charge	Access allowed
Sanjay Van	Reserved Forest	• Village commons (fields, grazing land and forests) • Heritage greens, Landscaped DDA Park (1970s) • Citizen restoration project (2010)	DDA, WWN	Open to recreational uses, citizens' plantation drives and educational excursions organized by WWN
Aravalli Bio-diversity Park	Not notified under forest laws. Orders to create conservation space by Supreme Court	• Village commons (forests and grazing land) • Mines and miners settlements (1960s) • Hotel Malls and office complex built (1996-2004) • Bio-diversity park (2004)	DDA, CEMDE	Recreational access restricted to visitors' trail, access for scientific studies and educational trips only with permission from CEMDE
Asola Bhatti Wildlife Sanctuary	Reserved forest Wildlife sanctuary	• Village commons (forests and pastures mostly used by Gujjar herdsmen) • Mines in Bhatti, settlements for miners, unauthorised colony of Sangam Vihar and farmhouses come up (1960s-70s) • Establishment and expansion of Asola Bhatti Wildlife Sanctuary (1986-1996)	Forest Department, ETF	Guided access allowed for nature education purposes on payment of fee at the nature education centre

What are the factors that enabled these diverse spaces to be claimed by discourses of conservation? One clear factor is the presence of a contestation and the particular claims of the competing discourses that participate. The particularities of each negotiation can lead to different problem definitions and different solutions. It might seem on the surface that conservation policy is top-down but as we have seen in this chapter significant interaction and bargaining was/is involved in the formation of these conservation units. Therefore, geographical spaces need to be understood as political spaces (Massey, 1992). The fact that certain environmental problems are defined and receive attention at a certain time is not contingent on solely an objective urgent need for conservation but the interaction of the existing situation with a contesting discourse and different actors (Hajer, 1995; Herzele, 2005). A certain aspect is defined as a political problem based on the narratives involved. Thus the moments of 'dislocation' when the regularity is broken are also moments of power struggles as "power is not simply in the discourse but in the performance of a conflict, in the particular ways that actors mobilize discourses and reconnect the previously unconnected" (Hajer & Versteeg, 2005, p. 182).

Despite a steady priority to conserve the Ridge seen in the master plan and the mandate of the Forest Department, the translation to concrete policy or action on conservation was missing and restoration was not even on the agenda of any existing policy or plan. The triggers for active conservation and restoration projects were seen taking the shape of a conflict over construction of hotels and malls in the Mahipalpur Ridge, the greening for heritage monuments followed by a citizen initiative in Sanjay Van and closing of mines, public interest litigation and judicial pressure in the Asola Bhatti area. These interactions led to the implantation in the legal and administrative set up, of the idea of the forest as a space of value to the city and in need of protection and restoration. The ensuing negotiations led to changes in the laws or stricter implementation of existing laws, it has legitimatised new forms of control regimes and authorised new actors like the ETF, WWN and CEMDE.

These conflicts regarding forests have had a distinct socio-political aspect with stark implications for actors involved. Discursive interactions involve inter-definition of actors (Callon, 1986), i.e. not just how actors look at environment but how they look at other actors looking at the environment (Blaikie, 1995). Blame, responsibility, possible and desirable solutions are all negotiated by various discourses (Blaikie, 1995; Cronon, 1992; Forsyth, 2003). It is for this reason that Hajer (1993, p. 45) terms the study of the political process as a 'mobilisation of bias'. The perception that some people and practices associated with them are either conservation compatible or environmentally destructive (Zimmerer, 2000) positions them as legitimate or illegitimate in their use of forests.

In argumentative discourse analysis, a discourse is considered dominant in a field when two conditions are met. The first is that of 'discourse structuration', i.e. "central actors are persuaded by, or forced to accept the rhetoric power" of the discourse and must formulate their negotiations in terms of the dominant discourse to have legitimacy in the domain (Hajer, 1993, p. 44). The second condition is of 'discourse institutionalisation' which entails solidification into institutional practice for example in official plans and policy and regulatory documents (Hajer, 1995, 2006). Meeting these two conditions, the conservation discourse can be said to be dominant in the case of the Ridge. However, policy and practice are not the direct reflection of any single discourse. Firstly, the dominant discourse is not automatically reflected in practice. The formation of the wildlife sanctuary or the designation of Sanjay Van as Reserved Forest did not lead to the immediate imposition of the conservation agenda. Secondly, the conservation discourse in itself is not a consistent whole. One storyline is that of the Ridge being 'natural heritage' (present in all environmentalist discourses around the Ridge and institutionalised in the three conservation spaces). This storyline is related to protection of urban forests which provide life-supporting eco-system services and a space for aesthetic experience of nature. It aspires to the conservation of this space against permanent loss to built environment but does not necessarily insist on removal of all other uses. The second storyline within this discourse is that of 'ecological crisis'. This is intimately related to the physical characteristics of the environment. The crisis of water scarcity (in each of the three cases) and desertification (especially in the case of Asola Bhatti) has provided the logic for demand of urgent and strong measures. The urgency has also provided an excuse for not having a more deliberative process as in the case of the immediate eviction of Lal Khet, Indira Nagar and Balbir Nagar. At other times the perceived urgency led to half-baked measures as in the case of the transfer of land to the Forest Department on paper without accounting for the existing uses or lack of resources to realise the sanctuary. The crisis storyline has had an enabling function for activists demanding conservation measures as well as for government agencies looking to implement these. Harvey (2000, p. 217) refers to forceful lobbying backed by an impression of impending destruction of environment as 'crisis environmentalism' which "helps legitimise all manner of actions irrespective of social or political consequences (and)[…] often sparks elitist, authoritarian impulses". On the other hand the prospect of the tangible problem of water scarcity has also provided a means of recruiting support from local urban residents for activists against the mall. A third storyline within the ecological frame is that of the Ridge as a biodiversity reserve forwarding an instrumentalist view of nature, i.e. nature for its own sake, and privileges the interests of biophysical nature in the given space (most strongly visible in the discourse of the CEMDE and embodied by the biodiversity park but present in all management agencies of conservation spaces). This storyline demands a strict separation of human presence and activity from the forest, though in practise it also demands active management of the ecosystem by a small group of experts.

In the conservation and restoration models institutionalised in the three spaces examined, the storylines have interacted within different contexts to produce particular hybrid socio-natures. Each of these has to varying extents, functioned to restrict of claims, uses or even presence of poor local residents positioned as illegitimate. The social, political and economic complexity of long used spaces in the Ridge has been reduced to a "a flat, colourless cartoon" (Cronon, 1995, p. 35) where the conservation is considered the non-negotiable goal. There has been little effort by environmentalists or by the policy making actors to attempt any evaluation of the various place-based understandings or uses of the Ridge, the forest is seen as a value in itself, above any possibilities for alternate ecological imaginaries (Peet & Watts, 1996) which account for some of the uses or claims presented by marginalised groups.

There is, however, also a social justice discourse that can be discerned. In the case of Lal Khet, this was incorporated into the struggle against the construction of the mall but in a supporting role to the main conservation discourse. In the case of Sanjay Colony, the claims and recruitment of the residents have been given some support by environmentalists like Ravi Agarwal and Iqbal Malik (Down To Earth, 1997). Yet, there is no integrated narrative, no vision of the bringing together of conservation with social justice in the dominant storylines. An attempted synthesis can be seen in the demands of Sanjay Colony residents of being recruited to reforest the sanctuary instead of being evicted, but this has not found any success in translating to policy since it is presented in terms of social justice claims rather than in terms of the conservation discourse. Moreover, the residents of the space are considered illegal and illegitimate according forest laws.

Actors forwarding a certain discourse and associated storylines must enrol other actors into a discourse coalition in order for the discourse to become dominant (Hajer, 1993, 1995, 2006) but this may change elements of the storylines (Herzele, 2005). The conservation discourse once accepted by the state can have new adversaries (such as slums) not originally targeted by the environmentalists. The implication of the social justice narrative not being well integrated with the main conservation agenda is that while environmentalists pay lip service to social issues concerning the Ridge, they focus on conservation issues and are not as active in contesting the outcomes as long as the main demand is met. On the other hand, those who articulate their claims mainly in terms of human rights or justice issues are not seen as valid actors even if they position themselves in terms of the conservation discourse. As Demeritt (1994, p. 163) notes, people who imagine the world through different discourses "would likely talk past one another because, quite literally, they speak different languages and use incommensurable metaphors". So while environmentalists protested the proposed landfill in Asola (Toxicslink, 2001), it was social activists who campaigned actively against the removal of miners settlements (Soni, 2006a, 2006b). The first cause was argued in terms of damage to the ecology of the area and was successful as the conservation discourse has already been institutionalised in the Asola Bhatti area.

In each case, value of the forest in the conservation discourse (intrinsic value as well as in terms of eco-system services like pollution abatement and water catchment, education and aesthetics) is scaled up to the level of the city so that local demands and use claims are seen as a small sacrifice in the face of larger urban good. The common metaphor for the Ridge, 'Lungs of the City', is one example of its presentation as a city level asset. Nature is externalised from people by means of a "rhetoric manoeuvre" that allows a few people (experts, government officials, scientists, and middle class environmentalists) to speak as "nature's representatives" (Braun, 1997, p. 25). This discourse is based on the assumption of an external, universal nature (Wilson, 1999) such that a 'connection of people with nature' in Sanjay Van, for example, is only possible after walls are erected, security staff employed and those involved in plantations are selected and legitimised to do so by the managing authorities with specific goals in mind. The forest is separated from its various relationships by means of a

> "discursive razor that defines included and excluded, relevant and irrelevant, empowered and disempowered [...] A powerful narrative reconstructs common sense to make contingent seem determined and the artificial seem natural" (Cronon, 1992, pp. 1349–1350).

The scale of negotiations through which the conservation discourse is translated into regulation and policy (legal and administrative) also is at the level of the city or even at national level, for example through enrolling the Lt. Governor, national media, the Centrally Empowered Committee of the Supreme Court and even the national parliament. On the other hand, the scalar reach of the local residents, especially those facing displacement and restrictions in access, is limited mostly to local politicians and slum Pradhans (headmen). Thus, more often than not, there is no interaction at all between the two discourses. This has to do with the nature of political participation in the city, a point developed in the next chapter.

Another reason for existence and maintenance of a disjunction between the two narratives has been the powerful position afforded to agencies with no accountability and the sole responsibilities of conservation and restoration, be it the CEMDE, ETF, WWN or the Forest Department. Neither is answerable to popular demands or protests, nor are they responsible for meeting needs in terms of housing, services, infrastructure etc. They do not have to engage in the non-forest aspects of the space. As a scientist from CEMDE put it in an interview, "they (slum dwellers) also have needs, but that is the MLA and DDAs responsibility. Not ours"[365]. Thus not only is the conservation discourse institutionalised, the separation between the conservation and other discourses is also institutionalised.

Biophysical elements of the landscape also have an impact. For instance, the need to re-introduce native vegetation was a legitimising factor for actors like the WWN and CEMDE. The semi-arid conditions of the city provided a means to push for afforestation by linking forests to more rainfall (Sinha, 2014) and a higher water table (EPCA, 1999). This also provided a means for mobilising actors and individuals who may not have otherwise been involved in lobbying for conservation such as residents around the Mahipalpur Ridge. The high bird population of Delhi has also been an argument for protecting forests in the city (Kalpavriksh, 1991; Prakash, 2013; Srishti, 1994). On the flip side, the absence of charismatic large mammal species and lack of visual appeal provided by the semi-arid open scrub forest has been seen as one reason for lack of public support to the Ridge as a conservation space (Mandal & Sinha, 2008).

To conclude, each of the current conservation and restoration models in Delhi is far from neutral and is a result and site of conflict ridden histories. Rather than being purely 'natural' entities, they must be regarded as hybrid social and ecological spaces in order to enable a holistic vision. The discourses and practices that are embodied in these spaces have been examined to expose the political nature of seemingly benign notions of conservation. This demonstrates the existence of layered hybrids related to the larger dynamics of the city that have been transformed from the community owned village commons to extractive landscape of mines to spaces of conservation and biodiversity, where remnants of previous uses are still inscribed in the form of subsistence uses like grazing and fuelwood collection, mining pits and miner's settlements.

365 Interview EX:3

8 UNEQUAL POLITICS: FORMATION AND USE OF CONSERVATION SPACES

To apply discourse analysis to politics, it is necessary to relate this analysis to the questions of power and exclusion (Hajer, 2006), which is also a defining objective of post-structural political ecology and urban political ecology (Blaikie, 1994, 1995; Heynen, Kaika, et al., 2006; Robbins, 2012). Moreover, the argument in this study has been that urban political ecology must be contextualised to understand the interaction of discourses. The previous chapter examined the legal, administrative, discursive and material dynamics of the Ridge through a micro-history of the formation of the three distinct conservation units that exist today. This exposed the inequalities inherent in these processes and their outcomes, a point that demands further elaboration, which is the aim of this chapter.

The chapter begins with an analysis of three specific groups of actors and associated discourses and practices that have been powerful in the formation of conservation spaces in the Ridge. Here, power is reflected in the extent of the ability of actors to control their own environment and the environment of others (Bryant, 1998; Bunker, 1985). The first subsection examines urban environmentalism around the Ridge, tracking similarities, exchanges and areas of difference with the dominant forms of environmentalism in India. The role of the state as multiple actors was explored in chapter 5; one of these actors is the Supreme Court which has been instrumental in institutionalising the conservation discourse in the Ridge. Therefore the role of the court in conservation conflicts, particularly through public interest[366] litigation is reviewed. This is followed by a critical look at the role of scientific experts, a group that is now in a position of considerable power through the institutionalisation of the biodiversity park model in the Ridge.

The second section of the chapter examines the implications of conservation projects envisioned and implemented through negotiation between powerful actors in terms of restriction of access for subsistence and livelihood uses and eviction of settlements of the urban poor. The chapter ends with an exploration of why, despite small and fragmentary resistance on the ground, there is no visible, effective challenge in terms of social justice demands by those affected negatively by the conservation agenda.

In short, this chapter presents the power struggle in the Ridge in terms of dominance and silences contained in the discursive and material formation of conservation spaces in Delhi in the context of larger socio-political factors that these contestations are based in.

366 As PIL is the main practice through which courts have intervened in the issue of conserving the Ridge

8.1 COMPARATIVE ANALYSIS OF ENVIRONMENTALISM AROUND THE RIDGE

The beginning of environmentalism in India is usually traced to the 1970s[367], marked by the Chipko movement in 1973 in the foothills of the Himalayas, where group of locals (predominantly women) hugged trees that were important for subsistence use to prevent government-licensed timber contractors from cutting them (Gadgil, 2001; Guha, 1989). The dominant form of environmentalism in India, has been labelled 'environmentalism of the poor' as socio-economic factors have been at the helm of environmental movements, leading to a intertwining of 'red' and 'green' issues (Baviskar, 2001; Gadgil & Guha, 1992, 1995; Gadgil, 2001; Guha & Martinez-Alier, 1997; Williams & Mawdsley, 2006a). As opposed to 'post-materialist' environmentalism, which is a product of desire for space for recreation and education, once basic material concerns have been fulfilled (Guha & Martinez-Alier, 1997; Guha, 1994), the 'environmentalism of the poor' focusses on conflicts over access and use of resources. This characterisation of the dominant form of Indian environmentalism rejects the idea that the poor are destructive to the environment (as they use it desperately for short-term gain due to lack of awareness and ecologically unsound priorities). Instead, it argues that though not interested in conservation for the same reasons as post-materialist environmentalists, the poor who depend on natural resources for survival and livelihood do respond to environmental destruction (Guha & Martinez-Alier, 1997). On the other hand, the colonial and later post-colonial state, captured by elite interests, diverts resources for industry, commercial extraction and large-scale agriculture, leading to environmental destruction. This estranges local subsistence users from the resources who then desperately use limited resources in a destructive way and eventually are dispossessed and turn into 'ecological refugees' as the urban poor. Thus, costs of metropolitan and elite consumption are passed on to the marginalised in the countryside through control of state power by powerful groups (Gadgil & Guha, 1995, p. 80).

Guha and Martinez-Allier (1997) have argued that this predominance of social justice issues sets the environmental movements in India (and the Global South in general) apart from the dominant form of environmentalism in the Global North where aesthetic and intrinsic values of environment are privileged. This binary thesis pits the idea of the wilderness (intrinsic value) based environmentalism firmly against the environmentalism of local land users "who have no recourse except direct action, resisting both state and outside exploiters through a variety of techniques" (Guha & Martinez-Alier, 1997, p. 5). Thus these move-

367 There were precursors to Indian environmentalism for example, in overt and covert challenges to colonial usurpation of multiple local uses of forest land in favour of state owned, 'scientifically' managed forests for timber extraction (Guha, 1989b; Rangan, 1997). But these were not articulated as environmental movements, rather protests demanding access and use rights in the face of a usurping state (Baviskar, 2001)

ments regard productive landscapes as central to their struggle as opposed to strains of environmentalism common in the Global North (and also practiced by the elite of the Global South), which are rooted "not in the politics of production but in forms of consumption" (Baviskar, 2005, p. 163). This narrative of two environmentalisms has been critiqued, particularly for its simplistic view of the monolithic, exploitative state (Rangan, 2000) and its essentialisation of local resource users as ecologically wise natives which engage in ideal forms of resource use (Baviskar, 1995; Rangan, 2000). However this approach to defining environmentalism in India remains powerful in capturing the central stage that social justice issues have had in major Indian environmental movements in rural areas as opposed to in the urban sphere.

The urban middle class[368] has been a protagonist in 'environmentalism of the poor' based in rural areas (Baviskar, 1995; Guha & Martinez-Alier, 1997; Guha, 1997; Rangarajan, 2001, 2003). Where a favourable alliance has been worked out, support from this group to ecological movements has provided struggling rural communities the resources of their education, connections with powerful state officials (Baviskar, 2005; Guha & Martinez-Alier, 1997) as well as sympathetic media coverage (Williams & Mawdsley, 2006a).

There are commonalities between the larger landscape of environmental movements in India and that of the urban forest conservation sphere in Delhi. This is could be linked to the fact that the first phase of environmentalism around the Delhi Ridge was connected to broader international and national environmental discourses. Issues and ideas from these scales were dispersed through the media that actively reported ecological conflict stories to a class characterised by its command of English (Gupta, 2000; Lakha, 2000). Secondly, there was direct exposure of some students involved in the issue of the Ridge, to environmental con-

368 Definition of the 'middle class' in India, is complex and deeply contested (Fernandes & Heller, 2006; Mawdsley, 2004). The term refers to a vast heterogeneous group in terms of occupation, income, education, cultural and religious backgrounds. In the absence of clear definitions and boundaries, estimates for how much of Indian populace is middle class, range from around 50 million to 300 million people out of a billion (Mawdsley, 2004). Different aspects of the middle class have been presented as central elements, including simple classification based on profession (Misra, 1961). Another classification alludes to their incorporation in global discourses and global cultural and economic exchanges (Gupta, 2000; Lakha, 2000). Apart from this affinity, they have been defined in terms of opposition by Nandy who posits that the lower middle class defines itself by opposition to slum dwellers (Nandy, 1998). Rather than the more usual liberal hegemonic pattern where the middle class seeks to gain the consent of the lower classes, the middle classes in India are said to be characterized by a conscious distancing from the lower classes (Fernandes & Heller, 2006). In line with Fernandes and Heller, who suggest defining the middle class as a 'class in practice' i.e. through their everyday politics that reproduce their positions of power (Fernandes & Heller, 2006, p. 496), for the purpose of this research I use the term 'middle class environmentalists' functionally to refer to a highly educated group which engages in environmental issues. They access the forest for recreational uses and not for productive or subsistence uses. Their resources include organisation capacity and possession of scarce occupational skills (made scarce by the requirement of legal sanction like educational degrees or credentials or social networks and gatekeeping) (Fernandes & Heller, 2006)

flicts including the Chipko Andolan; members of Kalparvriksh undertook study tours to sites of environmental conflicts[369] exposing them to environmental issues that were intimately connected with livelihood and subsistence aspects.

Gadgil and Guha (Gadgil & Guha, 1994; Gadgil, 2001) discern common strands of environmentalists in India, they categorize them as: wilderness enthusiasts, guardians of the sacred, community managers, technocratic conservationists, appropriate technologists, technocratic planners, crusading Gandhians, judicial activism and ecological Marxists. In the case of the Ridge, almost all of these strands (save that of the crusading Gandhians and ecological Marxists) can be discerned, at times in overlapping ways. The 'wilderness enthusiasts', who hold a more instrumentalist view of nature and pit themselves against the destruction of forests by 'development' by demanding implementation of protected areas, have played a key role in pushing the conservation agenda. Though it must be pointed out that unlike in Gagil and Guha's formulation where wilderness enthusiasts are marked by apathy for the poor who depend on forests, environmentalists in Delhi have not demanded exclusion of this group from the Ridge. This strand was central to the student groups that agitated against encroachments and mismanagement in the Ridge through the 1980s and then the 1990s. It remains the main motivation of the more recent push for conservation in the three sites examined in chapter 7. It is therefore central to the conservation discourse of the Ridge and is common to various storylines under this cluster. 'Judicial activism' has proven to be an effective means of pushing through legislation and enforcing implementation of forest laws by various state authorities responsible for the Ridge. Right from the public interest litigation filed by M.C. Mehta in 1985 in which the Supreme Court declared the Ridge must be maintained in its '*pristine glory*'[370] to the specific place-based conflicts in the Mahipalpur ridge as well as in the constant pressure of the National Green Tribunal in demanding finalisation of the boundaries of the Southern Ridge; judicial activism remains a key means for conservation in the face of both administrative transgression and apathy. The 'community management' strand, though present in the case of the people of Sanjay Colony who demand that they be employed in afforestation of the sanctuary they live in (Soni, 2006b), remains delegitimised in policy. 'Technocratic conservationists' like the CEMDE are in a position of power in the Ridge, especially in the biodiversity park as they have both financial and political resources through their access to sizable government funding and the support of the highest executive (the Lt. Governor) who heads the Delhi Biodiversity Trust. The 'appropriate technology' strand is not widely shared by actors but is seen in Sanjay Van, where it is embodied in the inclusion of scientists in WWN and in their claim to legitimacy through the scientific, stakeholder based reforestation and building of small-scale water harvesting structures. However as mentioned before, there remains a limited inclusion of claims and limited participation of 'community' even in this case. The 'guardians of the sacred' or community conservation of tracts of forest considered

369 Interview EV:4 and EV:8
370 MC Mehta vs. UOI & Ors Writ Petition (Civil) No. 4677/1985

sacred for a community, can be seen in the sacred grove of Mangar Bani in the Ridge near Delhi, in Haryana (see Box 4). This case is not discussed here as it is outside the purview of this study which concerns itself with the Ridge in Delhi.

Despite these similarities in parts, urban environmentalism presents a different picture from the dominant strand of rural environmental movements in India and cannot be captured by the 'environmentalism of the poor' thesis. A large section of the urban poor do not stand to lose long term livelihood in the absence of conservation efforts[371] and have not organised to protect their subsistence and access demands. Rather, the activities and needs of some of them can be directly disruptive to the conservation and expansion of forest land. Urban environmentalists in India, largely belonging to non-poor groups have been accused of being anti-poor in their disposition (Baviskar, 2011; Fernandes, 2004; Sharan, 2002; Truelove & Mawdsley, 2011). Rather than seeing the dichotomy as 'environmentalism of the poor' and environmentalism of the urban non-poor, Baviskar asserts that these are interlinked as the same urban environmentalists that ally with and provide support to environmental movements in the countryside, "have an interest in imagining heroic people (like tribal peasants and pastoralists) and pristine places *elsewhere*, so that their own patterns of resource use are not effected" (Baviskar, 2005, p. 174).

Academic literature analysing 'green' issues in Indian cities and Indian urban environmentalism is not yet extensive (exceptions include Baviskar, 2002, 2003a; D'Souza & Nagendra, 2011; Rademacher & Sivaramakrishnan, 2013a; Véron, 2006; Zérah & Landy, 2012; Zérah, 2007) but throws up certain common strands in environmental conflicts and outcomes in Indian cities. In terms of conceptualisation, the phrase 'bourgeois environmentalism' (Baviskar, 2003a, p. 90) has been used to capture struggles over environment in the city. Baviskar (Baviskar, 2002, 2003a, 2011) uses this term to describe the implications of a disproportionately powerful post-liberalisation urban elite that used the judiciary, media and other political resources at their disposal to demand the implementation of an environmental agenda at the expense of the poor. In spite, of the fact that marginalised classes provide for the consumption demands of the bourgeois environmentalists (through domestic labour or small scale industry, for example), their presence is deemed unaesthetic, unsafe and undesirable (Ghertner, 2011b). Central to the demand for a clean and green city are the aesthetic, leisure and safety concerns of the urban elite, and consequently urban 'beautification drives' are aimed at the removal of the urban poor and their settlements which are equated with pollution (Baviskar, 2002, 2003a; Bhan, 2009; Ghertner, 2011a). White-collar production, commerce, leisure and aesthetic consumption activities are demands of such environmentalist, directly pitted against the subsistence or production based needs of

371 Exceptions exist, in the case of the Ridge for example owners of cattle dairies and goat and pig farmers who use the Ridge for pasture do derive livelihood from it and would suffer if the forest would be built over. They do not seem to be a noticeable majority in any settlement studied around the three conservation units and it is hard to measure their numbers given that grazing in areas under forest laws is an illegal activity and is mostly covert

the urban poor. The demands of bourgeois environmentalists combine with the need of the state to create legible planned spaces and with the interests of commercial capital to shape urban environments in particular ways; while safe, hygienic, aesthetic spaces are valorised, the basic demands of a vast majority of city dwellers are actively pushed to the fringes (Baviskar, 2003a; Ghertner, 2011b).

The case of the Ridge however demonstrates that bourgeois environmentalism is not the only form of environmentalism that can be seen in Delhi, though implications for the urban poor are not necessarily very different in other forms. In the Mahipalpur Ridge for example, the malls and hotels would have represented an aspirational place of consumption and leisure as is seen desirable by bourgeois environmentalists, yet we see a long-drawn opposition against these by middle class activists. Moreover, the environmental justice demands of the poorer residents were also incorporated at a subordinate level (to support the main argument against the malls) in Mahipalpur when the concerns of social injustice were included into the discourse of the Ridge Bachao Andolan by including residents of the demolished slums[372] and later allying with the villagers of Mahipalpur to demand restoration of the Ridge as village commons (Singh, 2013a). Environmental organisations like Kalpavriksh and Srishti explicitly included social justice demands in their reports (Kalpavriksh, 1991; Srishti, 1994).

While the urban poor were not actively targeted by environmentalists, the key discourse of conservation had stark implications as regards access to and use of space. Thus, while the 'environmentalism of the poor', which incorporates a social justice argument at its helm does not characterise struggles around the Ridge, 'bourgeois environmentalism' does not suffice as a description of the contestation around the Ridge either. In each case, a variety of factors and motivations have enabled the conservation agenda, pitting environmentalists against an array of actors including the state and commercial interests. Yet there are certain common elements in the environmentalist groups concerned with the urban forests in Delhi. These commonalities contain insights to environmentalism in the city and highlight the preponderance and omissions of certain interests within them.

The first factor of commonality is the prevalence of middle class environmentalists; this is closely connected to the second factor, which is, the strategies used by these environmentalists to realise their demands. Studies have highlighted the prominence of the middle class and its interests in urban environmentalism in India (Baviskar, 2003a; Coelho & Raman, 2013; Mawdsley, 2004; Rademacher & Sivaramakrishnan, 2013b; Srivastava, 2009; Véron, 2006). Mawdsley has provided an analysis of the middle classes regarding environmental activism and points out that

> "middle classes exert a disproportionate influence in shaping the terms of public debate on environmental issues through their strong representation in the media, politics, scientific establishment, NGOs, bureaucracy, environmental institutions and the legal system" (Mawdsley, 2004, p. 81).

372 Interview SA:1

The motivation for the engagement of this group in environmental conservation lie, to some extent, in post-materiality or a shift towards 'green issues' accompanying a cultural shift related to increase in wealth (Ingelhart, 1995). This is akin to what Guha and Martinez-Alier call the 'full stomach environmentalism of North' where nature is seen as a valuable space for leisure and recreation (Guha & Martinez-Alier, 1997, p. xxi). This nostalgic yearning for romanticised 'pristine nature' is seen as connected to the total separation of 'nature' from daily lives of the middle classes in cities (Williams, 1973). The reasons for the rising interest of the middle class in environmental issues include the increased exposure of this group to such issues and a certain conception of human-nature relationships. Mawdsley lists influence of the media (television shows on the National Geographic Channel, and Programmes like Earth Matters on state run television), the incorporation by law of environment as a subject in curriculums at all levels in school and universities and the proliferation of eco-clubs in schools and colleges as avenues of such exposure (Mawdsley, 2004). Eco-clubs have been instrumental in the environmental engagement of students in the issue of the Ridge according to those who were part of these movements as students (Kothari & Rao, 1997; Srishti, 1994). In the case of the Delhi Ridge, bird-watching groups would be another avenue for middle-class enrolment into environmentalism. A member of Srishti remembers: (at the handing over of the Ridge to the DDA in 1992) "For those who had grown up walking in the forest or bird watching, this was both a death-knell and a call to action. The moment converted ordinary citizens into activists" (Agarwal, 2010, p. 49). More recent institutionally organised activities for environmental education in various parts of the Ridge, including guided tours by experts for schoolchildren (by the Bombay Natural History Society in Asola Bhatti Wildlife Sanctuary, WWN in Sanjay Van and CEMDE in the biodiversity park), may play a similar role, though it is too early to tell.

Further, the urban middle classes are also affected by environmental problems in cities and have therefore shown some initiative to engage with them for example through resident welfare associations (Fernandes & Heller, 2006; Tawa Lama-Rewal, 2007). Problems like pollution, water scarcities and high temperatures face direct implications related to the deforestation of the Ridge and have been consistently forwarded by environmentalists as reasons for the need for conservation. As mentioned in the previous chapter, all these issues and particularly that of water scarcity have been used as a means to enrol support in the case of the Ridge.

A third rationale for engagement in environmental issues is at the level of the individuals and their background in terms of "formative moments and personal characteristics, and their role in leading and stimulating others" (Mawdsley, 2004, p. 92). This role of specific individuals is common at the national scale as well, where certain prominent individuals can be discerned[373]. The role of conspicuous individuals is very noticeable in the case of the Ridge: for example the founder of

373 For example, Salim Ali associated with the Bombay Natural History Society, columnist M. Krishnan and E.P Gee (Rangarajan 2001; See chapter on 'Independednt India's Naturalists' p. 80–93)

WWN, who has vast quasi-legal powers in Sanjay Van solely based on his connection with the Lt. Governor. Various Lt. Governors of Delhi have also been inclined to conservation and sympathetic to environmentalist demands. In the case of the Mahipalpur Ridge, Vikram Soni and Diwan Singh were central figures and continue to work on environmental conservation in the Yamuna River. Even in the larger, city level movement for conservation of the Ridge, primatologist and convenor of the Joint NGO Forum, Iqbal Malik remains a well-known name. M.C. Mehta who filed a slew of public interest litigations demanding enforcement of environmental laws, including pushing for conservation in the Ridge, is another example as are certain Judges of the Supreme Court who are seen as more likely to rule in favour of environmental issues than others[374]. All these individuals are from the group of highly educated, well-to-do citizens. Thus certain individuals from within a larger middle class are prone to endorse and actively pursue the environmental agenda. A member of Kalpavriksh recalls:

> "Many of the people (in Kalpavriksh) came from the kind of environment where they had leftist or liberal parents who were quite keen to go out and look at issues like this. These were not rich kids, by and large it was an upper middle class thing, but very high in terms of cultural capital in the sense they were able to draw upon certain intellectual resources"[375].

Apart from the exposure to the environmental agenda and to conducive personal environments, there is the more theoretical point of recognising that individuals as actors have 'knowledgeability' and 'capability' to process choices within the given structures (Giddens, 1984). Individual strategy and understanding thus comes into play through actors making choices between different storylines available to them according to their 'discursive affinities' (Hajer, 1995, p. 67), which are rooted in larger structures. Therefore even if discourses remain a "kind of practice that belongs to collectives" (Diaz-Bone et al., 2008, p. 10), the question of agency cannot be brushed over if one is to recognise that certain individuals from a group may make discursive choices that others do not (Hajer, 1995; Long & Long, 1992) While socio-economic structures enable members of the middle class to concern themselves with environmental causes, the acceptance of their agency is necessary to explain why a few individuals are prominent in the debates around the Ridge in a situation marked by apathy of most others in their class.

This composition of environmentalist groups is related to structures of access to political resources. In what he terms 'middle-class culture of public life', Nandy suggests that "the entire ideology of the Indian state is so formatted and customized that it is bound to make more sense [...] and give political advantages" to the urban middle class (Nandy, 1998, p. 4). The post-liberalisation urban middle class had access to global ideas as well as new tools of interacting with the state to implement their visions, such as public interest litigation, through which they could challenge corrupt or inadequate government practices (Fernandes & Heller, 2006).

374 One example of such an individual in the judiciary is Justice Kuldip Singh often labelled the 'green judge' by the media (Baviskar, 2005, 2012)

375 Interview EV:4

Thus, closely linked to the middle class background of the environmentalists and attendant implication for access to social and political resources, are the strategies and practices they use in order to enrol other actors in discourse coalitions and affect changes in urban environments. While in the broader realm of Indian environmentalism, the presence of popular Gandhian means of protest including satyagrah (non-violent protest including sit down strikes and demonstrations), have been found to be common and effective (Baviskar, 2005; Gadgil & Guha, 1994). It is not so the case of the Ridge. There were attempts seen in the student marches held in the 1990s and demonstrations held against the building of the malls by RWAs, students and NGOs in the case of the Mahipalpur Ridge. These, however, cannot be characterised as large-scale Gandhian protests, the absence of which is linked to the limited mobilisation and support these movements could achieve, as noted by many environmentalists seeking to enrol other citizens on the issue[376] (Kothari & Rao, 1997; Srishti, 1994). This, as argued in the previous chapter, has to do with the dominance of the particular strands of conservation discourse forwarded by environmentalists that do not provide avenues for effective engagement with broader demands and privilege specific ways of knowing and seeing nature.

On the other hand, the strategies that have been used effectively in the struggles around the Ridge have further disenfranchised the poor by taking recourse to means that require social and political capital which the poor lack access to. Specifically, the use of media[377], directly petitioning higher executives like the Lt. Governor and at times even the Prime Minister (PM)[378] and public interest litigation which circumvents the political process. Other tools used by environmentalists to forward their agenda have been, presentations to influential agencies and individuals, nature education walks and lectures (Kothari & Rao, 1997). None of these are accessible to a broader base including those who live in informal settlements around the Ridge or access the space for livelihood and subsistence uses; these groups are therefore marginalised in the decision-making process because they cannot provide counter-narratives in these arenas.

As regards the positioning of other actors by environmentalists (an aspect of discourse structuration in terms of argumentative discourse), a review of environmentalism around the Ridge shows that the main targets of environmentalists have been various agencies of the state for neglect of plans and laws and at times collusion with commercial (hotels and malls in Mahipalpur Ridge) and elite private interests (farmhouses in Asola Bhatti Sanctuary). As a headline of a major

376 Interviews MEV:2, EV:5

377 In terms of environmental reporting, it has been found that both in content and standpoint, the national newspapers, contain reports on environment that mirror the instrumentalist view of the middle class (Chapman, 2000; Mawdsley, 2004)

378 Direct Petitioning to the Lt. Governor, it was discussed in the previous chapter, is what led to the legitimacy of WWN in Sanjay Van. As for Direct petitioning to the Prime Minister, Indira Gandhi halted construction of transmission towers on the Ridge on being petitioned by students after which the Lt. Governor declared certain areas as Protected Forests (Rangarajan, 2009)

national newspaper put it, the Ridge is seen under threat from "big money, corrupt systems" (Seth, 2006, n.p.). To summarise from earlier chapters, the Student groups and JNGO opposed the lack of implementation of forest laws in the Ridge, large scale encroachment including by government agencies and the conversion of forest land to landscaped parks for recreation. The opposition in the Mahipalpur Ridge was to the DDA and to its collusion with commercial interests that was allowing large-scale construction for elite consumption, and to some extent to the army that was also constructing in the area. The main opposition in Sanjay Van was to neglect and mismanagement by the DDA. Therefore a common strand in environmental struggles around the Ridge by environmentalists has been a critique of the practices of state agencies.

Ironically the result of these struggles has been the extension of control by state agencies like the DDA and Forest Department in these areas and the increase in the exclusionary power of these agencies through legislation and court decisions. Here the concept of 'anti-politics machine' developed by Ferguson[379] (1994) seems to capture this irony. He postulates that through the actions of NGOs, a 'particular sort of state power' is asserted, at the same time, they have a 'powerful depoliticising effect' as complex socio-political issues are subjected to technocratic and bureaucratic management. In such a case "what is extended is not the magnitude of the capabilities of "the state", but the extent and reach of a particular kind of state power" (Ferguson, 1994, p. 274). This valorisation of state control is also a result of the model inscribed in the conservation discourse, which sees 'pristine' landscapes as the ideal of conservation units. In keeping with this vision, environmentalists have seen the legal and coercive power of the state as essential to the production of conservation landscapes, thus empowering the very actor they positioned themselves against. Therefore, once the conservation is institutionalised, it empowers the state to impose conservation projects in a way that the urban poor face negative implications that were not anticipated by environmentalists.

As for the positioning of local users and inhabitants in discourses of the environmentalists, the opposition is not as clear. The Indian middle class in general, has been characterised by a marked indifference to a wider public good when their own interests are not at stake (Gupta, 2000; Varma, 1998). In conceptualising the large urban middle and elite population[380], Gadgil and Guha coined the term 'omnivores' to describe a group that draws resources from a wide area with no concern for its effect on the 'ecosystem people' who depend on local resources for livelihood and subsistence (Gadgil & Guha, 1995, p. 4). Their practices are thus detrimental to both the environment and social justice but they dominate the state which is in turn controlled by their interests (Gadgil & Guha, 1992). Despite this, they argue, there is a tendency of these classes to put the blame for environmental degradation on the poor (Gadgil, 2001). In studies of environmentalism in the

379 He provides this term through a study of development projects in Lesotho
380 They also include rural elite like large landholders as omnivores

urban context as well, this anti-poor tendency of middle class environmentalist has been highlighted (Baviskar, 2002; Coelho & Raman, 2013; Véron, 2006)

In contrast to the 'omnivore' formulation and that of 'bourgeois environmentalism' the struggles for conservation around the Ridge have not directly targeted informal settlements. On the contrary, many environmentalists have condemned the removal of such settlements being carried out in the name of environmental conservation. A leading member of Srishti, for example, questioned the viability of the Asola Sanctuary if the mining settlements inside it were removed as these residents, she argued, keep a check on illegal mining activity in the area (Down To Earth, 1997). The reports prepared by Srishti, Kalpavriksh and Vatavaran do not target informal settlements but a whole host of encroachments belonging to the state and religious institutions, such as: police stations, radio relay stations, wireless monitoring stations, military constructions, petrol pumps, schools, hospitals, temples, polo ground of the President etc. (Chauhan, 2011; Kothari & Rao, 1997; Malik, 1998; Srishti, 1994). The issue of removal of informal settlements has at times been a supporting argument to demand the removal of other undesirable encroachments in the Ridge for example the Delhi Polo Club (Kothari & Rao, 1997) and the malls in Mahipalpur[381].

Nevertheless, it has been argued in this study that the dominant discourse of conservation in these struggles side-lines the interests of the poor even while it does not directly target them, as environmentalists in Delhi have not actively engaged in the question of social justice related to conservation in the city, and the consequences born by poorer citizens, is inscribed in the lack of integration of the core conservation discourse with social issues.

To sum up, neither the concept of 'environmentalism of the poor' that is used to describe the dominant form of Indian environmentalism in general, nor the concept of 'bourgeois environmentalism' that conceptualises commonly visible urban environmental agendas in India capture environmentalism around the Ridge. There are certain commonalities in the unconnected environmentalisms of the Ridge, these being; primacy of middle class activists, use of similar strategies (including litigation and petitioning higher executives), a positioning against corruption and mismanagement by state agencies as well as commercial and private interests and an expressed but unsubstantiated sympathy for the urban poor connected directly to the Ridge. Commonalities also lie in the impact of these conflicts which have enabled certain modes of state control (including closing of access and expert control of the area for selected uses), removal of informal settlements and restriction of access to these spaces.

381 Onkareshwar, K., (2006. 9 August) Press release, Ridge Bachao Andolan

8.2 CRITICAL REVIEW OF PUBLIC INTEREST LITIGATION AS A DISCURSIVE PRACTICE

Through Public Interest Litigation (PIL), the Supreme Court of India has taken an active role in environmental governance since the 1990s (Bhushan, 2004; Rajamani, 2007). PIL was introduced following the end of a period of emergency rule[382] under Prime Minister Indira Gandhi, a period during which, the Supreme Court failed to uphold the fundamental rights guaranteed to citizens by the constitution. It flowed from the egalitarian values of certain senior judges who undertook the role of activists (Mate, 2013) as well as the larger institutional need for the court to restore some legitimacy after its erosion during emergency (Baxi, 1980; Bhuwania, 2014). In effect, the court expanded its own power and jurisdiction, in some cases taking over functions of the legislature and executive as well as having a large oversight over their matters (Bhuwania, 2014; Mate, 2013).

PIL provides a means for third party groups or individuals to move the court in public interest in order to increase participation, as it was considered too expensive and time-consuming for the poor to approach courts (Dias, 1994). It also allow for relaxation of formal legal proceedings, including pleading and evidentiary norms. Any citizen can move the court with a PIL without necessarily being the effected party. Further, the case can be initiated by informal means, by notifying a judge with a letter for example. In some cases judges have even initiated proceedings themselves based on newspaper reports[383]. The court can engage in proactive investigation by appointing fact-finding commissions or expert committees for example, to facilitate their decisions. Judges can order a wide range of remedial measures through PIL and can also demand compliance and reporting from government agencies on orders passed (Ghai & Cottrell, 2010; Mate, 2013; Rajamani, 2007).

The primary focus of PIL on civil liberties and rights expanded to include environmental concerns in the 1980s (Baviskar, 2012). This period coincided with the phase of coalition governments at the central level and the court, which earlier functioned as an agent of the central government (in bringing state and local governments in compliance), has since been actively confronting both the central and state governments in spheres like environmental policy (Baxi, 1985). Since economic liberalisation in the 1990s, the role of PIL in environmental decision-making has become clearly visible in raising concerns of the Indian middle classes, regarding government corruption, inefficiencies and lack of transparency to the judicial sphere (Ghai & Cottrell, 2010; Mate, 2013).

Courts in India hear a larger number of socio-political matters as compared to other countries; they function as political institutions through PIL (Gadbois, 1985)

382 The imposition of 'state of emergency' between 1975 and 1977 gave the power to the Prime Minister to rule by decree and saw widespread suspension of civil and political liberties in the country

383 This was the case for the first PIL case in India regarding the conditions of prisoners in Bihar (Baxi, 1980) as well as in the Yamuna pollution case in Delhi (Baviskar, 2012)

and have had a far reaching impact on a variety of issues notably including those related to environmental policy[384] (Dhavan, 2000; Dias, 1994; Mate, 2013). Through a series of decisions[385], the 'right to a wholesome environment' has been interpreted in various environmental PIL cases as part of the fundamental right to life[386] guaranteed to all citizens by the Indian constitution. This gives it more legal weight as well as enables petitioners to directly approach the Supreme Court[387] (Rajamani, 2007).

While the rationale for PIL was based on increasing access to the larger public and preventing human rights violations, such that it was commented that "the Supreme Court of India is finally becoming the Supreme Court of Indians" (Baxi, 1985, p. 289), certain criticisms have been raised against the use and results of PIL cases in terms of their representativeness and issues of access and participation as well as the effectiveness of their decisions (Rajamani, 2007). These bear out in the case of the Ridge. Firstly, results of judicial activism have been found to be highly unpredictable because they depend heavily on individual judges (Ghai & Cottrell, 2010; Rajamani, 2007). In the PIL against the hotels in the Mahipalpur Ridge, the decision given by the three judge bench in September 1996, was overturned later by a two judge bench in August 1997, including replacement of the EIAA which was instituted by the former to carry out environmental assessments in the whole city for all large projects. The second bench also retrospectively cleared the hotel that had been constructed, which according to the first bench needed to be subject to environmental clearance. This case-to-case ad-hoc decision making has prevented long-term integrated policy, in spheres where judicial intervention has been the dominant means of effecting policy, like the environment (Dias, 1994; Upadhyay, 2001). Judges have considerable power especially in PIL to "design innovative solutions, direct policy changes, catalyse law making, reprimand officials and enforce orders" (Rajamani, 2007, p. 294). Given the flexibility of PIL proceedings and the power to effect innovative solutions, there is ample space for the personal and class preferences of a judge to play a part (Ghai & Cottrell, 2010; Rajamani, 2007). Since judges are from the middle or elite classes, they are found to be more receptive to cases regarding demands of this group such as those of green spaces and cleanliness rather than hearing issues related to livelihood or low-cost housing (Rajamani, 2007; Véron, 2006).

In some cases, the effects of judicial activism have been remarkably anti-poor (Baviskar, 2003a; Ghertner, 2011b) and the tendency of court decisions to be detrimental to marginal groups has increased since liberalisation in the 1990s

384 The other areas where PIL has made noticeable impact are those of corruption and judicial appointments (Dhavan, 2000)

385 These decisions were in the cases of Subhash Kumar vs State of Bihar (1992) 1SCC 598, M.C. Mehta vs. Union of India(1992)3 SCC 256 at 257, Virender Gaur vs. State of Haryana (1995) 2 SCC 577 (Sharma, 2008)

386 Article 21 of the Constitution of India

387 Normally one has to approach lower courts from the district to state level before appealing to the Supreme Court, this is not the case where the issue involves a fundamental right guaranteed in the Constitution of India

(Bhushan, 2004). An illustration is provided by the case of the Yamuna Pushta, where an informal settlement of almost 350,000 urban poor was evicted on grounds of being the reason for the pollution of the river. In the following series of evictions by the end of 2006, the riverbed was cleared of informal settlements (Ghertner, 2011a). It has been pointed out since, that while the court removed squatters in order to control pollution without any evidence for how much pollution they were actually responsible for[388], they ignored the source of massive pollution from industrial effluents and domestic sewage[389] (Baviskar, 2011b; Ghertner, 2011a; Misra, 2010). On the other hand large-scale constructions like a massive temple complex on the riverbed was legalised in retrospect (Srivastava, 2009). Another such decision based on a PIL petition, had small industrial units removed from the city in order to control pollution at substantial costs to members of the urban working class (Baviskar, 2003a). In the case of the Ridge, court orders to create a Reserved Forest in response to the PIL filed by M.C. Mehta led to the removal of miner's settlements.

Judges are not only more likely to pick issues of interest to the middle class (Cottrell, 1992; Friedman, 2008); they are also more likely to pass decisions favourable to the petitioners in such cases (Rajamani, 2007). It has been found on a review of environmental movements that in most of the cases where the Supreme Court (and central government) intervened in favour of conservation, it has favoured protectionist style of conservation, thereby disregarding issues of access and control (Lele, 2007; Rosencranz et al., 2007).

Judges tend not to research ground realities in-depth, but often rely on the suggestions of petitioners, leading to certain kinds of views to dominate the proceedings (Ghai & Cottrell, 2010). Thus the environmental discourse in the courts is framed in a particular way. There is also the risk of the better articulated and presented side dominating the proceedings (Mate, 2013). Moreover certain 'modes of argumentation' like rational-legal arguments receive more preference than social ones (Rajamani, 2007, p. 302). Cases are expected to be well supported by documentation and so English speaking middle and upper classes are more likely to move the court (Dembowski, 2001). Further, since there is a plethora of PIL cases to be heard, constant lobbying is required to get attention from the judges, leaving the interests of marginalised groups (where they are included) to be routed through activists whose own priority is better reflected in the PIL than that of the marginalised groups (Véron, 2006). Thus the needs for legal literacy, "court and media savviness" and devotion of considerable time by the plaintiff, limit participation by the urban poor (Rajamani, 2007, p. 303). Of the PIL cases filed regarding the conservation of the Ridge, the first was filed by activist lawyer M.C. Mehta and the later PIL petitions in the struggle against the malls and hotels were backed by professional lawyers (the coordinator of the RBA, being a lawyer himself).

388 A study by an NGO later put forward that only 0.5 per cent of the effluent discharge into the river could be attributed to the Pushta dwellers (Ghertner, 2011a)

389 An estimated amount of 1,789 million liters of untreated water from the city

The court also has limited access to experts from various fields, judges often depend on government experts, which diminishes the diversity in views presented to the them (Dias, 1994). Although environmental conflicts have many layers, not all affected parties are presented in court or are even approached or heard by the fact-finding commissions or expert committees formed to advise the court (Rajamani, 2007; Rosencranz et al., 2007). To begin with, the court forms the committee on a discretionary basis and no guidelines are laid down for their functioning, this leads to public participation being entirely subjective to the members of each committee (Rajamani, 2007). Thus the recommendations given to the court may not be representative of all interests or contestations and only certain aspects of environmental problems are considered. The aspects of social justice are further excluded from the discussion because their priorities are considered of private rather than public interest (Baviskar, 2003a, 2012; Véron, 2006). While the causes of natural heritage and biodiversity are presented as related to public interest, the concerns with habitation, sanitation and pastureland are considered in the realm of private resources, disqualifying them from receiving a sympathetic hearing. The series of demolitions carried out in response to the M.C. Mehta case were done without consultations, the settlers of Asola Bhatti Sanctuary were not a part of the deliberations for example (Soni, 2006b). In the Mahipalpur Ridge, even though social activists demanded information about the petition to the CEC, none was provided to them. None of the informal settlements was a party to the CEC's hearings[390].

The effectiveness of court decisions has also been questioned given that they often lay down unrealistic deadlines to accomplish directions and orders, resulting in complications due to reactive actions by officials and bureaucrats. This also leads to officials spending too many resources including much of their time, simply responding to the court (Rajamani, 2007; Rosencranz & Lele, 2008). Within the Forest Department in Delhi, the orders and directives of the court demand considerable attention and short deadlines lead to either unaccomplished tasks or hurried measures like handing over of unsubstantiated documentation on forest boundaries[391].

It has been found that there is a tendency in environmental cases for the court to rule against small business and small development projects and lower tier government agencies but not against larger business and commercial ventures and large infrastructure projects (Dias, 1994). It is also a pattern that where environmental claims are pitted against large projects, the court has usually decided in favour of the latter, often disregarding the suggestions of expert committees appointed by the court itself[392] (Bhushan, 2004). In the cases against the hotels and

390 Letter to CEC : With reference to news reports about NGOs petitioning it for declaring Lal Khet area ridge, to draw attention to available statutory protections, PIL being heard by High Court, etc. by MPISG, 14th July 2004

391 Interview FD:1 and MoEFD:1

392 Examples of these include the Narmada case where the court ruled in favour of the large dam project against the rights of those whose homes and livelihoods would be submerged (Bhushan, 2004)

the malls in Mahipalpur Ridge, the expert committees (EPCA and MoEF expert committees respectively[393]) observed that the area under question was of ecological value, despite this in both cases large-scale construction was legalised retrospectively justified on the basis of the investment already made by private parties in complicity with the DDA.

Judges are insulated from the political process and more likely to order unpopular measures than local governments (which are implicated in electoral politics) would undertake on their own volition (Dias, 1994). They are also more likely to view environmental cases one-dimensionally when they are posed in terms of ecology and the only opposition is from marginalised groups (Bhushan, 2004). The institution of special judicial bodies like the CEC, with a specific agenda to promote forest conservation, further cements the uni-dimensional perspective taken on board during proceedings. The CEC is furthermore nominated by and responsible only to the SC, making it improbable to be responsive to popular demands (Rosencranz & Lele, 2008). Local government bodies thus often use court orders to carry out actions that would otherwise be politically difficult (Bhushan, 2004). The prompt removal by the DDA of the Lal Khet settlement after the court ordered areas outside a randomly marked 'constraint area' be set aside for conservation, was once such instance. The target of the petitioners had been large-scale construction and not informal settlements which the DDA acted against.

To summarise, trends show that the court is more likely to decide in favour of large projects like malls and hotels as against both environmental and social issues. However when the issue of environmental conservation is pitted not against large projects but solely against social justice concerns, it is the latter that are sacrificed (Bhushan, 2004; Dias, 1994). These trends are evident in cases related to the Ridge. Environmentalists have used PIL to enrol the court into the coalition for the conservation discourse in order to fight the corruption and mismanagement by state agencies and the takeover of forest space by private commercial interests with some success. Often times the result of the PIL leads to new actors and implications that the environmentalists had not bargained for (as in the case of Mahipalpur Ridge, where a biodiversity park was set up under strict scientific control rather than the demand of the RBA and CPQLW to let the forest remain as it was and one hotel and malls were allowed to continue). This confirms the assertion of argumentative discourse analysis that the enrolment of new actors can change the outcome through negotiation within a coalition as well as legitimise new actors (Hajer, 1995; Herzele, 2005).

The judiciary can innovate broad scale decisions compared to local governments and can be effective in checking corruption, abuse and apathy by both state bureaucracies and political elements (Dembowski, 2001). However, PIL is not without pitfalls in terms of participation, access and the trends in outcomes which have been shown to neglect the interests of the marginalised and have led to unfa-

393 EPCA (2000) and Report of the MoEF cited in T.N. Godavarman Thirumulpad vs Union of India and Ors. Writ Petition (civil) 202 of 1995. Orders dated 17 October, 2006

vourable results for those who use the forest for habitation, subsistence or livelihood.

8.3 SCIENTIFIC EXPERTS AS POLITICAL ACTORS

Scientific experts are often presented as providing an "accurate and politically neutral assessment of environmental problems" (Forsyth, 2003, p. 12). On the contrary, political ecologists have pointed out that expert knowledge is socially and politically constructed and has social and political implications (Forsyth, 2003, 2011; Jasanoff & Wynne, 1998; Nadasdy, 2011). This is not to say that scientists like ecologists and conservation biologists, who are often involved in conservation efforts, do not provide useful and necessary explanations for environmental problems, but expert knowledge alone, isolated from all other contextual claims and considerations can be inaccurate and inadequate (Forsyth, 2003, 2011; Lachmund, 2004; Nadasdy, 2011; Socolow, 1976). With the position of power afforded to institutionalised scientist managers in the Ridge, through the establishment and extension of the biodiversity park model across Delhi, the role of professional scientists in conservation has been established and is set to grow in spatial terms. Much has been written about science and research as expressions of power and politics (Forsyth, 2003; Jasanoff & Wynne, 1998; Leach & Mearns, 1996; Stott & Sullivan, 2000; Walker, 2007). In this case, two specific points are important to note: Firstly, scientific choices are laden with the scientists' values and political choices (Jasanoff & Wynne, 1998; Lele & Norgaard, 2003). Secondly, privileging experts and expertise as a way of defining a problem and possible solutions, forecloses contestation and side-lines other demands (Baviskar, 2003b; Lele & Norgaard, 2003; Nadasdy, 2011).

Scientists themselves draw on a social position of power from their middle class backgrounds, institutional backing, access to and legitimacy in policy debates (Guha, 1997; Socolow, 1976). Along with the social and institutional relations of the expert bodies, scientific conservation models also carry certain value judgments (Forsyth, 2003; Lele & Norgaard, 2003). Based in their institutional and disciplinary backgrounds, scientific explanations embody a certain conception of nature-society relationship (Blaikie & Brookefield, 1987; Escobar, 1998; Forsyth, 2008a; Latour, 2004; Nadasdy, 2011). The notion of environment in scientific discourses, like that of conservation biology, is often based on the concept of external nature i.e., nature is seen as an independent entity with its own set of laws, knowable only to a set of experts (Latour, 2004). This separation "authorises certain 'disinterested' voices, the resource manager, the ecologist or nature's 'defender', to speak as nature's representative" (Braun, 1997, p. 25).

This concept of the environment in the scientific discourse is not just a representation of what human-nature relationships are, but also what they ought to be. These sometimes "buried normative elements" (Jasanoff, 2010, p. 248) shape ideas about what the resources under question are, how they are to be used and by whom (Nadasdy, 2011). These normative choices are reflected in the issue-framing and delimitation of possible solutions, as well as in the identification of who can participate in these framings and solutions (Forsyth, 2011). Scientific evidence is produced with a goal in mind, this goal is socially defined (Socolow, 1976) and directly affects what kind of information is collected and which actors are considered in what roles (Blaikie, 1999; Forsyth, 2003, 2011). The assumptions, choices and management techniques of experts can directly bear upon the conservation unit and have diverse impacts on various groups (Stott & Sullivan, 2000). At times these normative choices are simply thinly veiled prejudices against certain groups and in favour of others, for example a senior scientist in the biodiversity park stated:

> "We have changed the quality of the people who come here, the composition of people and for what purpose they are coming here. It is a big change in terms of thinking. We have removed gamblers and drinking people. Now embassy people and elite class people can come here; walkers, bird watchers, nature lovers. This is one successful aspect of the project"[394].

The discourse of scientific conservation carries embedded boundaries for what can and cannot be deliberated and delimits policy options serving as a precursor to policy outcomes where the role of scientists in policy making is institutionalised (Forsyth, 2003; Hajer & Versteeg, 2005; Jasanoff, 1996; Litfin, 1994). The view of local users as short-sighted and destructive is often displayed by conservationists, including scientists involved in conservation efforts, therefore creating inviolate spaces, land without users (or only specific kinds of users), is seen as the only viable solution (Forsyth, 2011; Guha, 1997; Neumann, 1998a). In none of the three conservation units was any attempt made to understand the existing resource use and its effect on the area before all subsistence and livelihood uses were summarily banned by the imposition of forest laws. The tendency to assume that human resource use in conservation areas is negative remains dominant in Indian conservation models (Baviskar, 2003b; Rangarajan, 2003); 'use' is often referred to as 'disturbance' and therefore as leading to 'degradation' (Lele & Norgaard, 2003, p. 167). This universalisation of environmental explanations, which are then used as scientific law, such as all human activity needs to be banned to protect and promote biodiversity, has been shown to be faulty by several place-based studies which have highlighted that context specific, participatory science needs to preclude decision-making (Fairhead & Leach, 1996; Forsyth, 2003; Guha, 1997; Leach & Mearns, 1996). In the case of the Greater Himalayan National Park, for example, the fact that grazing is detrimental to the landscape and biodiversity was taken as self-evident along with the legal sanction of the Wildlife (Protection) Act. It was found four years after the notification (leading to prohibition of graz-

394 Interview EX:2

ing and other activities), that the level of grazing that had taken place had little or no negative impact on the ecosystem (Baviskar, 2003b)[395].

Conservation biologists and ecologists also tend to favour the interests of biophysical elements of the landscape over human stakeholders (Guha, 1997; Jalais, 2010; Lachmund, 2004; Saberwal, 1996). In the biodiversity park, the scientist in-charge is quite clear, that when he says that he is concerned for the security of the residents of the park, the residents mean birds, animals and plants which have been curated carefully by the CEMDE team[396]. They also see the presence of certain species of flora and fauna as preferable to others, for example in the case of the biodiversity park, the replanting of native species provides the logic for transformation of the landscape, requiring restriction of access and control of movement to enable sapling growth. Scientists in the park have applied themselves to the goal of creating butterfly and orchid conservatories (which none of the actors demanded) but not to ascertaining, if certain amount of firewood can be extracted from the area while meeting conservation goals. This selective use of selective kinds of science is linked to the normative assumptions inherent in the conservation model. Where human use is seen as acceptable or tolerable, it encompasses specific kinds of recreation and educational activities. Like the biotic environment of the park, which is being subject to careful selection and elimination by the scientists, the social composition of people and the uses they engage in is also subjected to control. The park authorities express concerns regarding "how do you take out people? What is the composition of people coming? For what purpose are they coming?"[397]. The section of the population whose subsistence and livelihood concerns are negated, bear the brunt of the newly applied conservation practices. Thus scientific decisions have led to unequal outcomes for different sections of society as scientists have engaged in the reconstitution of the space in biotic terms (by removing exotic species and creating plantations to green the area) as well as social terms (by allowing only certain uses by certain citizens).

Environmental problems have been defined as 'wicked problems' in that they defy definite formulations and resolutions due to complex, open-ended and dynamic elements (Rittel & Webber, 1973, p. 160). Within any chosen framework, the variables and units are not pre-determined (Lele & Norgaard, 2003). The certainty on which scientific models are based is achieved by excluding certain experiences and claims from consideration (Funtowicz & Ravetz, 1993). Through an application of scientific conservation, complex socio-political problems are dealt

395 Studies have shown that similar universalised assumptions have been erroneously applied across geographical and spatial scales, only to be disproven by place-based studies (Bebbington & Batterbury, 2001; Calder & Alyward, 2002; Fairhead & Leach, 1996; Forsyth, 2003). Forsyth uses the term 'environmental orthodoxies' to speak of such widely institutionalized conceptions of environmental degradation and change that are treated as facts but are often shown by field research to be biophysically inaccurate and responsible for unnecessarily restrictive policies. They are in effect received wisdom accepted as fact, not an outcome of local scientific study (Forsyth, 2003, pp. 36–37)

396 Interview EX:2

397 Interview EX:2(b)

with using technocratic and bureaucratic measures (Bryant, 2000). The idea of forests as 'wilderness' which dominates conservation research (Forsyth, 2011; Rangarajan, 2003) precludes arguments from any present livelihood, cultural or other use claims (Neumann, 1998). The expert bodies are also insulated from demands of users as they are not accountable to political pressures from them. Thus while scientists have a privileged position in environmental decision making, especially in regard to certain issues or landscapes, they do not have to communicate with those whose interests they do not represent (Ludwig, Mangel, & Haddad, 2001), allowing for decisions that may impact certain groups negatively. This one-sided conservation view makes possible the negation of historical and existing uses, for example, grazing in Asola Bhatti Sanctuary is considered illegal and herders, including traditional Gujar herders (which are noted as using the Southern Ridge even in the Gazetteer of 1883-84), are considered criminal. At the same time, there is an imposition of ahistorical aspects on the landscape, like captured monkeys in large numbers and fruit trees for feeding them, from those quarters which can influence policy (in this case orders from the SC). Thus while the politically-fraught nature of the landscape is not addressed by scientific conservation models, science is used to enable "production and imposition of one nature/society and erasure of others" (Nadasdy, 2011, p. 132) which in itself is a political act.

All actors do not privilege the same values in the landscape, for example, the middle class users of the biodiversity park see its value in recreational terms while residents of local informal settlements see the habitation, subsistence (sanitation, firewood) and livelihood (pasture) uses as important as well[398]. The scientist in-charge of the park is aware of this as he states that it is difficult to convince local users to respect the rules as: "they see fuelwood, they see fodder, they see space to play"[399] (as opposed to viewing the biodiversity park as a forest). In instances where urban landscapes have been treated differently from less anthoropogenically affected areas, it has enabled innovative solutions to include non-expert demands along with scientific conservation. In the case of Südgelände Nature Park in Berlin, for example, a long negotiation between scientists and lay citizens has provided possibilities for a model of conserving existing flora and fauna while incorporating some demands of use and experience of the space by citizens (Lachmund, 2004).

The criticism listed here does not call for the elimination of scientific explanation in environmental problems but rather, in line with other political ecologists, calls for a broadening of the frame by including different kinds of data and actors rather than brushing over the social and political elements (Escobar, 1998; Forsyth, 2011; Jasanoff & Wynne, 1998; Latour, 2004; Ludwig et al., 2001; Nadasdy, 2011; Richmond et al., 2007). In short, ecological and biophysical explanations are important but not enough. Normative and social justice aspects need to be considered along with scientific explanations in defining the goals and

398 Field Notes FN-ADBP
399 Interview EX:2

means of conservation (Forsyth, 2011; Ludwig et al., 2001). Science itself will have to be interdisciplinary and socially engaged (Lele & Norgaard, 2003) in order to move away from the emblem of the 'the authoritarian biologist' (Guha, 1997) and engage in providing ecologically as well as socially viable and desirable solutions.

8.4 IMPLICATIONS FOR THE MARGINALISED

The previous section discussed how certain elements enabled the material and discursive establishment of conservation as the dominant discourse in the Ridge. The three conservation units in the Ridge follow, to varying degrees, the preservationist agenda, which is the mainstay of conservation in India and is based on the idea that

> "the only way to conserve wildlife effectively in our complex socioeconomic and demographic milieu, is by investing in strong legislation, creating inviolate spaces and enforcing the law strictly with a regimented state force" (Gubbi, 2010, p. 61).

The application of such an agenda, in areas located in a highly populated city with heavy pressures on land and long histories of anthropological use, has unequal implications for various social groups. It has been suggested by urban political ecology studies in the Global North that those parts of a city inhabited by marginalised groups bear negative environmental change while the more economically and politically powerful areas often see an enhanced quality of environment (Dooling et al., 2006; Heynen, 2006; Swyngedouw & Heynen, 2003). However in cities like Delhi which contain mixed socio-economic profiles in the same area (Baud et al., 2008), changes in environmental characteristics in any locality may be of benefit to some and detriment to others who have different priorities and needs.

On the surface, conservation is a moral agenda of protecting important biodiversity and ecosystem functions, neutral in its social effects and carried out in the name of public good (Bryant, 2000; Sundberg, 2003). A closer look at three bounded spaces of conservation in the Delhi Ridge reveals that the burden of eviction (or possibility of eviction in the future) is borne disproportionately by informal settlements that were located in the area due to availability of affordable space in the urban fringe or due to earlier state-supported productive uses of the Ridge for mining (in Asola Bhatti and Mahipalpur Ridge). On the other hand, commercial constructions like malls and hotels as well as army constructions have been allowed to remain. The access patterns have also been subjected to tighter control in these spaces by the authorities that seek to enable only certain recreational and educational uses of the forest and eliminate other subsistence and livelihood uses.

8.4.1 Access to the Forest for Subsistence and Livelihood

The establishment of protected conservation areas leads to displacement of varying levels (Cernea, 2006). It has been argued that even where there is no physical displacement, involuntary restriction of access counts as displacement, this would include loss of access to use of land and resources, foreclosure of rights to future use and loss of non-consumptive uses like cultural and recreational uses (Adams & Hutton, 2007). The Ridge is accessed by the inhabitants of surrounding informal settlements and urban villages for various purposes, these include fuelwood collection, fodder for cattle, goats and pigs, use for sanitation purposes where settlements have inadequate facilities, and for recreation. The lack of implementation of forest laws in large parts of the Ridge allowed these uses to continue even after the notification of the forest spaces as Reserved or Protected Forest[400]. Access to this space is being increasingly restricted due to the pressure on state agencies from courts, environmentalists and scientist managers for stricter implementation of forest laws.

The Mahipalpur Ridge provided fuelwood to some residents of informal settlements of Bhawar Singh camp and Kusumpur Pahadi[401] and pasture for the cows from the dairies in Kusumpur Pahadi[402] that supply milk to the nearby houses[403]. Pigs graze at the edges of the biodiversity park next to Kusumpur Pahadi. Many residents use the space for sanitary purposes as three out of four public toilets are not functioning in these settlements and few people have private facilities in their homes[404]. The closing of the boundaries by park authorities was not accompanied by a renewal of toilet facilities and the practice of using the forest space for sanitation continued[405]. Fuelwood is still collected covertly with risk of impoundment of tools and fines. Increasingly strict security has led women from Bhawar Singh Camp to collect wood in the army-controlled part of the forest rather than in the biodiversity park[406]. The dairy owners of Kusumpur found themselves walled out of the forest in 2012 but there were certain openings in the fence which they could gain access through. With the building of a wall in place of the wire fence in early 2013 (the wall was being constructed in March), they found themselves without options for pasture. Ten out of fifteen dairies had closed and the five remaining

400 Interviews SVEV:1, DDA:2, SVFO, EX:5, ETF:1 and FD:1

401 There is no clarity on how many of them collected fuelwood as answers were varied given the covert nature of the activity

402 A resident and local leader in Kusumpur estimated they have 100–150 cows but the dairies refused to provide a number saying only that they are taking their cows back to the village since there is no access to pasture anymore. Interviews KPLS:2 and FN-KP (b)

403 Field notes FN-KP, Interview KPLS:2

404 Field notes FN-KP, Interview KPLS:2

405 The cost of constructing a toilet ranges from 15–30,000 rupees and not all can afford this. The public toilets are also short of water facilities rendering them unhygienic: residents reported that the manager of the toilet was selling the water it received for 6 rupees a can (30 litres). Interview KPLS:2 and Field notes FN-KP

406 Field notes FN-BS

dairy owners spoke about returning to their villages in Haryana with their cattle[407]. A group of Gujar women from dairy owning families expressed their frustration and suggested that this 'surrounding' of the settlement by walls was a strategy to get rid of them so that they would leave without having to be re-settled by the government[408]. Though this may or may not be a factual claim, it points to the antagonism that exists between the park authorities and some residents of the surrounding informal settlements. Residents from both settlements referred to the DDA having 'captured' the land ("kabza kar liya hai"), some spoke of them having bought it[409]. Neither of the settlements was consulted when the park was proposed or designed, the residents came to know of it only after it was established. A local political worker in Kusumpur Pahadi was invited to the inauguration of the biodiversity park (he is associated with the political party from which the leader inaugurating the park hailed), yet he had heard nothing about the biodiversity park before the inauguration as no announcement was made and none of the deliberations that led to the establishment of the park, included these settlements[410].

Sanjay Van included forest, pasture and farmland where crops such as wheat, millets, sorghum, chickpeas and later vegetables were grown[411]. The land was taken over by the DDA in 1962 as part of land acquisition for the first Master Plan and the owners of farmland were compensated monetarily, some also received jobs with the DDA[412]. The area continued to be used as village commons for fuelwood and pasture until 2010[413], when the forest renewal and the attendant boundary enforcement were undertaken by DDA along with WWN[414]. Pigs still graze at the edges of Mehrauli Ward No. 2 and Kishangarh; the two urban villages that the authorities in Sanjay Van identify as 'problem areas'[415] since uncertainties of land ownership have prevented boundaries from being built in some places along these settlements. Although cattle and goat grazing ended due to stricter security and boundary walls in most areas, fuelwood collection continues [416].

Asola Bhatti Sanctuary is a water scarce area and has historically been used as a grazing ground rather than farmland (Gazetteers Organisation, 1999). Mining, including illegal mining has mostly been curbed after the establishment of the ETF in the Sanctuary[417]. Given the lack of boundary walls due to unsettled boundaries, grazing of cattle and goats continues along with fuelwood collection

407 Field notes FN-KP (b)
408 Field notes FN-KP (b)
409 Field notes FN-KP and FN-BS
410 Interview KPLS:2
411 Interview SVFO
412 Interview SVFO
413 Interviews DDA:4 and SVFO
414 Interviews DDA: 4, SVFO and SVEV: 1, SEV:1 (b).
415 Interviews SVEV:1, DDA:4 and SVFO
416 Field notes FN-KG and FN-MRW2, Interviews SVEV:1, DDA:4, SVFO
417 Interviews ETF:1

from surrounding villages and informal settlements[418]. The restriction of access in the sanctuary is not as strict as in the other two parts studied, given that this covers a much larger area in the peri-urban region with disputed boundaries, rendering effective policing difficult by the staff strapped Forest Department and ETF.

In all three areas the authorities are using hired security guards (including ETF staff in the case of the sanctuary), impounding of cattle and tools (scythes used for firewood collection), and threat of fines to discourage these uses that are seen as detrimental to tree plantation and creation of forested landscapes[419].

The theory of access developed by Ribot and Peluso defines access as "the ability to benefit from things, including material objects, persons, institutions, and symbols"[420] (Ribot & Peluso, 2003, p. 153). Therefore, issues of access entail questions of who does (and who does not) get to "use what, in what ways, and when (i.e. in what circumstances)" (Neale, 1998, p. 48). Access remains a dynamic concept as relationships between different groups of people and resources are different at various historical and geographical scales (Neumann, 1998; Ribot & Peluso, 2003). The interaction of discourses, changing legal framework, conflict and change in social relations can lead to changes in access patterns (Ribot & Peluso, 2003). This is seen in the case of the Ridge where access possibilities have changed historically as the site of the forest and the socio-political relationships it embodies has been altered, from being fields, village commons and mines to conservation areas. Spatially too, the three conservation units are subject to stricter control by the authorities involved than the larger Ridge landscape and within these three spaces, the sanctuary remains harder to wall and separate from use than the other two conservation units.

Access is placed within power structures of societies as certain institutions or individuals control resources and others must secure access through them (Peluso, 1992). Access is different from property rights as it is based on ability to benefit from resources rather than an enforceable right to do so based in law or custom. The flow of benefit is controlled by certain institutions and socio-political relations as well as discursive strategies, some of these could be seen as illegitimate by some or all parts of society and some are remnants of earlier legitimised practices and discourses (Ribot & Peluso, 2003).

Under this theory of access, social action is divided into: access control or the "ability to mediate others access" (ibid., p.158); access maintenance referring to the "expending of resource and power to keep a particular sort of resource access open" (ibid., p.159) and thirdly, gaining access, means through which access is

418 Filed notes FN-SC, Interviews ETF:1 and FD:1

419 Interviews FD:1, EX:3, DDA:4, SVFO, Field notes FN-BS and FN-KP

420 Ribot and Peluso draw this theory from the existing theories of property rights with important difference in terms of enforceability of claims as well as flow of benefits. A person or group may or may not have property rights over the resources they derive benefits from. On the other hand, those who have property rights over a certain resource may not have access to it if they cannot derive value or benefit from it for example, if they do not have the capital or technology to exploit the resource they have property rights over (Ribot & Peluso, 2003, pp. 157–160)

established. Access can be controlled by law, custom and convention as well as discursive manipulations (ibid., p. 158-159). Ribbot and Peluso further identify certain 'mechanisms of access based in socio-political structures through which groups or individuals control, gain or maintain access to resources (ibid., p. 161). These can be both legal and illegal, the former are mechanisms sanctioned by law, custom or convention as opposed to the later. These mechanisms are overlapping and heuristic (ibid.).

The first such example of an access mechanism is technology. As a control mechanism, this would include walls, which both physically keep people out as well as signify intent to exclude people from access (Fortmann, 1995; Ribot & Peluso, 2003). In the case of the Ridge it would also be possible to use technology to control access in non-coercive ways for example by providing cooking fuel or even the wood of the removed Prosopis juliflora which is known to burn well (Gold, 1999) to reduce fuelwood dependence or by building adequate and functioning sanitation facilities in the informal settlements to negate the need for using the forest for sanitation purposes.

Another such mechanism is that of capital in terms of finances (Ribot & Peluso, 2003). As a mechanism of control, it can be used in order to acquire technology to control access, like walls, surveillance cameras. It is also used to hire labour to guard resources. It could also be used to grant access to certain actors, for example through grant money to certain institutions like the CEMDE, ETF etc. As a means of gaining access, it could be used in legal ways like the payment of fees at the educational centre at the Asola Bhatti Sanctuary in order to receive an educational tour. It could also be used in illegal ways through payment of bribes.

Access to labour and labour opportunities (Ribot & Peluso, 2003) as a mechanism of control includes hiring of guards as well as staff for plantations and other afforestation activities. As a mechanism to gain access, it has been attempted by the residents of Sanjay Colony through their suggestion that they be allowed to remain on forest land and residents from their settlement be hired to engage in reforestation efforts.

Access to knowledge as a mechanism of control has been discussed in previous sections. Ribot and Peluso define knowledge as "beliefs, ideological control, discursive practices and negotiated systems of meaning" (Ribot & Peluso, 2003, p. 169) This could be in symbolic terms through certain degrees and titles as well as through higher education, training and access to certain powerful discourses that shape frames of access (Forsyth, 2011). Certain groups like experts can claim access on the basis of scientific knowledge required to effect scientific conservation measures (CEMDE, WWN) or planned reforestation measures (ETF). Knowledge as a mechanism to gain access is also evident in the discourse of the Sanjay Colony residents that present tribal earth working knowledge as a claim to remain on forest land. Environmentalists can prevent or secure access of other actors through their ability to engage in legal arguments, use their organising capacity and social networks, to enable certain control patterns through courts and state agencies.

Access to authority as a mechanism of access, can enable access control through law and influence on legal and policy processes and through access to individuals and authorities who make and implement laws (Ribot & Peluso, 2003). This is seen in direct petitioning and lobbying and well as in use of PIL by environmentalists as well as participation of some individuals in the RMB. Experts whose role in forest management is institutionalised have direct access to policy and authority figures. Informal settlements and urban villages access authority by depending on political patronage of local councillors and politicians looking for votes (Ribot, 1995; Robbins, 2000).

Another mechanism is seen in social identity i.e. membership of a certain group or community (Moore, 1986; Ribot & Peluso, 2003). Example of positive effect on access through social identity would include middle class recreational users like bird watchers and 'nature lovers' who are seen as legitimate and desirable elements in conservation spaces (Mawdsley et al., 2009). Social identity is also used to exclude users from protected areas, especially those who engage in extractive uses like fuelwood collection or grazing (Adams & Hutton, 2007; Peluso, 1992). These are seen by conservation authorities as 'criminal people'[421] and as resource controllers they seek to exclude such groups.

Mechanisms to control access are concentrated in the hands of the state agencies and allied expert bodies (ETF, WWN and CEMDE). These resource controlling actors interact with actors that have the discursive, socio-political and legal means to manipulate the resource control through them like middle class environmentalists and courts. The mechanisms to gain and maintain access for resource users especially those engaged in extractive or productive uses are limited through both their social position as urban poor and their position in the conservation discourse as human disturbance to forest area. They do not have access to capital, technology and authority. Moreover, due their ascribed position in the dominant discourse of conservation as detrimental elements to forest conservation, their access to labour opportunities, favourable social identity and an acceptance of their knowledge is curtailed. This unequal access to both the forest as well as to sites of discursive struggle is reflective of the larger inequalities in the social and structure of the city within which these competing views of the landscape are situated; i.e. the Ridge as a site of productive space for practical use as opposed to a site of aesthetic consumption, biodiversity conservation and ecological services.

8.4.2 Green Evictions in the Ridge[422]

A specific kind of access, access to the Ridge as a space for habitation, is discussed in this subsection as linked to the larger urban process of relocation of urban poor from within inner city limits. The first wave of environmentalism ac-

421 Interviews ETF:1, FD:1 and EX:2

422 The term 'Green Evictions' from title of: Ghertner, 2011a

companied by the M.C. Mehta PIL, led to a series of demolitions of structures of all kinds in the Ridge. Environmentalists had to point out that most of the initial demolitions were informal settlements and pushed for the same treatment for other private or government structures that had come up on forest land (Kothari & Rao, 1997). The case of the Mahipalpur Ridge was a stark example of such difference in treatment, where the axe of eviction for conservation fell on the inhabitants of informal settlements of Lal Khet, while other large constructions were allowed. The pattern was repeated in the Asola Bhatti Sanctuary where after the eviction of mining settlements, media reports had to point out the enormous illegal farmhouses on Ridge land, leading to court orders for marking boundaries and demolishing defaulting structures. In each case, though it was not the urban poor who were targeted by the environmentalists, the consequences of conservation were born by this section with the loss of their homes. Linked to the elements of the middle class environmental discourse, the background and functioning of PIL and the role of expert control of these spaces, the larger modification of Delhi's social cartography provides the context for this result.

This alacrity shown by state agencies either on their own or acting on court orders, to evict poor settlers, is part of the urban social reconstitution through which state agencies in large cities in India have been relocating informal settlements to the fringes (Baviskar, 2003a; Bhan, 2009; Coelho & Raman, 2013; Ghertner, 2011a). Since the early 1990s, there has been a noted drive in Indian cities to clear public lands of informal settlement and 'clean up' urban space for the use by 'proper citizens'[423] (Chatterjee, 2004, p. 131). Moreover, there have been marked instances of such evictions being presented as required to protect and promote urban environmental quality (Baviskar, 2003a, 2011; Coelho & Raman, 2013; Ghertner, 2011a; Véron, 2006).

Linked to the politics of neo-liberalisation, with its need to create a "legible and attractive cityscape" (Truelove & Mawdsley, 2011, p. 407), urban authorities have combined the promotion of large scale infrastructure and real estate development like highways, metro rails and shopping malls with the removal of slums in a bid to attract investments (Ghertner, 2011a; Srivastava, 2009). In Delhi in particular, the Master Plan, 2021 declared aspirations to turn Delhi into a 'world class city' which centres on recreating the city in the image of a global city and a hub of investment, for which slum removal is seen as necessary (Baviskar, 2003a). Yet these 'worlding'[424] aspirations (Roy, 2011c, p. 6) of planning agencies like the DDA and certain groups of citizens, are mediated by 'actually exist-

423 'Proper citizens' as a rhetoric term refers to non-poor citizens who are also consumers targeted by the state as opposed to those poor citizens who illegally encroach on public land. They have also been referred to more specifically as 'tax paying citizens (Lemanski & Tawa Lama-Rewal, 2013). It has been pointed out however, that both state agencies and non-poor residents also engage in illegalities of corruption and contraventions of master plans (Ramanathan, 2006; Roy, 2011b; Truelove & Mawdsley, 2011).

424 Roy uses the Term 'worlding' to denote various models of urbanism and strategies used by state agencies or non-state groups to integrate a city into global networks of economic exchange and creating 'world class cities' (Roy, 2011c)

ing urbanisms' (Shatkin, 2011, p. 79) i.e. local planning bodies, in their attempts to implement plans have to engage with ground realities which are

> "rooted in alternative social dynamics (informality, violence, alternative cultural, and social visions, vote-bank politics), that resist worlding practice. These existing urbanisms are manifest in a variety of appropriations of space and social behaviour that contravene master planning: appropriations of urban public space for housing and economic activity by both the poor and the wealthy; the performance of activities in public space deemed inappropriate by planners; and the existence of power centres outside the state"(Shatkin, 2011, pp. 79–80).

Local planning bodies must interact with these existing realities that challenge the aims and legitimacy of the plans through various means. Therefore the results of neo-liberal planning and globalisation are varied socially and spatially as they are a result of this interaction explaining both the flexibility of planning and the inconsistency of results (Roy, 2011c).

The existence and growth of the large informal sector in Delhi, including informal settlements[425] and the vast array of services its residents provide to the city was not accidental but linked intimately with the building of the planned city under the DDA and its master plans. It was the "inadequate, ineffective and inequitable" nature of state planning that led to, and continues to foster such informalities (Truelove & Mawdsley, 2011, p. 410). Baviskar elucidates that the building of the planned Delhi needed workers for whom there were vastly inadequate housing provisions and facilities, leading to the establishment of informal settlements (Baviskar, 2003a). The residents of these settlements resort to political patronage of local politicians looking for votes as well as illegal means like bribes, in order to carve out an insecure niche in the city (Benjamin, 2008; Chatterjee, 1998). The state in many cases tolerated these informalities (Baviskar, 2003a; Truelove & Mawdsley, 2011) and through the 1970s and 1980s, cities in India saw the "emergence of an entire substructure of paralegal arrangements, created or at least recognised by governmental authorities, for the integration of low wage labouring and service population into the public life of the city" (Chatterjee, 2004, p. 137). In the period after independence from colonial rule, under the developmental nationalist state, the poor were seen as having certain right to the city as citizens and it was seen as the responsibility of the state to incorporate them in the national development project (Bhan, 2009). After the end of the emergency period, which was replete with atrocities including large scale slum removal, the government under the new Prime Minister V.P. Singh, gave slum dwellers in Delhi tokens[426]. This was in a bid to enumerate and identify residents of these settlements to attempt a pro-poor policy of slum up-gradation and resettlement, though the government did not last long enough for a change to materialise (Ramanathan, 2006).

425 It is estimated that there are 643 JJ clusters (slum clusters) in Delhi containing four lakh households and housing a population of twenty lakhs (GNCTD, 2014)

426 Known as V.P. Singh tokens, these are used to prove that the holder has been residing in a slum since (or before) 1990 which is the cut-off date for eligibility for rehabilitation (Ramanathan, 2006)

In the post 1990 period when India went through a wave of economic reforms moving towards to a more neo-liberal economic and political stance, a shift is noted in the drive to evict residents of informal settlers in Delhi. Firstly, the logic of clearance is as mentioned before, closely linked to the aspirations of state agencies to attract national and international capital and consumers (Baviskar, 2003a; Fernandes, 2004; Truelove & Mawdsley, 2011). Large, highly visible modern infrastructure projects that are seen as reflective of the neo-liberal ambition, further draw limited funds away from important state investments in public goods like housing and public services (Ghertner, 2010). Further, the demand from "aspirational middle class (and elite) consumer citizens" (Bhan, 2009, p. 141) for better housing, recreational and commercial areas have encouraged the DDA to convert public lands to lucrative real estate by removing squatters (Truelove & Mawdsley, 2011). There has been a marked increase in slum removal in this phase, further accelerating in the mid-2000s. It has been estimated that between 1990 and 2003, 51,461 informal houses were demolished in Delhi, whereas, between 2004 and 2007 alone, at least 45,000 homes were demolished (Bhan, 2009). Moreover, fewer than 25 per cent of the households evicted in this latter time were given resettlement options (Hazards Centre, 2007).

Secondly, the push for clearances since the 1990s has often come from the middle class 'bourgeois environmentalists' (Baviskar, 2003a) along with the active role of the judiciary through PIL, in pursuing a healthy environment for 'public good', which ironically has been detrimental to the urban poor (Baviskar, 2003a; Bhan, 2009; Chatterjee, 2004; Dembowski, 2001; Fernandes, 2004; Ghertner, 2011b; Ramanathan, 2006; Véron, 2006). The increasing role of the middle classes in pushing the implementation of the law and holding state agencies accountable, positions them as ideal citizens and upholders of law, the flip side is that the informal practises of the poor are increasingly vouched in terms of illegality and criminality (Baviskar, 2003a; Truelove & Mawdsley, 2011). Through a review of judicial rulings on the question of slum evictions, Ramanathan shows, that earlier decisions on evictions were tempered with a concern for resettlement rights as well as placed of blame and responsibility for the situation on the failures of the state to provide adequate housing facilities, which the poor deserved as citizens. This framing of the urban poor as desperate citizens who have been failed by inadequate state provision, was replaced by one of illegality of slums and criminality of residents of these settlements as encroachers on public land (Ramanathan, 2006). In February 2000, in response to a PIL[427], the Supreme Court ruled that Delhi as the capital of India "should be a showpiece" and this was prevented by the presence of slums on "large areas of public land [...] usurped for private use free of cost" resettlement of such households on alternative sites by the government would be akin to "rewarding a pickpocket" (Ramanathan, 2006, pp. 3194–3195). This case, therefore, firmly placed in judicial and larger public spheres the picture of slum dwellers as illegal and framed the responsibility of state agencies to remove such hindrances to a hygienic and aesthetic city, under-

427 Almitra Patel vs Union of India Writ Petition (Civil) No. 888 (1996)

cutting the rights of large sections of Delhi's population (Bhan, 2009; Ramanathan, 2006)

These discourses of illegality and a 'world class city' have come together to create a situation where, through a series of decisions in the early 2000s, courts have been ordering demolitions based purely on the aesthetic quality of a settlement (Ghertner, 2010, 2011b). Ghertner has revealed that even authorised, legal settlements have been removed on the basis of being a 'nuisance'[428] due to unaesthetic appearance (Ghertner, 2010). This shift in practise enabled by the courts, allow circumvention of the complications and costs associated with slum removal. Before 2000, the responsibilities of removing slums rested with the state agencies that owned the land and had to be preceded by surveys, notification, resettlement plans etc. The costs of relocation, the ambiguity of land ownership, the absence of clear documentation, corrupt and evasive practices of slum residents and lower bureaucracy made slum removal complicated and time-consuming (Benjamin, 2008; Ghertner, 2011b; Ramanathan, 2006). Reducing the decision to the visual quality of the space flattens historical and legal considerations and allows for demolition without the earlier "calculative practices" like mapping and enumeration that had to be carried out by municipal bodies before relocation and demolition (Ghertner, 2010, p. 185). This has made it possible to skip the technical and political challenges of slum removal as well as made it difficult for residents of such settlements to prevent demolition or argue for better terms through documentary evidence of citizenship and state support like identity cards and proof of residence (Ghertner, 2010). At the same time, the aesthetic argument has also been used to argue for the planned nature of other unauthorised constructions. In the case against the mall in Mahipalpur Ridge the DDA

> "defended the project in the Court for being "planned" and thus legal because of the involvement of professional builders, its high-quality construction, and its strategic function in boosting Delhi's architectural profile. Showing graphic models and architectural blueprints of the proposed development, emphasizing the project's 300 million USD price tag, and describing the mall as a "world-class" commercial complex, the DDA suggested that the visual appearance of the future mall was in itself enough to confirm the project's planned-ness" (Ghertner, 2011b, p. 279).

As we know, the court allowed for the mall to remain despite the expert committee declaring it in contravention to the plan and harmful to the environment.

The discourse of conservation once established in policy, adds another layer of illegality to the informal settlements on forest land, even where they were settled while the land was still being used for productive use with state sanction (or through state agencies themselves) as in the case of the miners' settlements in Mahipalpur Ridge and Asola Bhatti Sanctuary. The residents of Bhawar Singh

428 Legally "nuisance refers to an 'offense to the sense of sight, smell or hearing" (Jain, 2005, p. 97)

camp near the biodiversity park, were included in the in-situ rehabilitation[429] scheme of the DDA and a big inauguration was reportedly held in the settlement by Delhi's Minister for Urban Development in 2009 just before elections. When the promised up-gradation did not materialise a resident filed an query[430] and was informed that rehabilitation for Bhawar Singh Camp was not possible since it is on forest land (Munshi, 2014).

Even though the benefits of conservation accrue on a larger scale, the costs are borne in the immediate locale of the conservation area (Balmford & Whitten, 2003) and indeed even at the ground level, the costs are borne disproportionately by the poor since: "scapegoating the poor has, over the past two decades, become a part of the official discourse on salvaging urban ecologies" (Coelho & Raman, 2013, p. 147). The conservation discourse has led to evictions in the Ridge, at times through the active orders of the court (in the sanctuary) and at other times through the selective targeting of informal settlements by state agencies (in Mahipalpur Ridge), ostensibly to promote the goal of conservation. This outcome is linked to the larger processes of urban reconfiguration in Delhi within which these discursive interactions are placed and to which they are linked.

8.5 THE SUBORDINATE SOCIAL JUSTICE DISCOURSE

This chapter has presented the ways in which participation in the negotiation for the redefining of the Ridge has remained tilted against the urban poor who access the space for various purposes. The discourse of conservation which has become inscribed in the Ridge is shared by a very small group of actors. The broader citizenry of Delhi, even in the locality of the conservation units, is largely unaware of the ideas of biodiversity and conservation. Often during field work for this study, when a local resident was asked if they ever visited the sanctuary or biodiversity park they would not even recognise the term. A journalist reporting on the Asola Bhatti Sanctuary met similar results as he states: "If you ask villagers and shopkeepers at Tughlakhabad about the whereabouts of the Asola-Bhatti Wildlife Sanctuary, either they shake their head in bewilderment or give you blank stares" (Victor, 2003, n.p.). What the users from local informal settlements do report is that their access has been severely restricted and in some cases, the insecurity of their residence enhanced by the government taking over control of the land.

429 In-situ rehabilitation is one of the measures of slum improvement in the Master Plan 2021 (the other two being relocation and environmental improvement of slums). In 2008 DDA announced the 'In-Situ Slum Rehabilitation Scheme', under which slums are redeveloped and upgraded into flats in the same plot of land on which they are located, by a private developer who gets the rights to the remaining land. Twenty one clusters in four pockets including Bhawar Singh camp were declared eligible for this up-gradation the first of which (Kathputli Colony) is being carried out presently (Banda, Vaidya, & Adler, 2013)

430 This was done under the Right to Information Act of 2005 under which citizens can ask the government to respond to demands for specific information through submitting written applications

Resistance to this imposition plays out in different ways. In Sanjay Colony, the resistance took the form of poplar protests to prevent demolition coupled with the "strategic essentialisation" (Brosius, 1999, p. 280) of its population as traditional earth workers to argue for a long term inclusion in the afforestation projects. In other parts has taken the form of more dispersed, less articulated protests through what Scott calls "everyday forms resistance" (Scott, 1985, p. 29) which refers to piecemeal protests on a small scale; for example, squatting on state forests, setting fires to plantations, collecting fuelwood by theft, disobeying security guards. Scott calls these the "weapons of the weak" that stop short of open, organised political activity that is the "preserve of the middle classes and intelligentsia" (Scott, 1985, p. xv). This form of resistance is common where the fear of retribution prevents large scale organised protest, in the case of the informal settlements this could include removal of the settlement due to pressure from the managing agency of the conservation unit for opposing conservation efforts. Such a concern was expressed by a local leader in Kusumpur Pahadi who feared that if the dairies continued to graze their cows, the park authorities could demand removal of the settlement to get rid of the problem[431]. Scott also points out that since people rarely want to draw attention to such resistance and so resistance is hard to document, but these resistances are an assertion of the actors' perceived right to shape the environment (Scott, 1985). In the case of the biodiversity park, fights between security guards and women collecting firewood are common[432]. Tensions between park scientists and locals sometimes result in scuffles when locals are caught in the park collecting fuelwood or with cattle[433]. Setting fire to the plantations is also an act of protest that locals have used on more than one occasion against the park authorities to register their protest [434].

However, the contesting discourses evident through these protests at the ground level, do not directly confront the conservation agenda at the policy formulation level. The conservation discourse is institutionalised and implemented on the ground after which, interests of the various groups of people are bent to fit the situation or they are sought to be removed from the area altogether. There is little conversation between those who can effectively negotiate on the formulation of conservation policy (environmentalists, certain state agencies, the Lt. Governor, courts) and those who are concerned with access to the forest for livelihood, subsistence and habitation.

One factor is the moral dimension of the conservation discourse coupled with the discourse of illegality associated with informal settlements in India. It has been noted that most conservation agendas contain "moral assumptions and assertions" (Bryant, 2000, p. 677) and solutions to perceived conservation problems are often presented in moral terms, precluding the consideration of the costs paid by certain groups for the implementation of conservation in an area (Bryant, 2000;

431 Interview KPLS:2
432 Filed notes FN-ABDP
433 Interview EX:3
434 Interviews EX:2 and EX:3

Neumann, 1998; Wilson, 1999). Where actors supporting the conservation discourse have the power to influence policy, this moral authority can translate to legal and institutionalised forms of access and limitations (Sundberg, 2003). Environmentalists have stressed the immorality of state agencies in not implementing forest laws. The courts, environmentalists and state agencies have stressed the ecological services the forest provides the city to add an element of normative need to the conservation discourse. On the other hand the local users have been presented as criminal people with no regard for the importance of the forest[435] by managing authorities. Residents of informal settlements are already presented in the judicial and official spheres in terms of illegality and criminality (Ramanathan, 2006), in the vicinity of conservation areas an added layer of illegality exists as they are not considered, 'environmental subjects' i.e. those whose "come to care about, act in relation to, and think about their actions in terms of something they identify as 'the environment[436]'" (Agrawal, 2005, p. 162) and therefore cannot be allowed in a space that has been set aside for conservation.

In conservation debates referring to rural landscapes in India, urban middle class conservationists have over recent years, largely come to support the view of local people as essential stakeholders that are seen as carriers of local knowledge which the state lacks (Rangarajan, 2003). They have provided support to this essentialised 'other' that is seen as ecologically wise in environmental movements in the countryside (Baviskar, 2005). This is closely linked to the 'environmentalism of the poor' idea based on division of publics into 'ecosystem people' and 'omnivores' (Guha & Martinez-Alier, 1997). This division is not clear in urban areas where local resources are not the main source of livelihood for the bulk of the urban poor, who inhabit varying degrees of omnivorism. These groups of users have not been able to present a moral claim to access and use the Ridge in most cases and where they have (Sanjay Colony), the dominant discourse of illegality has thwarted their demands so far. An activist who worked with Sanjay Colony residents, summarises the frustration faced when trying to position the residents as holders of local ecological knowledge in the context of negotiations in the sanctuary:

> "The Bhatti Mines case is not about an inevitable conflict between the human rights of a working-class community living within the confines of a sanctuary, and the legitimate concerns of the state government, bound by law to protect the sanctuary. It is about other things like unaccountable governance, class prejudice, and contempt for social and spiritual dignity of simple, hardworking people. Also, about fabricating evidence of a spectacular increase of forest cover, to forward political ambitions of the leadership and justify claims for bringing more land under the state government's control. Above all, it is about the alienation of the

435 Interviews EX:2, EX:3, DDA:4, ETF:1

436 Environmental subjects "care about the environment" (Agrawal, 2005, p. 162), even if it is due to their interest being linked to the protection of the forest. According to Agarwal, this care for the environment is not pre-given but come about as a result of certain experiences (in Agarwal's study, as a result of new participatory forms of forest regulation)

dominant Anglophone elite from cultural traditions and knowledge systems of their desi[437] subjects. Ecologists know the value of ethno-science, and insist that regeneration of degraded eco-systems is not possible without the involvement of local communities, especially rural women. But the officialdom will have none of that. Rural migrants coming to work in the city as labourers become instantly categorised as nondescript 'urban poor', without cultural identities of their own, fit only for getting 'shifted out' when the spaces which they occupy are required for some more prestigious purpose" (Soni, 2006b, n.p.).

The second factor that removes the conservation discourse from contestation is the nature of political participation in Delhi. It has been stated that unlike in countries of the Global North, a really inclusive public sphere does not exist in India (Williams & Mawdsley, 2006b). Political participation or "activity that is intended to or has the consequence of affecting either directly or indirectly government action" (Verba, Scholzman, & Brady, 1995, p. 9), is splintered in India and this has implications for its outcomes (Harriss, 2006).

The concept of civil society which can be defined as follows, is limited and does not capture the larger sphere of politics in India

"social sphere outside the state and market institutions, in which people come together, as equal rights-bearing citizens, on a voluntary basis, and may engage in deliberative action in addressing public problems" (Harriss, 2006, p. 441),

Partha Chatterjee thus provided the distinction between 'civil society'[438] and what he calls 'political society' [439] to describe the different relationships of groups and governments in post-colonial, non-Western societies that are characterised by high levels of poverty and inequality (Chatterjee, 1998, 2001, 2004). He noted that only a small number of 'citizens' have the cultural and social means to access civil society and yet the state and its legal and coercive power has been able to reach almost all of its population (Chatterjee, 2001, 2004). To understand this, Chatterjee presents the concept of 'political society' to describe popular politics of the marginalised, lying "between civil society and the state" (Chatterjee, 2001, p. 173). Political society is the realm of 'populations' i.e. if 'citizen' carries the "ethical connotation of participation in the sovereignty of the state", concept of 'population' means nothing more than a 'target' for the government policy (Chatterjee, 2004, pp. 34–35). Since "most of the inhabitants of India are only tenuously, and even then ambiguously and contextually, rights-bearing citizens in

437 Desi refers to someone who belongs to the Indian subcontinent, used here in the sense of native or 'of the soil'

438 Chatterjee defines civil society as "characteristic institutions of modern associational life originating in Western societies that are based on equality, autonomy, freedom of entry and exit, contract, deliberative procedures of decision making, recognized rights and duties of members, and other such principles" (Chatterjee, 2001, p. 172)

439 These categories of 'citizen and population' 'civil and political society' have been criticised for being too binary and not always being discreet entities in reality (Menon, 2010). Nonetheless, scholars have found these useful concepts and remind us that these categories should be treated as 'ideal types' for conceptual use and not as airtight representations of fact (Routray, 2014)

the sense imagined by the constitution" (Chatterjee, 2004, p. 34) they rely on legal and para-legal negotiations to make claims (Routray, 2014).

Populations are to be categorised and enumerated by the state but lack "basic material and cultural prerequisites of membership of civil society", they are thus denied the "normative status of the virtuous citizen" (Chatterjee, 1998, p. 210). In studying the politics of slums in Calcutta, Chatterjee notes that given their illegal status, the state cannot recognise them; yet they cannot be ignored given the importance of their numbers in democratic politics (Chatterjee, 1998). The struggle of such populations is therefore in the realm of political society where

> "claims and benefits can be negotiated between governmental agencies responsible for administering welfare and groups of population that count according to calculations of political efficacy" (Chatterjee, 1998, p. 282).

The characteristic features of political society are very different from the state society relationship of rights embodied in the constitution. These are: the political society is often based on violation of law (squatters, poachers, users of forest produce in Reserved Forests for example), despite this they demand welfare from the government, these demands are made as collective demands rather than in terms of individuals due to their lack of claim of citizenship (in symoblic terms) and finally, the state and civil society sees them as populations deserving welfare and not as rightful citizens (Chatterjee, 2001). This fosters dependency on political parties, lower bureaucracy and local headmen. Moreover the working of political society is contingent upon factors such as the ruling party at the time, the vote-bank dynamics of the area, the demographic strength to affect electoral result, the power of other local groups, the judicial disposition, the value of the land they are settled on etc. (Routray, 2014).

John Harris has operationalised this separation in the urban context through empirical research on political participation in cities including Delhi (Harriss, 2005, 2006, 2007). He finds that the capacities of groups and individuals to influence government action and policy varies, as do the priorities and demands (Harriss, 2005) and this has implications for the extent and content of the citizenship of these groups[440] (Harriss, 2006; Zérah & Landy, 2012). The modes of engagement with politics available to groups, have an impact on the access to officials as well as the terms of engagement (Harriss, 2006). He finds that the poor[441] residents of slums often try to solve their problems in groups (as opposed to individually taking up a cause and then trying to rally people as we have seen in the case of WWN and CPQLW) but these groups are not organised and are largely local neighbourhood communities. The norm is for local groups to approach political parties through local representatives like the slum headman (Pradhan), who in turn is used by parties to ensure a good electoral turnout. Party politics is also largely oriented to garnering votes from the 'political society'. Wealthier, more

440 Here citizenship is understood as constituted by rights that can be claimed from the state (Chatterjee, 2004)

441 Harris equates the urban poor with informal working class largely coinciding with residents of slums (Harriss, 2006)

educated citizens tend to show relatively little participation in electoral party politics, whereas civil society association in Indian cities largely composed by these groups (Harriss, 2005, 2006). These modes of political participation, therefore, represent different kinds of relationships with the state, ranging from a 'direct relationship' (including direct petitioning and legal action), to a 'brokered relationship' (relationships with political parties and patrons) or a 'confrontational relationship' (demonstrations) (Harriss, 2006, p. 452).

The concerns prioritised by different groups are also diverse. Communities in informal settlements understandably tend to prioritise issues of water supply, sanitation, electricity, livelihood, security of habitat (Baviskar, 2003a; Bhan, 2009; Harriss, 2005). Due to the relative inaccessibility of the state to these groups, their chances of having their demands met by the government remain weak (Harriss, 2006). On the other hand, middle class activists have largely restricted themselves to the immediate concerns of their surroundings rather than espousing larger causes of social justice. They have stressed, "roads rather than public transport; garbage and pollution, rather than public housing; mosquitoes and public toilets rather than public health" (Nair, 2005, p. 336).

Even where causes of the urban poor have been advocated by civil society, the urban poor themselves have largely been excluded as active participants (Harriss, 2007). Civil society therefore has largely excluded representation of the urban poor and has used newly available ways to influence the state (for example PIL), thereby circumventing the messiness of popular politics to meet some of its demands (Harriss, 2006, 2007; Nair, 2005). On the other hand, there has been little organisation of the urban poor or working classes in most Indian cities including Delhi (Baviskar, 2002; Fernandes & Heller, 2006) and therefore brokerage politics is often the only option available to these groups (Nair, 2005).

To summarise, "civil society is the site of middle class activism while the poor have politics" (Harriss, 2007, p. 2717). This results in a situation where the social justice discourse has little opportunity to interact with the conservation discourse. While the environmentalists can forward their demands and opposition through access to the government or through the courts (even at times entering into co-management bodies like the RMB), the contestation of the local marginalised users are usually presented to their slum headman who may or may not present it to the political representative, who in turn can do little to challenge the already negotiated and institutionalised conservation discourse backed by the courts. On the ground this disjuncture produces a situation where local users are told ex post facto that they live on conservation territory and thus will be removed or access will henceforth be blocked.

8.6 CONCLUSION

Since discursive contestations that shape the Ridge are set in wider socio-political dynamics, the outcomes of these contestations are tied to those unequal dynamics. Actors forward certain demands through particular discourses but the outcome is mediated through the interests and discourses of other actors as well as the historical and political landscape that empowers certain actors and disempowers others. Thus while, environmentalists have argued for conservation, the models of conservation that have resulted from their negotiation have varied according to context and the particular set of actors involved. Moreover, while environmentalists around the Ridge have not engaged in anti-poor rhetoric, the burden of restriction of access and loss of homes and livelihood is borne by the poor in the vicinity of the Ridge who in turn have not been able to participate in the conservation debate due their position in the dominant discourse and the nature of political participation in Delhi.

9 DISCOURSES, STORYLINES AND METABOLISM OF THE RIDGE

This study argued that the Delhi Ridge is a socially constructed socio-nature hybrid. To this end, a historical overview of loss of forest cover and afforestation efforts was provided in chapter 4 as a background. The claim was developed through detailed examination of contestations involved in shaping the Ridge through the following chapters. To take the example of Asola Bhatti Sanctuary; lying in a water scarce area, the landscape was historically used as collectively owned, consolidated pastureland rather than privately owned farmland (Kaul, 1990). This enabled the appropriation of a large expanse of land by state authorities, as the area was incorporated into urban development processes. The presence of resources for the construction industry led to the exploitation of the land with state sanction and to the demand for labourers who settled in the area (Soni, 2000). Availability of land in the urban fringe and the comparative lack of regulation by state authorities (Schenk, 2005), led to the establishment of settlements of various sections of society in and around the sanctuary over the years. The legitimacy these heterogeneous interests and uses was challenged through discursive framing of the area as ecologically important and the institutionalisation of this framing through judicial orders and implementation of the Wildlife (Protection) Act. The area was thereby vested with the Forest Department for the 'creation of a Reserved Forest' under orders from the Supreme Court, entailing supply of manure, tree saplings, water, walls, labour for afforestation and guarding, monkeys and fruits and vegetables to feed them, along with capital input to secure these in order to fulfil this established aim. Similarly, in the other two sites, the previous uses and discourses were undermined by the framing of the Ridge as a space for conservation, leading to legitimisation of new actors with the power to redefine the landscape and the interaction of different groups with it. The tendency of the conservation discourse has been to present the Ridge as a natural environment under threat from the processes of urbanisation, therefore requiring of certain kinds of legal, scientific and administrative control to attempt "purification" of the hybrid i.e. while elements of 'nature' and 'culture' are inseparable in hybrids, powerful discourses present them as belonging to one pole (in this case that of nature) rather than the other (Latour, 1993, p. 10).

It does not suffice to establish that these 'natural' spaces are socially constructed. The aim of this study, based in the agendas of political ecology and urban political ecology, has been to examine how these constructions are undertaken and what power dynamics are at play, i.e. an exposition of the politics of formation of socio-nature hybrids (Blaikie, 1995; Forsyth, 2003; Heynen, Kaika, et al., 2006; Robbins, 2012; Zimmer, 2010). By reconstructing the development and establishment of the conservation discourse around the Ridge, this study confirmed the assertion that hybrids are products of historical processes and are "malleable and intermediate" (Gandy, 2006, p. 64) and therefore, the processes through which they are formed are constantly struggled over (Keil & Boudreau, 2006). This struggle is captured by the concept of metabolism which entails both physical and discursive flows related to the creation and contestation of hybrids (Swyngedouw, 2006; Zimmer, 2010). Since metabolic processes are set in existing socio-political conditions, the resulting hybrids benefit some to the detriment of others (Heynen, 2006; Swyngedouw, 2006). Domination over the metabolic processes by certain groups and discourse defines how hybrids are formed and used, who uses them and to what extent (Keil & Boudreau, 2006; Swyngedouw, 2006). Further, relations between sections of society and between society and environment are dialectic and re-enforcing; while metabolism takes place within existing socio-political conditions, those performing environmental practices seen as legitimate are empowered (Robbins & Sharp, 2006) and conversely, those whose practices are delegitimised, are disempowered, thus reproducing social inequalities.

This legitimacy of certain ways of interacting with the environment (including securing benefits from and participation in shaping of the hybrid) is produced and negotiated through discursive interactions that contest to position claims and actors (Hajer, 1995; Zimmer, 2010). This study used argumentative discourse analysis (Hajer & Versteeg, 2005; Hajer, 1993, 1995, 2006) to examine the discourses of various actors involved in the creation and contestation of hybrid conservation spaces. To understand the formation and reproduction of complex, overlapping and at times contradictory discourses, it is important to look at the 'moment of dislocation' (Hajer & Versteeg, 2005, p. 182) (the point where a problem is defined by certain actors) and the following argumentative interaction of discourses (Hajer, 1995, 2006). To this end, the various seemingly unconnected struggles to conserve the Ridge were analysed. The study established that at present, the conservation discourse is the dominant discourse in the case of the Ridge, as central actors must refer to it in order to have any credibility in the concerned domain i.e., the discourse has achieved the condition of 'discourse structuration'. Conservation has also been institutionalised in certain institutional and organisational practices, embodied in official policy documents (notifications, plans, and court orders for example) and achieved 'discourse institutionalisation' (Hajer, 1995, pp. 61–62, 2006, p. 70). However the conservation discourse does not encompass the claims and demands of all actors and a subordinate social justice discourse can also be discerned, which is not reflected in policy and has not been able to successfully contest the dominant discourse.

The concept of storylines was used to explain how complex environmental issues are deliberated upon by actors and a finite definition of the problem is arrived at in order to negotiate the solutions. Actors do not engage in the details of the storyline, but these are rather used as metaphors once they are ritualised (as they are used by central actors with regularity) (Hajer, 1995, 2006). These storylines help overcome fragmentation of information, interests and viewpoints but also create a discursive order by placing ideas of blame, responsibility and urgency on various actors (Hajer, 1993, 1995). Within the dominant conservation discourse, there are several common overlapping storylines which can be read. These storylines have been present in the discourse of environmentalists since the 1980s and once the discourse achieved dominance, they have been shared by the main empowered actors (environmentalists, state agencies and scientific experts).

The first storyline is one that poses the Ridge as natural heritage. The demand of the storyline is that there should be no conversion of the forest land to other uses, especially built-up area or landscaped, horticultural parks. This storyline privileges natural heritage over cultural heritage as seen in the case of Sanjay Van where old religious sites are seen as hampering the forest regeneration efforts. Moreover, it obfuscated the long history of the Ridge as a site of habitation and use and its inter-connectedness with the social and economic lives of the surrounding inhabitants. This creates the ironic situation where traditional Gujar herdsmen of Asola Bhatti area and the religious sites of Sanjay Van are positioned as criminal, even though the idea of the Ridge as pristine forest was not established until more recent times[442]. Historical evidence shows that the protection and afforestation of the Ridge in modern times, was initiated by colonial officers for the sanitary and aesthetic concerns of colonial settlers and since then, the Ridge has been sought to be cleared of claims from subsistence and livelihood related uses in the name of public heritage (Mann & Sehrawat, 2009). The second storyline within the conservation discourse is that of ecological crisis. This is related to the framing of the importance of the Ridge in terms of the ecological services it provides the city, including serving as a water catchment area, reducing pollution, preventing desertification and lowering temperatures. This storyline has been used by environmentalists to enrol support and by state agencies and experts to justify their policies. The third storyline presents the Ridge as an important biodiversity reserve, privileging non-human elements (fauna and flora) above most human interests. According to this storyline, the Ridge must be protected from 'human disturbance' for intrinsic purposes. This particular storyline was behind the suggestion of Srishti and Kalpavriksh that core areas must be maintained in the forests where no human use would be allowed (Kalpavriksh, 1991; Srishti, 1994). Since 2004, the biodiversity storyline has empowered scientist managers in the biodiversity park in particular and in the Ridge in general (as the scientists of CEMDE advise the Forest Department and ETF in the sanctuary, and WWN also bases its legitimacy on scientific afforestation) to 'speak for nature' (Braun, 1997, p. 11). These scientists have been authorised to engage in active restoration and conservation projects that entail high levels of control on space and legitimacy to select biophysical as well as use and access elements of the forest.

While three main storylines of the conservation discourse; 'natural heritage', 'ecological crises' and 'biodiversity reserve', are common across the Ridge, they do not result in the same practices across actors such that, once operationalised in conservation models, one finds a diversity of results. Hajer stresses that the power

442 As it may be recalled, parts of the Southern Ridge were legally protected only beginning in 1986 and Sanjay Van was declared Reserved Forest in 1994

of storylines in influencing policy and regulation does not come from their consistency or their basis in factual evidence, but from their 'multi-interpretability' (Hajer, 1995, p. 63). The natural heritage storyline, for example, has seen solidification in a stakeholder afforestation model in Sanjay Van as WWN sees it as a means to connect citizens to their heritage. In the other areas however, such an attempt has not taken place. The biodiversity of the three bounded spaces is also differently conceived. In the Asola Bhatti Sanctuary, the main aim of the Forest Department and ETF was to increase green cover and Prosopis juliflora was planted to achieve quick results, while court orders led to the introduction of a large number of rhesus monkeys into the sanctuary. In the biodiversity park, Prosopis juliflora is being actively removed and replaced, while in Sanjay Van, saplings of other indigenous trees are being planted alongside the existing invasive species. Different elements of the 'ecological crisis' storylines are stressed in different areas depending on their context. In the Mahipalpur Ridge, it was the issue of water scarcity faced by local residents and the role of the Ridge as an important catchment area that was used by environmentalists as a means to enrol support against construction in the Ridge. In the sanctuary, the prospect of desertification is highlighted along with ground water recharge given that it lies in the leeward region of the Aravalli, on the outskirts of the city in a heavily mined area.

This variation in results can be explained by the contextual differences in the three sites as well as through the particularities of the interaction of discourses that are embodied in the outcome. This argumentative interaction which is central to the formation and change of discourses, is captured by the concept of discourse coalitions which are a set of storylines, the actors who forward them and the associated practices in which the discourse is based (Hajer, 1995, 2006; Herzele, 2005). Actors must enlist other actors into a discourse coalition in order for a discourse and its associated storylines to become dominant (Hajer, 1995). However, as this research demonstrates, the enrolment of other actors also changes elements of the discourse, as each actor in the coalition is drawing on various storylines available to them and does not necessarily have the same aims and interests. Moreover, the interactions of actors in the coalition can legitimise new actors such as scientists and military units (CEMDE and ETF) and place blame on actors not targeted by initial demands for conservation, such as informal settlements.

In the case of the Ridge, environmentalists have tried to enrol state agencies in order to achieve discourse institutionalisation. However, as is evident from the different models of conservation in the three conservation units, the particularities of the negotiations has produced a varied set of results. Thus, looking at structural aspects alone would not explain why the same structures result in different outcomes. The DDA for example, presented the Ridge as a space for recreation and horticultural practices in the MPD 1962 (DDA, 1962). This shifted to a discourse of conservation and maintaining the 'ecological balance' in the MPD 2001 (DDA, 1990) and MPD 2021 (DDA, 2007). However, the authority continued to carry out horticultural practices in Reserved Forest of Sanjay Van until 2010 and did not consider Mahipalpur Ridge ecologically important or legally protected and therefore open to construction. Following contestation by environmentalists, who managed limited mobilisation of citizens but enrolled key executives like the Lt. Governor and the Prime Minister through direct petitioning and the Supreme Court through PIL; the DDA was added to the conservation discourse coalition. On the ground, the same authority presently manages Sanjay Van and Aravalli Biodiversity Park in very different ways in collaboration with CEMDE and WWN, which subscribe to different philosophies of management, with the former being far more deep-ecological and the latter attempting a stakeholder-based project of afforestation.

The temporal and spatial incoherence in the discourses and practises of the DDA have been traced by this study as being based in the variety of contestations (from other state agencies as well as non-state actors) the authority has faced in the context of the Ridge, in addition to its role in planning and restructuring urban land in the neo-liberal context. The agencies of the state, therefore, have not been actors representing a stable set of interests and discourses, as a more structural approach would suggest. Rather, they have functioned as members of the discourse coalitions responding to conflict and interaction with other discourses within the given structures.

It is this interaction of actors and associated storylines that contributes to the outcome of environmentalism being detrimental to the interests of the poor once the conservation discourse attains dominance in a certain area, since interaction can change elements of the discourse and define new problems (Hajer, 2006). This outcome is based in contextual political economic (structural) elements of inequality in city and linked to the associated discourses and storylines of the actors enrolled into the discourse coalition (including the DDA, Forest Department, ETF, WWN the Supreme Court and CEMDE). The demands of conserving the forest by environmentalists have targeted the corrupt (supporting private and commercial interests in the Ridge), inappropriate (horticultural practices) and at times, inadequate actions of the state in preventing the conversion of forest land. However, the institutionalisation of the conservation discourse has to be mediated through state agencies and has led to the empowerment of state agencies to impose models of conservation that disregard competing claims from inhabitants of informal settlements and urban villages. This is linked to the discourse of Illegality of the informal settlements of the economically marginalised and the discourse of a 'world class city' that the conservation discourse interacts with within the discourse coalition.

Once a discourse achieves the condition of discourse structuration, actors must present their claims and challenges in terms of the dominant discourse to have credibility in a given domain (Hajer, 1993, 1995). Therefore, those who forward their claims in terms of social justice, are not considered legitimate actors in conservation areas. Within the social justice discourse there are several storylines, evident in various parts of the Ridge. The first is that which claims that the inhospitable Ridge has been settled and made habitable by the residents of the informal settlements such as Sanjay Colony and Sangam Vihar, many of whom settled in the area before these areas had been marked as legally protected forests. The second storyline sees the Ridge as a village commons and cultural heritage as is evident in the Sanjay Van, where earlier associations with religious structures and other uses of space continue in the face of opposition from authorities. A third storyline is that of injustice, that demands that other encroachments not be allowed in the areas from where the informal settlements have been evicted, as the residents of Lal Khet argued as a part of the RBA[443]. A particular storyline within the social justice discourse that attempts a synthesis with the conservation discourse is presented by the residents of Sanjay Colony in the sanctuary. This storyline forwards the claim of tribal knowledge of earth working and water conservation, which can be used for afforestation efforts by employing the residents of the colony rather than evicting them. However these demands have not been given any consideration by state agencies, since residents of informal colonies are positioned as illegal and their presence on areas deemed forest land adds another layer of illegality precluding any negotiation with these groups.

In tracing the reasons for the subordination of the social justice discourse with regard to the Ridge, it was argued that they lie at the level of the positioning of local users in the conservation discourse and secondly, in the practices involved achieving discourse institutionalisation. Despite the fact that the diverse groups of environmentalists have brought up concerns of social justice, none of them have actively pushed for the inclusion of such concerns in conservation policy. Once the main demand for conservation is met, the side-effects of eviction and restriction of access fail to draw the same level of involvement from environmentalists. This limited integration of the conservation and social justice demands in the discourse of the environmentalists, morphs into a complete disjunction between the two in policy due to the interactions within the discourse coalition as described above. The institutionalised conservation discourse (as seen in the application of forest laws and creation of biodiversity parks) positions users of forests for purposes of habitation, subsistence and livelihood as illegal and detrimental to conservation efforts. Thus, within this discursive structure, the claims of marginalised 'illegal' local users bear no weight in spaces set aside for preservation of natural heritage, biodiversity and ecological services for the benefit of the larger public. Secondly, environmentalists have used PIL and petitioned higher executives with some success to translate their demands to policy. Some of them have access to the state through membership in the RMB. These practices limit inclusion of a broader base of claims in the debate as marginalised users do not have the social and political resources to effectively approach higher officials or participate successfully in litigation. While the contestation for forest conservation takes place in the realm of 'civil society', the marginalised local users can only put up counter claims at the level of 'political society' by approaching local headmen and political leaders (Chatterjee, 2001, 2004). The main bodies involved in legitimising and implementing the conservation agenda such as the CEMDE and the Supreme Court (Including the CEC) are impervious to political pressure and therefore beyond the reach of political society. Therefore, access of marginalised users to the negotiations related to formation of conservation spaces remains

443 These storylines are not limited to the examples provided and are overlapping

limited. Moreover once the conservation discourse is dominant, their position as illegal and undesirable elements is reinforced as their practices are further delegitimised within the context of conservation efforts. On the other hand, the practices of the managing agencies (including DDA, Forest Department, CEMDE, ETF and WWN) re-enforce their positions of power as actors performing afforestation and protection measures in the Ridge, backed by legal and institutional authority to do so.

In other words, the processes of formation of hybrids are unequal as dominant discourses (conservation) overlay subaltern ones (social justice) and certain groups (middle class environmentalists, courts, state agencies and scientific experts) can exert greater control upon these processes and the resulting hybrids have unequal benefits for various actors (Swyngedouw & Heynen, 2003). This flow of discourses, practices and materials is summarised in the figure below:

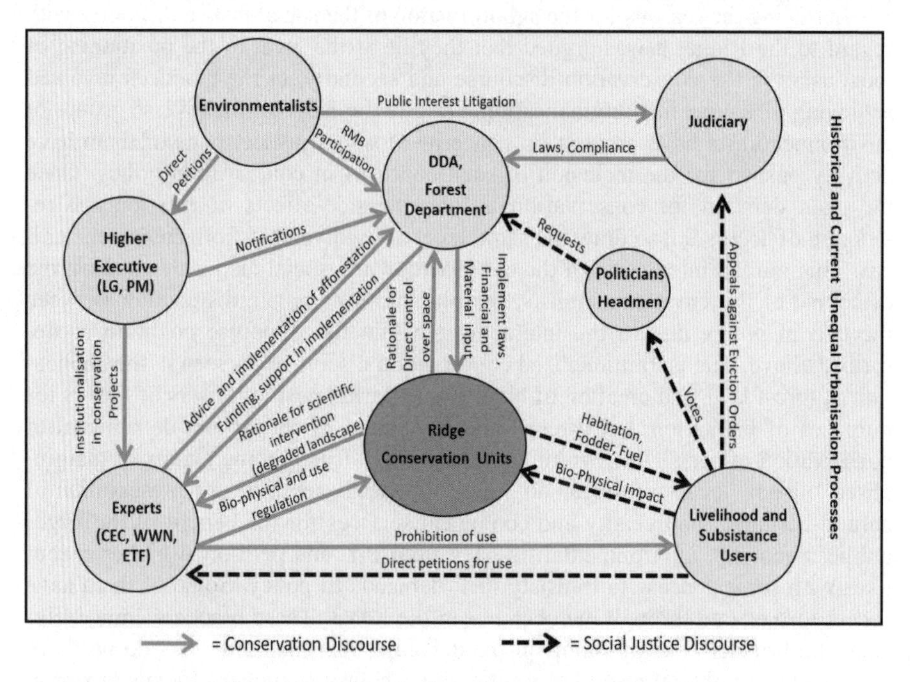

Figure 3: Metabolism of Conservation Units in the Ridge

In the dominant conservation discourse the only legitimatised flows out of the forest are related to nature education as defined by experts and ecosystem services such as air purification. The material and discursive flows are controlled by experts and state agencies which have the legitimacy and power to construct forests in discursive as well as material terms including form (boundaries) and content (species of flora and fauna and kinds of uses). These decisions are in themselves not free of contextual socio-political dimensions and can be influenced by demands from certain actors like other state agencies especially the court and higher executives as well as environmentalists. On the other hand, the social justice discourse presents the forest as a space infused with politics of access and claims over benefit from flows of material such as firewood, pasture and land for habitation. As seen in the figure, those who use the Ridge for fuel, fodder, habitation, sanitation etc., have a direct impact on the Ridge. However, they have limited access and legitimacy in the negotiations with other central actors who forward the conservation discourse.

The silencing of the social justice discourse in policy and its separation from the dominant conservation discourse has implications not only for the marginalised communities effected by conservation, but also for the outcomes of the conservation projects that face constant challenges in implementation of forest laws and require increasing levels of legal and physical protection.

10 CONCLUSION

10.1 SUMMARISATION OF MAIN ARGUMENTS

It was argued in the previous chapters that the diverse forms of conservation spaces in the Delhi Ridge have been formed within the historical and political specificities of the city through conflict and negotiation involving a variety of actors with varying levels of socio-political power. This unequal negotiation has led to the production of different kinds of forest spaces (with varying biophysical elements, legal and administrative aspects and access possibilities) and to the reproduction of socio-economic inequalities. These processes have had negative consequences for marginalised populations living in the vicinity of the forest, as well as for the viability of the conservation units in achieving their stated aims.

As regards actors examined, it was contended firstly, that the state, though central to formulation and implementation of environmental policy, is not a single actor. Rather, it comprises agencies with differing motivations and at times opposing mandates. Further, state agencies interact between themselves and with non-state actors and this interaction is reflected in their discourses and policies. Secondly, environmentalists concerned with the Ridge have employed diverse means to influence agencies of the state and successfully enrolled them into a coalition of the conservation discourse. The outcome of environmental activism however, has often had negative consequences for the marginalised due to the way the conservation discourse is framed and the nature of interactions that occur between the demand for conservation from environmentalists and the formalisation of conservation in legal and physical terms. Thirdly, marginalised populations in the vicinity of the conservation spaces, especially those residing in informal settlements, are directly affected by restriction of access and in some cases, eviction. This group is not part of the negotiations for the formation of conservation units and does not have the political resources to effectively counter the models of conservation imposed in their area. This is due to their positioning in the conservation discourse as well as the structural fragmentation in the nature of political participation in Delhi.

To capture the nuances of socio-environmental transformation presented in this research, the insistence of post-structural political ecology on context-specific detailed analysis through a study of discourses of various groups, were adhered to and urban political ecology concepts of hybrids and metabolism were drawn upon to illustrate the case. The limitations of urban political ecology, based in its structural legacy can be overcome by the epistemology and expansiveness of situated, post-structuralist political ecology. This study has therefore attempted to deal with the criticism of both political ecology and urban political ecology in being geographically and methodologically constricted, by demonstrating that the methodologies, agendas and concepts of the two are complementary rather than exclusive and restrictive. Through an analysis based in 'post-structural urban political ecology' in the context of the case, the study calls for a critical envisioning of environmental issues as being implicated in the socio-political, economic and historical context; implying that solutions must aim at intervention at the level of social aspirations, urban development processes and related inequalities rather than ignoring these issues by regarding them as separate from the legal, administrative and scientific debates on conservation.

10.2 POST-STRUCTURAL CONTEXTUAL URBAN POLITICAL ECOLOGY

This research fulfils the broad manifesto of political ecology and urban political ecology presented in chapter 2. To summarise, the social construction of the Ridge was examined critically through discourses and practices involved in shaping, re-shaping and contesting the forest as a hybrid socio-natural space. This was done by analysing the positions of multiple actors from micro-level politics to broader policy formulations within the historical and political context of urban development in Delhi. Various relationships among groups of society and between different groups of society and the Ridge were examined to unveil the power dynamics embedded therein.

To expand the explanatory potential and analytical framework of political ecology and urban political ecology in order to study environmental transformations in a city of the Global South, post-structural methodologies were used along with contextual literature on political participation, environmentalism and the state's role in mediating socio-environmental relationships.

In tracing the contestations regarding the Ridge, the need for such a contextual understanding was evident through the variety of outcomes within the same structural background. Firstly, the state has managed the Ridge in different ways both temporally and spatially depending on the contestation faced within the state and from certain non-state actors. Secondly, the opposition to the practices of the state has also played out differently from rights based protests (Sanjay Colony) to middle class mobilisation and litigation to fragmentary, covert practices of protests like setting fire to plantations and illegal fuelwood collection. Thirdly, within the discourses of middle class environmentalism, the state has been positioned differently as seen in the collaborative approach in Sanjay Van as opposed to the confrontational approach in Mahipalpur. Analysing actors and their interaction through discourse analysis illuminated how these various outcomes have come about.

Both the actor approach (Long & Long, 1992) used in this research and argumentative discourses analysis (Hajer, 1993, 1995) draw on the structuration theory of Giddens and Foucauldian discourse theory which acknowledge the effect of structures but point out that they cannot be understood without the role of agency to account for the dynamism within powerful structural elements (Foucault, 1969; Giddens, 1984). Further, both agree that the role of structures and agency must be seen in context as power is based in social interaction and must be constantly negotiated (Feindt & Oels, 2005; Giddens, 1984). Therefore broader inequalities within which discourses are produced and contested are important to explain why certain discourses are rendered powerful as against others (Foucault, 1980; Latour, 1986). Studies on fragmentation in political participation, background and role of state and other powerful actors like scientists and middle class environmentalists and their practices along with meta-representations of illegality and 'world class city' linked to neo-liberal agenda, provided this context in the present study. To summarise: socio-environmental interactions in the case of the Ridge were found to be temporally and spatially specific but display certain identifiable interlinked patterns. These patterns are accounted for by a) the broader context of the neo-liberal shift evident in the DDAs practices and the associated weakening of marginalised groups in demanding rights from the state, b) the fragmentation in political participation and differential access to authority and discursive practices that have proven to be successful such as PIL, c) the particular interplay of discourses and the position of actors therein that reinforce the legitimacy of certain actors and delegitimise others.

Through this contextualisation, the study evaded "unreconstructed pluralism" (Blaikie, 1999, p. 114) by relating discourses and practises to the larger process of unequal development to understand the power relations reflected in the environmental outcomes.

10.3 SOME OPTIMISTIC REFLECTIONS

While socio-economic and political inequalities have considerable impact on the participation of various actors in the formation and use of conservation spaces, there has been a degree of dynamism in the ways in which this negotiation has taken place displaying the possibilities for actors' agency playing a part in the outcome. As discourse analysts have pointed out, new storylines can change the way in which a contestation occurs and the ways in which the problem is defined (Fischer, 2003; Hajer, 1995; Herzele, 2005). This provides some room for optimism regarding the possibilities of finding ways to meet both social and environmental justice goals, though this is tempered by the obstacles in realisation within the context of unequal urban politics.

Environmentalists have been instrumental in establishing the discourse of conservation and setting the Ridge apart from the category of landscaped urban greens. So far, various groups of environmentalists have mostly reacted to perceived threats to the forest from other actors, especially the DDA. This confirms the assertion of argumentative discourse analysis framework that the genesis of change in discourse lies in the presence of a contestation (Hajer, 1995). However, this confrontation has more often than not, been reactionary, leading to 'crisis environmentalism' (Harvey, 2000, p. 217), whereby strategic compromises have been made to secure forest land from conversion to other uses. As a result, environmentalists concerned with the Ridge have not been able to substantiate their concern for social justice linked to conservation. Where a less confrontational stance has been evident, as in the case of Sanjay Van, it has made the collaboration of environmentalists with state agencies possible, resulting in alternative models. This points to the possibility of deliberating goals and means of conservation with the state, where actors have the requisite political and social connections. There is a possibility that if conservation demands were proactive rather than reactive, there would be a better chance for the integration of the two discourses. In building a broader narrative of the needs for conservation and the ways in which conservation goals can be achieved, environmental activists would be able to attempt the mobilisation of a larger support base for their cause as well. Since the mid-1990s, environmentalists have relied on PIL which, as described in chapter 8, has certain pitfalls in terms of participation and outcomes. A wider constituency for conservation needs to be built in order for conservation demands to have resonance in policy without depending on judicial enforcement for setting the terms of resolution.

State agencies and expert bodies associated with them have the power and legitimacy to try models of urban conservation that are suited to the context of Delhi. While implementation of forest laws prevents, to a large extent, the further conversion of forest land to built environment, these laws also provide space for negotiation of the exact forms of management of forests. It has been pointed out that wilderness based preservationist conservation will meet little success in heavily populated urban areas, as it stresses the separation of human and forest areas (Dearborn & Kark, 2010; Rosenzweig, 2003b). One possible way for more pragmatic management of urban forests would be an acceptance of 'reconciliation ecology' (Rosenzweig, 2003a, p. 194) which accepts that natural environments in urban areas can be "valuable without being pristine" (Dearborn & Kark, 2010, p. 6), and attempts ways to reconcile certain conservation goals with human demands. Motivations for conservation have to be explicitly negotiated between various actors and stakeholders and a considerable degree of pragmatism would be called for as different actors have varying interests. A necessary step would be to understand the usage of the forest space and produce (fuelwood and fodder) to negotiate which of these demands are compatible with the conservation goals that have been deliberated upon.

10.4 SUGGESTIONS FOR FURTHER RESEARCH

It was pointed out that the environmentalists in the case of the Ridge, do not clearly confirm the anti-poor tendencies visible in the dominant strain of urban environmentalism in India. A comparative analysis of environmental engagement in other cases, including that of environmentalism related to the Yamuna river in Delhi, may help illuminate patterns of similarity and difference which could lead to a better understanding of urban environmentalism in India.

The role of conspicuous individuals within the middle class was referred to in chapter 8 (section 8.1). An ethnographic study of place-based environmental understandings of such individuals would be a step towards understanding why certain figures dominate environmental activism and are more prone to take up environmental issues than others.

As mentioned in the previous section, a detailed empirical study on the varied uses and aspirations related to urban forests held by different actors would provide some understanding of the effect of the practices of various users on the forest along with providing a starting point to negotiate the goals of conservation in the city.

REFERENCES

Abebe, T. (1994). Growth Performance of Some Multipurpose Trees and Shrubs in the Semiarid Areas of Southern Ethiopia. *Agroforestry Systems*, *26*(3), 237–248.

Adams, W., & Hutton, J. (2007). People, Parks and Poverty: Political ecology and biodiversity conservation. *Conservation and Society*, *5*(2), 147–183.

Adams, W., & Mulligan, M. (2002). *Decolonizing Nature: Strategies for Conservation in a Post-colonial Era William Mark Adams, Martin Mulligan*. Earthscan.

Adas, M. (1989). *Machines as the Measure of Men: Science, Technology, and Ideologies of Western Dominance*. Ithaca: Cornell University Press.

Agar, M. (1996). *Professional Stranger: An Informal Introduction To Ethnography* (2nd ed.). Academic Press.

Agarwal, A. (1997). The politics of development and conservation: legacies of colonialism. *Peace and Change*, *22*(4), 463–482.

Agarwal, R. (2010). Fight for a forest. *Seminar*, *613*, 48–52.

Agrawal, A. (2005). Environmentality: Community, Intimate Government, and the Making of Environmental Subjects in Kumaon, India. *Current Anthropology*, *46*(2), 161–190.

AMDA Bulletin. (2011). *Newsletter of Association of Municipalities and Development Authorities (July–September)*, *1*(3).

Anderson, D., & Grove, R. (1987). *Conservation in Africa: People, Policies and Practice*. Cambridge: Cambridge university press.

Angelo, H., & Wachsmuth, D. (2014). Urbanizing Urban Political Ecology: A Critique of Methodological Cityism. *International Journal of Urban and Regional Research (Early Online Version)*. doi:10.1111/14.

Arabindoo, P. (2005). Examining the peri-urban interface as a constructed primordialism. In V. Dupont (Ed.), *Peri-urban dynamics: Population, Habiat and environement on the peripheries of large Indian metropolisis A review of concepts and general issues. CSH Occassional paper No. 14* (pp. 39–74). New Delhi: CSH.

Armitage, D., de Loë, R., & Plummer, R. (2012). Environmental governance and its implications for conservation practice. *Conservation Letters*, *5*(4), 245–255.

Ashok, S. (2014, April 29). Asola Sanctuary encroachments Razed. The Hindu. New Delhi. Retrieved from http://www.thehindu.com/todays-paper/tp-national/tp-newdelhi/asola-sanctuary-encroachments-razed/article5958287.ece (01.05.14).

Ashraf, J. (2004). *Historical ecology of India*. New Delhi: Sunrise Publications.

Asian Age. (2006, April 27). RWAs against Malls in Ridge forest areas. New Delhi.

Bailey, S., & Bryant, R. L. (2005). *Third World Political Ecology: An Introduction* (p. 256). London: Routledge.

Bajpai, R. (2007, August 17). A twist in the mall tale. *Hindustan Times*. New Delhi.

Balmford, A., & Whitten, T. (2003). Who Should Pay For Tropical Conservation and How Could the Costs Be Met? *Oryx*, *37*, 238–250.

Banda, S., Vaidya, Y., & Adler, D. (2013). *In-Situ Slum Rehabilitation Scheme*. New Delhi.

Barnes, T., & Duncan, J. (Eds.). (1992). *Writing Worlds: Discourse text and metaphor in representation of landscape*. London: Routledge.

Basset, J. (1998). The political ecology of peasant herder conflicts in nothern Ivory coast. *Annals of the Association of American Geographers*, *78*, 453–72.

Batra, L. (2010). Out of Sight out of mind: Slum dweller in “world class” Delhi. In *Finding Delhi: Loss and renewal in the megacity* (pp. 16–16). New Delhi: Penguin India.

Batterbury, S., Forsyth, T., & Thomson, K. (1997). Environmental transformations in developing countries: Hybrid research and democratic policy. *Geographical Journal*, *163*(2), 126–132.

Baud, I., Sridharan, N., & Pfeffer, K. (2008). Mapping Urban Poverty for Local Governance in an Indian Mega-City: Case of Delhi. *Urban Studies*, *45*, 1385–1412.

Bauer, M. (2000). Classical content analysis: A review. In M. Bauer & G. Gaskell (Eds.), *Qualitative Researching with Text, Image and Sound: A Practical Handbook for Social Research* (pp. 131–151). London: Sage.

Baviskar, A. (1995). *In the Belly of the River: Tribal Conflicts over Development in the Narmada Valley*. Delhi: Oxford University Press.

Baviskar, A. (2001). Environmental Movements in India : The South forges its own discourse. In J. Nickum (Ed.), *Environmental management, poverty reduction and sustainable regional development* (pp. 93–108). Westport, Connecticut: Greenwood Press.

Baviskar, A. (2002). The Politics of the City. *Seminar*, *516*, 40–42.

Baviskar, A. (2003a). Between Violence and Desire: Space, power, and identity in the making of metropolitan Delhi. *International Social Science Journal*, (Baviskar 2001), 89–98.

Baviskar, A. (2003b). States communities and conservation. In V. K. Saberwal & M. Rangarajan (Eds.), *Battles over nature: Science and the politics of conservation* (pp. 267–297). New Delhi: Permanent Black.

Baviskar, A. (2005). Red in Tooth and Claw: Looking for Class in Struggles over Nature. In R. Ray & M. F. Katzenstein (Eds.), *Social Movements in India: Poverty, Power and Politics* (pp. 161–178). Oxford, UK: Rowman and Littlefield Publishers.

Baviskar, A. (2010). Indian environmental politics: An interview. *Transforming Cultures eJournal*, *5*(1).

Baviskar, A. (2011). What the Eye Does Not See: The Yamuna in the imagination of Delhi. *Economic and Political Weekly*, *XLVI*(50), 45–53.

Baviskar, A. (2012). Public Interests and Private Compromises. In J. Eckert, B. Donahoe, C. Strümpell, & Z. Ö. Biner (Eds.), *Law Against the State: Ethnographic forays into the laws transformation* (pp. 171–201). Cambridge: Cambridge University press.

Baviskar, A. (2013,July 1). The Sacred grove. *Outlook Traveller*. Retrieved from: http://www.outlooktraveller.com/trips/The-sacred-grove-1005241#5539 (22.22.13)

Baxi, U. (1980). *The Indian Supreme Court and Politics*. Lucknow: Eastern Book Company.

Baxi, U. (1985). Taking Suffering seriously: Social action litigation in Supreme Court of India. In V. R. Krishna Iyer, R. Dhavan, & K. Salman (Eds.), *Judges and the Judicial Power: Essays in honour of Justice V.R. Krishna Iyer* (pp. 289–315). London and Bombay: Sweet, Maxwell and Tripathi.

Bebbington, A. J., & Batterbury, S. P. J. (2001). Transnational livelihoods and landscapes: Political ecologies of globalization. *Cultural Geographies*, *8*(4), 369–380.

Beck, U. (1999). *World risk society*. Oxford: Blackwells Beier.

Benjamin, S. (2008). Occupancy Urbanism: Radicalizing politics and economy beyond policy and programs. *International Journal of Urban and Regional Research*, *32*(3), 719–729.

Bentinck, J. V. (2000). *Unruly urbanisation on Delhi's fringe changing patterns of land use and livelihood (PhD. Thesis)*. Univeristy of Groningen. Retrieved from dissertations.ub.rug.nl/FILES/faculties/rw/2000/j.v.bentinck/thesis.pdf (02.06.13).

Bhan, G. (2006, November). Does No One Care that I've No Place Here? *Tehelka*. Retrieved from www.hindu.com/2004/07/06/stories/2004070609450400.htm (09.04.14).

Bhan, G. (2009). "This is no longer the city I once knew": Evictions, the urban poor and the right to the city in millennial Delhi. *Environment & Urbanization*, *21*(1), 127–142.

Bhaskar, R. (1975). *A Realist Theory of Science*. London: Routledge.

Bhatnagar, G. V. (2013, January 27). The Monkey Menace. *The Hindu*. New Delhi. Retrieved from www.thehindu.com/todays-paper/tp-national/tp-newdelhi/the-monkeymenace/article4349713.ece (23.05.14).

Bhushan, P. (2004). Supreme Court and PIL: Changing Perspectives under Liberalisation. *Economic and Political Weekly*, *39*(18), 1770–1774.

Bhuwania, A. (2014). Courting the People The Rise of Public Interest Litigation in Post-Emergency India. *Comparative Studies of South Asia, Africa and the Middle East*, *34*(2), 314–335.

Billig, M. (1989). The argumentative nature of holding strong views: A case study. *European Journal of Social Psychology*, *19*(3), 203–223.

Blaikie, P. (1985). *The Political Ecology of Soil Erosion in Developing countries*. London: Longman.

Blaikie, P. (1994). *Political ecology in the 1990s: An evolving view of nature and society. CASID Distinguished Speaker Series, NO. 13.* (p. 33). Michigan State University.

Blaikie, P. (1995). Changing environments or Changing Views? A political ecology for developing countries. *Geography*, *80*(348), 203–214.

Blaikie, P. (1999). A Review of Political Ecology Issues , Epistemology and Analytical Narratives. *Zeitschrift Für Wirtschaftsgeographie*, *43*, 131–147.

Blaikie, P. (2000). Development, post-, anti-, and populist: A critical review. *Environment and Planning A*, *32*(6), 1033–1050.

Blaikie, P. (2008). Epilogue: Towards a future for political ecology that works. *Geoforum*, *39*(2), 765–772.

Blaikie, P., & Brookefield, H. (1987). *Land degradation and society*. London: Methuen.

Bove, P. (1995). Discourse. In F. Lentricchia & T. McLaughlin (Eds.), *Critical Terms for Literary Study* (2nd ed., pp. 50–65). Chicago: University of Chicago Press.

Braun, B. W. (1997). Buried Epistemologies: The Politics of Nature in (Post)colonial British Columbia. *Annals of the Association of American Geographers*, *87*(1), 3–31.

Brosius, J. P. (1999). Analyses and Interventions: Anthropological Engagements with Environmentalism. *Current Anthropology*, *40*(3), 277–309.

Bryant, R. L. (1991). Putting politics first: The political ecology of sustainable development. *Global Ecology and Biogeography Letters*, *1*(6), 164–166.

Bryant, R. L. (1992). Political ecology: An emerging research agenda in Third-World studies. *Political Geography*, *11*(1), 12–36.

Bryant, R. L. (1997). Beyond the impasse: The power of political ecology in Third World environmental research. *Area*, *29*(1), 5–19.

Bryant, R. L. (1998). Power, knowledge and political ecology in the Third World: A review. *Progress in Physical Geography*, *22*(1), 79–94.

Bryant, R. L. (1999). A political ecology for developing countries? Progress and paradox in the evolution of a research field. *Zeitschrift Fu¨r Wirtschaftsgeographie 43(3–4), Special Issue, Politische O¨ Kologie: Neue Perspektiven in Der Geographischen Umweltforschung*, 148–157.

Bryant, R. L. (2000). Politicized Moral Geographies Debating Biodiversity Conservation and Ancestral Domain in the Philippines. *Political Geography*, *19*, 673–705.

Bryant, R. L., & Parnwell, M. (1996). *Environmental Change in South-East Asia: People, Politics and Sustainable Development*. London: Routledge.

Buncombe, A. (2005, October 25). City at wits' end over "Simian menace." *The Newzealand Herald*. Retrieved from www.nzherald.co.nz/world/news/article.cfm?c_id=2&objectid=10471905 (23.05.14).

Bunker, S. G. (1980). The Impact of deforestation on peasant communities in the Medio Amazonas of Brazil. *Studies in Third-World Societies*, *13*, 45–60.

Bunker, S. G. (1982). The cost of modernity inappropriate bureaucracy, inequality, and development program failure in the Brazilian Amazo. *Journal of Developing Areas*, *16*, 573–596.

Bunker, S. G. (1985). *Underdeveloping the Amazon: Exraction, unequal exchange and the failiure of the modern sate*. Urbana,IL: University of Illinois Press.

Calder, I., & Alyward, B. (2002). *Forests Floods: Perspectives on watershed management and interagted Flood management*. Rome and Newcastle: FAO, United Nations and University of New Castle.

Callon, M. (1986). Some elements of a sociology of translation: Domestication of the scallops and the fishermen of St. Brieuc Bay. In J. Law (Ed.), *Power, Action and Belief. A New Sociology of Knowledge?* (pp. 196–223.). London: Routledge & Kegan Paul.

Castells, M. (1977). *The Urban question: A Marxist approach*. Cambridge Mas.: MIT.

Castells, M. (1978). *City,Class and Power*. London: MacMillan.

Castree, N. (2002). False Antitheses? Marxism, Nature and Actor-Networks Greening the Geographical Left. *Antipode, Volume 34*(1), 111–146.

Castree, N. (2003). Environmental Issues: Relational Ontologies and Hybrid Politics. *Progress in Human Geography*, *27*(2), 203–211.

Castree, N., Demeritt, D., Liverman, D., & Rhoads, B. (Eds.). (2009). *A companion to environmental geography*. Sussex: Blackwell.

CEC. (2004). *Report regarding preservation of environment and biodiversity in the land extending from south–west of Mehrauli to Masudpur and north of Vasant Vihar (Application 331).* New Delh: Central Empowereed Committee, Supreme Court of India.

Census of India. (2011). Minisry of Home Affairs. New Delhi: Government of India.

Cernea, M. M. (2006). Population Displacement in Protected Areas: A redefination of concepts in conservation politics. *Policy Mattters*, *14*, 8–26.

Chakarbarty, D. (2007). *Provincialising Europe:Post colonial thought and historical difference*. Princeton: Princeton University Press.

Champion, H. G., & Seth, S. K. (1968). *A Revised Survey of the Forest Types of India.* New Delhi: Government of India.

Chandrasekaran, S., Saraswathy, K., Saravanan, S., Kamaladhasan, N., & Nagendran, N. A. (2104). Impact of Prosopis juliflora on Nesting Success of Breeding Wetland Birds at Vettangudi Bird Sanctuary, South India. *Current Science*, *106*(5), 676–678.

Chapman, G. (2000). “Other Cultures”, “Other Environments” and the Mass Media. In Earthscan (Ed.), *Thr Daily Globe: Environmental Change, Public and Media* (pp. 127–150). London.

Chatterjee, P. (1998). Community in the East. *Economic and Political Weekly*, *XXXIII*(6), 277–282.

Chatterjee, P. (2001). Civil and Political Society in Postcolonial Democracies. In S. Kaviraj & S. Khilnan (Eds.), *Civil Society: History and Possibilities* (pp. 165–178). Cambridge: Cambridge University press.

Chatterjee, P. (2004). *The Politics of the Governed.* New York: Columbia University Press.

Chaturvedi, B. (2010). Introduction. In *Finding Delhi: Loss and renewal in the megacity*. New Delhi: Penguin India.

Chauhan, A. (2011). *Status of Sanjay Van, A preliminary report (November 2010-May 2011)*. New Delhi: Srishti

Chauhan, C. (2012, February 10). Rare bird spotted after decades. *Hindustan Times*. New Delhi. Retrieved from http://www.hindustantimes.com/newdelhi/rare-bird-spotted-after-decades/article1-809276.aspx (09.06.14).

Chopra, J. (2013, December 1). ’Green Soldiers Mark 31 years of Eco-Service. *The Pioneer*. Dehradun. Retrieved from www.dailypioneer.com/print.php?printFOR=storydetail&story_url_key=green-soldiers-mark-31-years-of-eco-service§ion_url_key=state-editions (13.05.14).

Coelho, K., & Raman, N. (2013). From the Frying Pan to the Flood Plain: Negotiaing land water and fire in Chennai’s development. In A. Rademacher & K. Sivaramakrishnan (Eds.), *Ecologies of Urbanism in India:metropolitan civility and sustainability* (pp. 145–168). Hong Kong: Hong Kong University Press.

Cosgrove, D. E. (1984). *Social formation and symbolic landscape* (2nd ed.). Madison: University of Wisconsin Press.

Cottrell, J. (1992). Courts and Accountability:Public Interest Litigation in the Indian High Courts. *Third World Legal Studies*, *11*(8), 199–211.

Creswell, J. W. (2009). *Research Design: Qualitative, Quantitative, and Mixed Methods Approaches (3 ed.)*. Thousand Oaks, USA: Sage Publications.

Cronon, W. (1992). A Place for Stories : Nature , History , and Narrative. *The Journal of American History*, 78(4), 1347–1376.

Cronon, W. (1995). Introduction: In Search of Nature. In W. Cronon (Ed.), *Uncommon Ground: Rethinking the Human Place in Nature* (pp. 23–68). New York: WW Norton & Company.

D'Souza, R., & Nagendra, H. (2011). Changes in Public Commons as a Consequesce of Urbanisation: The Agara Lake in Bangalore, India. *Environmental Management*, *47*(5), 840–845.

Dainik Jagran. (2006, July 9). Vasant Kunj mall site par hungama (Ruckus on Vasant Kunj mall site). New Delhi.

Dalrymple, W. (1997). *The Last Mughal, The Fall of a Dynasty Delhi, 1857*. New York: Knopf.

Dasgupta, R., Baru, R. V, Deshpande, M., & Mohanty, A. (2010). Location and Deprivation Towards an understanding of the relationship between area effects & school health. *USRN Working Paper Series 2*, *1*(2). Retrieved from www.srtt.org/institutional_grants/pdf/Location _Deprivation.pdf (12.04.14).

Dash, D. (2006, June 15). When Efforts Bear Fruit. *Times of India*. New Delhi.

Dash, D. (2012, January 14). Mega Tourism Nod Pushes Sacred Woods To The Brink. *Times of India*. New Delhi. Retrieved from http://timesofindia.indiatimes.com/city/delhi/Mega-tourism-nod-pushes-sacred-woods-to-the-brink/articleshow/11480346.cms (02.01.15)

Dastisdar, A. (2007, January 1). Now Nobel Laurate Opposes VK Malls. *Hindustan Times*. New Delhi.

Davies, B., & Harre, R. (1990). Positioning: The Discursive Production of Selves. *Journal for the Theory of Social Behaviour*, *20*(1), 43–63.

Davis, R. (2003). Development Eco-Logies: Power and Change in Arturo Escobar's Political Ecology. *Studies in Political Economy*, *70*, 153–172.

DDA. (1962). *Delhi Master Plan 1962*. New Delhi: Delhi Development Authority.

DDA. (1990). *Master plan for Delhi 2001 What will be Delhi in 2001: Delhi Master plan August, 1990 (Revised and updated) As published in the Gazette of India, Part II of Section 3, Sub Section (ii), Extraordinary, dated 1.8.1990, S. No 437, Notification no. 606 (E) d*. Delhi Development Authority, New Delhi: Akaank Publications, Delhi Editions.

DDA. (2007). *Master Plan for Delhi: With the perspective for the year 2021 The Gazette of India, Extraordinary, Part II–Section 3 Subsection (ii) No. 125 (Magha 18, 1928) vide S.O. 141-(E). Incorporates Gazette notifications of amendments/modifications up to October 2*. Delhi Development Authority, New Delhi. Retrieved from https://dda.org.in/ddanew/pdf/Planning/ reprint mpd2021.pdf (05.03.13).

DDA. (2013). *Sanjay Van Arravali City Forest*. Retrieved from www.dda.org.in/tendernotices _docs/jan13/Website_SANJAYVAN (2).doc. (03.12.12).

Dearborn, D. C., & Kark, S. (2010). Motivations for conserving urban biodiversity. *Conservation Biology : The Journal of the Society for Conservation Biology*, *24*(2), 432–40.

Delhi Development Act. (1957). Number 61 of 1952, Government of India, New Delhi. Retrieved from http://dda.org.in/tendernotices_docs/march09/DDA_conduct_disciplanary_and _appel_ Regulation1999.pdf (13.04.13).

Dembowski, H. (2001). *Taking the State to Court in Metropolitan India: Public Interest Litigation and the Public Sphere in India*. Asia House. Retrrieved from, www.asienhaus.de/public /archiv/taking_the_state_to_court.pdf (14.11.14)

Demeritt, D. (1994). The nature of metaphors in cultural geography and environmental history. *Progress in Human Geography*, *18*(2), 163–185.

Demeritt, D. (2002). What is the "social construction of nature"? A typology and sympathetic critique. *Progress in Human Geography*, *26*(6), 767–790.

Dennis, J. M., & Wolman, A. (1959). 1955-56 Infectious Hepatitis Epidemic in Delhi, India [with Discussion]. *Journal (American Water Works Association)*, *51*(10), 1288–1298.

Denzin, N., & Lincoln, Y. (Eds.). (1994). Entering the field of qualitative research. In *Handbook of Qualitative Research* (pp. 1–17). London: Sage.

DeWalt, K., & DeWalt, B. (Eds.). (2002). *Participant Observation: A Guide for Fieldworkers*. Walnutcreek C.A.: AltaMira Press.

Dhavan. (2000). Judges and Indian Democracy: The Lesser Evil? In *Transforming India* (p. 322). Delhi: Oxford University Press.

Dhawale, M. (2010). *Narratives of the environment of Delhi*. New Delhi: Indian National Trust for Art and Cultural Heritage (INTACH).

Dias, A. (1994). Judicial Activism in the Development and enforcement of Environmental Law: Some Comparative Insights from the Indian Experience. *Journal of Environmental Law*, *6*(2), 243–262.

Diaz-Bone, A. R. (2006). Die interpretative Analytik als methodologische Position. In B. Kerchner & S. Schneider (Eds.), *Foucault: Diskursanalyse der Politik—Eine Einführung*. (pp. 68–84). Wiesbaden: VS-Verlag für Sozialwissenschaften.

Diaz-Bone, A. R., Bührmann, A. D., Rodríguez, E. G., Schneider, W., Kendall, G., & Tirado, F. (2008). The Field of Foucaultian Discourse Analysis : Structures, Developments and Perspectives Analysis in the Social Sciences / Diskursanalyse in den Sozialwissenschaften G, *33*(1), 7–28.

Diwedi, S. (2006, January 8). DDA ne Ridge ki zamin bech dali (DDA has gone and sold Ridge land). *Rashtriya Sahara*. New Delhi, Lucknow, Gorakpur.

Dogra, C. S. (2013, February 8). Aravallis being gobbled up by land developers. *The Hindu*. New Delhi. Retrieved from http://www.thehindu.com/news/national/aravallis-being-gobbled-up-by -land-developers/article4390369.ece (03.01.15).

Dooling, S., Simon, G., & Yocom, K. (2006). Place-based urban ecology: A century of park planning in Seattle. *Urban Ecosystems*, 9(4), 299–321.

Down To Earth. (1997, May 15). Delhi Ridge: A lifeline in danger. *Down to Earth*. Retrieved from http://www.downtoearth.org.in/node/23699 (12.04.14).

Dupont, V. (2005). Peri-urban dynamics: population, habitat and environment on the peripheries of large Indian metropolises: An introduction. In V. Dupont (Ed.), *Peri-urban Dynamics: Population, Habitat and Environment on the Peripheries of Large Indian Metropolises a Review of Concepts and General issues. CSH Occassional paper No. 14* (pp. 3–20). New Delhi: CSH.

Dupont, V. (2011). The Dream of Delhi as a Global City. *International Journal of Urban and Regional Research*, *35*(3), 533–554.

Eden, S. (2000). Environmental issues: Sustainable progress. *Progress in Human Geography*, *24*(1), 111–8.

Ekta. (2014). Dominance of Prosopis Juliflora (SW.) D.C In The Delhi Ridge Forest of the Semi-Arid Region, India. *The Biobrio: An International Quarterly Journal of Life Sciences*, *1*(1), 41– 44.

Engels, F. (1845). *The conditions of the working class in England*. Oxford: Oxford World's Classics (Reissue edition 28 May 2009).

EPCA. (1999). *International Hotels Complex Vasant Vihar, Report*. Environment Pollution (Prevention and Control) Authority, New Delhi.

EPCA. (2000). *Report on constraint area and M/S Unision Hotels Ltd., Vasant Kunj*. Environment Pollution (Prevention and Control) Authority, New Delhi.

EPW (2007). Green Courts: Turf Wars. *Economic and Political Weekly*, *XLII*(14), 1232.

Escobar, A. (1995). *Encountering Development: The Making and Unmaking of the Third World*. Princeton University Press.

Escobar, A. (1996). Constructing Nature: Elements for a Post Structural Political Ecology. In R. Peet & M. Watts (Eds.), *Liberation Ecologies:Environment, Development,Social Movements* (pp. 46–68). London: Routledge.

Escobar, A. (1998). Whose Knowledge , Whose nature ? Biodiversity , Conservation , and the Political Ecology of Social Movements. *Journal of Political Ecology*, *5*, 53–82.

Escobar, A. (1999). After Nature: Steps to an Antiessentialist Political Ecology. *Current Anthropology*, *40*(1), 1–30.

Evans, P. (1989). Predatory, Development and Other Apparatuses: A comparative political economy perspective on the Third World state. *Sociological Forum*, *4*(4), 561–87.

Fairclough, N. (1992). *Discourse and Social Change*. Malden, MA: Blackwell.

Fairclough, N. (2012). Critical Discourse Analysis. *International Advances in Engineering and Technology*, *7*(July), 452–487.

Fairhead, J., & Leach, M. (1996). *Misreading the African landscape: Society and ecology in a forest-savanna mosaic*. Cambridge: Cambridge university press.

Feindt, P. H., & Oels, A. (2005). Does Discourse Matter? Discourse Analysis in Environmental Policy Making. *Journal of Environmental Policy & Planning*, *7*(3), 161–173.

Ferguson, J. (1994). *The Anti-Politics Machine: "Development", Depoliticization and Bureaucratic Power in Lesotho*. Minneapolis: University of Minnesota Press.

Fernandes, L. (2004). The Politics of Forgetting: Class politics, state power and the restructuring of urban space in India. *Urban Studies*, *41*(12), 2415–2430.

Fernandes, L., & Heller, P. (2006). Hegemonic Aspirations. *Critical Asian Studies*, *38*(4), 495–522.

Fischer, F. (2003). Beyond Empiricism: Policy Analysis as Deliberative Practice. In M. Hajer & H. Wagenaar. (Eds.), *Deliberative Policy Analysis: Understanding Governance in the Network Society.* (pp. 209–227). Cambridge Mas.: Cambridge University press.

Fischer, F., & Forester, J. (Eds.). (1993). *The Argumentative Turn in Policy Analysis and Planning*. Duke University Press.

Follmann, A. (2014). Urban mega-projects for a "world-class" riverfront: The interplay of informality, flexibility and exceptionality along the Yamuna in Delhi, India. *Habitat International*, *45*(3), 213–222.

Forsyth, T. (2003). *Critical Political Ecology: The Politics of Environmental Science*. London: Routledge.

Forsyth, T. (2008a). Critical Realism and Political Ecology. In A. Stainer & G. Lopez (Eds.), *After Postmodernism: Critical Realism?* (pp. 146–154). London: Athlone Press.

Forsyth, T. (2008b). Political Ecology and the Epistemology of Social Justice. *Geoforum*, *39*(2), 756–764.

Forsyth, T. (2011). Politicising Environmental Explaination: What Political Ecology can learn form sociology and philosophy of science. In M. J. Goldman, P. Nadasdy, & M. D. Turner (Eds.), *Knowing Nature: Conversations at the intersection of Political Ecology and Science Studies* (pp. 1–31). Chicago: University of Chicago Press.

Fortmann, L. (1995). Talking Claims: Discursive Strategies in Contesting Property. *World Development*, *23*, 1053–1063.

Foucault, M. (1968). Politics and the study of discourse. In *The Foucault Effect: Studies in governmentality, 1991* (pp. 53–72). London: Harvester Wheatsheaf.

Foucault, M. (1969). *The archaeology of knowledge*. New York: Pantheon Books.

Foucault, M. (1973). *The order of things*. New York: Vintage.

Foucault, M. (1976). *History of sexuality*. New York: Vintage Randomhouse.

Foucault, M. (1979). *Discipline and punish: The birth of the prison* (Vol. 1979, p. 333). Sydney: Feral Publications.

Foucault, M. (1980). *Power/Knowledge: Selected Interviews and other Writings 1972–1977*. London: Harvester Press.

Foucault, M. (1981). The order of discourse. In R. Young (Ed.), *Untying the text: a post-structural anthology* (pp. 48– 78). Boston: Routledge & Kegan Paul.

Friedman, S. (2008). *Human Rights Transformed: Positive rights and positive duties*. Oxford: Oxford University Press.

Frykenberg, R. E. (1994). The Study of Delhi: An Historical Introduction. In R. E. Frykenberg (Ed.), *Delhi Through the Ages: Selected Essays in Urban History, Culture and Society* (pp. 1–18). New Delhi: Oxford.

FSI. (2013). *India State of the Forests Report 2013.* Forest Survey of India, Ministry of Environment and Forests. Dehradun: Government of India

Fuller, C. J., & Bénéï, V. (2000). *The everyday state and society in modern India*. New Delhi: Social Science Press : Distributed by D.K. Publishers and Distributors.

Funtowicz, S., & Ravetz, J. R. (1993). Science for the Post-Normal Age. *Futures*, *25*, 735–755.

Furnivall, J. S. (1956). *Colonial policy and practice: A comparative study of Burma and Netherlands India.* Cambridge: Cambridge university press.

Fuss, D. (1989). *Essentially Speaking: Feminism, Nature & Difference* (p. 144). New York: Routledge.

Gadbois, G. H. (1985). The Supreme Court of India as a Political Institution. In R. Dhavan (Ed.), *Judges and Judicial Power* (pp. 250, 257). Bombay, London: Sweet, Maxwell and Tripathi.

Gadgil, M. (1991). Deforestation: Problems and Prospects. In R. A.S. (Ed.), *History of Forestry in India* (pp. 13–85). New Delhi: Indus Publishing.

Gadgil, M. (2001). *Ecological Journeys; The Science and Politics of Conservation in India*. New Delhi: Permanent Black.

Gadgil, M., & Guha, R. (1992). *This Fissured Land: An Ecological History of India*. Berkley: University of California Press.

Gadgil, M., & Guha, R. (1994). Ecological Conflicts and the Environmental Movement in India. *Development and Change*, *25*(1), 101–136.

Gadgil, M., & Guha, R. (1995). *Ecology and Equity: The Use and Abuse of Nature in Contemporary India.* London: Routledge.

Gandy, M. (1996). Crumbling land: The Postmodernity Debate and the Analysis of Environmental Problems. *Progress in Human Geography*, *20*(1), 23–40.

Gandy, M. (2002). *Concrete and Clay: Reworking Nature in New york*. Cambridge Mas.: MIT.

Gandy, M. (2006). Urban nature and the ecological imagination. In N. Heynen, M. Kaika, & E. Swyngedouw (Eds.), *In the Nature of Cities: Urban Political Ecology and the Politics of Urban Metabolism Questioning cities series* (pp. 62–72). New York: Routledge.

Gandy, M. (2012). Queer ecology: Nature, sexuality and heterotopic alliances. *Environment and Planning D: Society and Space*, *30*(4), 727–47.

Ganguli, U. (1975). *A Guide to the Birds of Delhi Area*. New Delhi: Indian Council of Agricultural Research.

Gaskell, G. (2000). Individual and Group Interviewing. In *Qualitative Researhing with Text, Image and Sound.* (pp. 38–56). New Delhi: Sage.

Gazetteers Organisation. (1999). *Reprint of Delhi District Gazetteer, Gazetteers Organisation Revenue Department, Haryana 1883-84*. Chandigarh.

Geertz, C. (1973). *The Intepretation of Cultures*. New York: Basic Books.

Geological Survey of India. (2006). *Geological and Geomorphological Mapping of Part of NCT Delhi for Seismic Microzonation*. New Delhi.

Ghai, Y., & Cottrell, J. (2010). Rule of Law and Access to Justice. In Y. Ghai & J. Cottrell (Eds.), *Marginalised Communities and Access to Justice* (pp. 1–22). New York: Routledge.

Ghertner, A. (2010). Calculating Without Numbers: Aesthetic Governmentality in Delhi's Slums. *Economy and Society*, *39*(2), 185–217.

Ghertner, A. (2011a). Green evictions: Environmental discourses of a "slum-free" Delhi. In R. Peet, P. Robbins, & M. Watts (Eds.), *Global Political Ecology* (pp. 145–165). London: Routledge.

Ghertner, A. (2011b). Rule by aesthetics: World-class city making in Delhi. In A. Roy & A. Ong (Eds.), *Worlding Cities: Asian Experiments and the Art of Being Global* (pp. 279–306). Oxford: Blackwell.

Ghosal, A. (2014, April 28). Govt presses on with demolition at Asola Bhatti, house owners cry foul. *Indian Expres*. New Delhi. Retrieved from http://indianexpress.com/article /cities/delhi/govt-presses-on-with-demolition-at-asola-bhatti-house-owners-cry-foul/2/ (21.06.14).

Giddens, A. (1982). *Profiles and Critiques in Social Theory*. London: MacMillan.

Giddens, A. (1984). *The Constitution of Society: Outline of the Theory of Structuration*. Cambridge UK: Polity Press.

Giddens, A. (1987). *Social Theory and Modern Sociology*. Stanford: Stanford University Press.

Gill, R. (2000). Discourse Analysis. In M. Bauer & G. Gaskell (Eds.), *Qualitative research with text image and sound: A practicle Handbook* (pp. 172–190). London: Sage.

Glasbergen, P. (1998). The Question of Environmental Governance. In G. P. (Ed.), *Co-operative environmental governance: Public private agreements as a policy strategy*. Dordrecht: Kluwer Academic Publishers.

GNCTD. (2014). *Economic survey of Delhi 2012-2013*. Government of National Capital Territory of Delhi. New Delhi.

Gold, A. G. (1999). From Wild Pigs to Foreign Trees: Oral Histories of Environmental Change in Rajasthan. In S. T. Madsen (Ed.), *State, Society and the Environment in South Asia* (pp. 20–58). Richmond: Curzon Press.

Greening Delhi Action Plan 2007-8. (2007). Department of Forests and Wildlife, Government of National Capital Territiry of Delhi. New Delhi.

Gregory, D., & Walford, R. (Eds.). (1989). *New horizons in human geography*. London: MacMillan.

Guba, E., & Lincoln, Y. (1994). Competing paradigms in qualitative research. In *Handbook of qualitative research* (pp. 105–117). Thousand Oaks, CA: Sage.

Gubbi, S. (2010). Making Governance Effective. *Seminar*, *613*, 61–65.

Guha, R. (1989a). Radical American Environmentalism and Wilderness Preservation: A Third World Critique. *Environmental Ethics*, *11*(1), 71–83.

Guha, R. (1989b). *The unquiet woods: Ecological change and peasan resistence in the Himalaya*. New Delhi: Oxford University Press.

Guha, R. (1997). The Authoritarian Biologist and the Arrogance of Anti-Humanism: Wildlife conservation in the Third World. *The Ecologist*, *21*, 14–20.

Guha, R., & Martinez-Alier, J. (1997). *Varieties of Environmentalism: Essays North and South*. London: Earthscan.

Guha, S. (1999). *Environment and Ethnicity in india 1200–1991*. Cambridge: Cambridge University Press.

Gupta, D. (2000). *Mistaken Modernity: India Between Worlds*. New Delhi: Harper Collins.

Gupta, N. (1986). Delhi and Its Hinterland: The ninteenth and twenteeth centuries. In R. E. Frykenberg (Ed.), *Delhi through the ages: Selected essays in Urban history, culture and society*. Oxford, UK: Oxford University Press.

Gupta, N. (2010). Delhi's History as reflected in its toponomy. In M. Dayal (Ed.), *Celebrating Delhi* (pp. 95–109). UK: Penguin.

Gupta, P. D., & Puri, S. (2005). *Private Provision of Public Services in Unauthorised Colonies A Case Study of Sangam Vihar. Working Paper, Centre for Civil Society*. New Delhi. Retrieved from http://ccs.in/internship_papers/2005/12. Private Provision of Public Services.pdf (15.07.14)

Hajer, M. (1993). Discourse Coalitions and the Institutionalisation of Practice, The Case of Acid Rain in Britain. In F. Fischer & J. Forester (Eds.), *The Argumentative Turn in Policy and Planning* (pp. 43–76). Durham: Duke University Press.

Hajer, M. (1995). *The Politics of Environmental Discourse: Ecological modernisation and the policy process*. Oxford University Press.

Hajer, M. (2006). Doing Discourse Analysis: Coalitions, practices, meaning. In M. van den Brink & T. Metze (Eds.), *Words matter in policy and planning. Discourse theory and method in the social sciences* (pp. 65–74). Utrecht: Netherlands Graduate School of Urban and Regional Research.

Hajer, M., & Versteeg, W. (2005). A decade of discourse analysis of environmental politics: Achievements, challenges, perspectives. *Journal of Environmental Policy & Planning*, *7*(3), 175–184.

Haraway, D. (1991). *Simians, Cyborgs and Women: The Reinvention of Nature.* London: Free Association Press.

Harré, R. (1993). *Social Being*. Oxford: Blackwell.

Harriss, J. (2005). Political Participation, Representation and the Urban Poor: Findings from Research in Delhi. *Economic and Political Weekly*, *40*(11), 1041–1054.

Harriss, J. (2006). Middle Class Activism and the Politics of the Informal Working Class: A Perspective on Class Relations and Civil Society in Indian Cities. *Critical Asian Studies*, *38*(4), 445–465.

Harriss, J. (2007). Antinomies of Empowerment: Observations on Civil Society, Politics and Urban Governance in India. *Economic and Political Weekly*, *42*(26), 2716–2724.

Harvey, D. (1993). The Nature of Environment: The dialectics of social and environmental change. In R. Miliband & L. Panitich (Eds.), *Real problems, False solutions* (pp. 1–51). London: Merlin Press.

Harvey, D. (2000). *Spaces of Hope*. Calafornia: University of California Press.

Hazards Centre. (2003). *A People's housing policy: A Case Study of Delhi*. New Delhi. Retrieved from http://www.hazardscentre.com/hazards_publications/pdf/urban_governance_resistance/people_housing_policy.pdf (19.05.14).

Hazards Centre. (2007). *A fact finding report on the ecivtion and resettlement process in Delhi*. New Delhi. Retrieved from http://www.downtoearth.org.in/node/23699 (15.05.14).

Hazra, A. K. (2002). *History of Conflict over Forests in India: A Market Based Resolution*. Working Paper Series, Julian L. Simon Centre for Policy Research.

Hearn, G. R. (1906). *The Seven Cities of Delhi*. Delhi: W. Thacker and Company.

Herndl, C., & Brown, C. (1996). *Green Culture: Environmental Rhetoric in Contemporary America.* Madison: University of Wisconsin Press.

Herod, A. (2005). Gender Issues in the Use of Interviewing as a Research Method. *The Professional Geographer*, *45*(3), 305–317.

Herzele, A. Van. (2005). *A tree on your doorstep, a forest in your mind: Greenspace planning and the interplay between discourse, physical conditions and practice*. Wageningen Universiteit and Research Centre.

Heynen, N. (2006). Green Urban Political Ecologies: Toward a better understanding of inner-city environmental change. *Environment and Planning A*, *38*(3), 499–516.

Heynen, N. (2013). Urban political ecology I: The urban century. *Progress in Human Geography*, *38*(4), 598–604.

Heynen, N., Kaika, M., & Swyngedouw, E. (2006). Urban Political Ecology: Politicising the production of urban natures. In N. Heynen, M. Kaika, & E. Swyngedouw (Eds.), *In the Nature of Cities Urban Political Ecology and the Politics of Urban Metabolism* (pp. 1–19). New York: Routledge.

Heynen, N., Perkins, H. A., & Roy, P. (2006). The Political Ecology of Uneven Urban Green Space: The Impact of Political Economy on Race and Ethnicity in Producing Environmental Inequality in Milwaukee. *Urban Affairs Review*, *42*(1), 3–25.

HIC-HLRN. (2006). *Threat of Demolition of 4,500 homes, and the Forced Eviction of more than 25,000 people in Bhagirath Nagar Village, Bhatti Mines, India, Habitat International*

Coalition Housing and land rights network Case IND-FE 300806. Retirieved from, www.hlrn.org/img/cases/UA-IND-FE%20300806.doc (16.05.14)
Hocking, D. (Ed.). (1993). *Trees for Drylands*. New Delhi: Oxford and IBH Publishing.
Holifield, R. (2009). Actor-Network Theory as a Critical Approach to Environmental Justice: A Case against Synthesis with Urban Political Ecology. *Antipode*, *41*(4), 637–658.
HT. (1991, January 11). Environmental plan for Bhatti Mines. *Hindustan Times*. New Delhi.
HT. (1993, May 22). Move to expand Asola sanctuary. *Hindustan Times*. New Delhi.
HT. (2006a, June 18). Protesters seek PM help. *Hindustan Times*. New Delhi.
HT. (2006b, June 20). Developers, Activists Clash at DCs office. *Hindustan Times*. New Delhi.
HT. (2006c, August 8). Vasant Kunj Malls: Ministry Suggests Penalty. *Hindustan Times*. New Delhi.
HT. (2013). RWA to save South Delhi ridge. *Hindustan Times*. New Delhi.Retrieved from www.hindustantimes.com/newdelhi/rwa-to-save-south-delhi-ridge/article1-1009692.aspx (19.06.14).
HT Estates. (2006, February 11). Ridge Rejoinder. Hindustan Times. New Delhi.
Hyde, W. F., Newman, D. H., & Sedjo, A. (1991). *Forests, economics and policy analysis: an overview*. Washington D.C.: The World bank.
IFDC. (2014). Dr. Norman Borlaug Father of the green revolution. Retrieved from www.ifdc.org/Infographics/Dr-Norman-Borlaug-Father-of-the-Green-Revolution/ (13.03.14).
Indian Express. (1999, January 28). Look for animals in this sanctuary and you will find trucks. New Delhi.
Indian Express. (2005, February 3). No Mall at Ridge, say residents. *Indian Express*. New Delhi.
Indian Express. (2010, January 20). Unauthorised Colonies: Civic Bodies to Make Layout Plans. *Indian Express*. Retrieved from http://archive.indianexpress.com/news/unauthorised-colonies-civic-bodies-to-make-layout-plans/569400/ (22.02.14).
Ingelhart, R. (1995). Public Suport for Environmental Protection: Objective problems and subjective values in 45 societies. *Political Science and Politics*, *28*(1), 57–72.
INTACH. (1998). *Blueprint for Water Augmentation in Delhi*. New Delhi: INTACH
Jain, A. (2014, November 12). NGT Notice to Centre, Delhi Government. *The Hindu*. New Delhi. Retrieved from www.thehindu.com/news/cities/Delhi/ngt-notice-to-centre-delhi-govt/article6589476.ece (14.11.2014).
Jain, A. K. (2005). *Law and Environment*. Delhi: Ascent.
Jain, R. (1994, December 15). Fenced out. *Down To Earth*. New Delhi.
Jalais, A. (2010). *Forest of Tigers People Politics and Environment in the Sunderbans*. New Delhi: Routledge.
Jamatia, H. (2013, January 30). Catchers face extinction as simians run riot in city. *Hindustan Times*. New Delhi. Retrieved from http://www.hindustantimes.com/newdelhi/catchers-face-extinction-as-simians-run-riot-in-city/article1-1004226.aspx (02.02.13).
Jargowsky, P. A. (2005). Comparative Metropolitan Development 21. In V. Dupont (Ed.), *Peri-urban dynamics: Population, Habitat and environment on the peripheries of large Indian metropolisie: A review of concepts and general issues. CSH Occassional paper No. 14* (pp. 21–38). New Delhi.
Jasanoff, S. (1996). Beyond Epistemology: Relativism and Engagement in the Politics of Science. *Social Studies of Science*, *26*(2), 393–418.
Jasanoff, S. (2010). A New Climate for Society. *Theory, Culture & Society*, *27*(2–3), 233–253.
Jasanoff, S., & Martello, M. (Eds.). (2004). *Earthly Politics: Local and Global in Environmental Governance*. Cambridge Mas.: MIT Press.
Jasanoff, S., & Wynne, B. (1998). Science and Decision Making. In S. Raynor & E. L. Malone (Eds.), *Human Choice and Climate Change, Volume one: The societal framework* (pp. 1–87). Columbus, Ohio: Batelle Institute.
Jha, S. K. (2001, May 10). Mining Mafia Takes Govt. for Ride. *The Pioneer*. New Delhi.

Jumani, U. (2006). *Empowering Society: An Analysis of Business, Government and Social Development Approaches to Empowerment* (p. 272). Foundation Books.

Jyoti, A. (2006, March 17). DDA Told to Shut Down VK Malls. *Asian Age*.

Kaika, M. (2006). The political ecology of water scarcity: The 1989–1991 Athenian drought. In N. Heynen, M. Kaika, & E. Swyngedouw (Eds.), *In the nature of cities: Urban political ecology and the politics of urban metabolism* (pp. 157–172). New York: Routledge.

Kalpavriksh. (1991). Delhi Ridge: Decline and Consevation. New Delhi: Kalpavriksh

Kalpavriksh. (2002). Landfill proposal for Asola WLS. *Protected Area Update:News and Information from Protected Areas in India and South Asia, 34–35*, 7.

Kasarda, J. D., & Crenshaw, E. M. (1991). Third World urbanization: Dimensions, theories, and determinants. *Annual Review of Sociology*, *17* (1991), 467–501.

Kaul, M. C. (1990). Common Property Resources in Delhi: With Special Reference to the Bisagama Cluster. In *Paper submitted for "Designing Sustainability on the Commons" as a part of the First Annual Meeting of the International Association for the Study of Common Property, September 27–30, 1990, Durham, North Carolina*. Retrieved from: http://dlc.dlib.indiana.edu/dlc/bitstream/handle/10535/2309/Common_Property_Resources_in_Delhi__With_Special_reference_to_Bisagama_Cluster.pdf?sequence=1 (23.04.14)

Keil, & Boudreau. (2006). Metropolitics and metabolics: Rolling out environmentalismin Toronto. In N. Heynen, M. Kaika, & E. Swyngedouw (Eds.), *In the nature of cities* (pp. 40–61). New York: Routledge.

Keil, R. (2003). Urban Political Ecology 1. *Urban Geography*, *24*(8), 723–738.

Keil, R. (2005). Progress Report—Urban Political Ecology. *Urban Geography*, *26*(7), 640–651.

King, N., & Horrocks, C. (2010). *Interviews in Qualitative Research*. London: Sage.

Kothari, A., & Rao, S. (1997). How are we Managing? Saving Delhi's Natural Ecosystems: A Model of Citizen Participation. *Ecosystem Health*, *3*(2), 123–126.

Kothari, A., Suri, S., & Singh, N. (1995). Conservation in India: A New Direction. *Economic and Political Weekly*, *30* (43), 2755–2766.

Krishen, P. (2006). *Trees of Delhi: A Field Guide*. New Delhi: Dorling Kindersley (India) Pvt Ltd.

Krishen, P. (2010). Avenue Trees in Lutyen's Delhi. In M. Dayal (Ed.), *Celebrating Delhi* (pp. 76–93). UK: Penguin.

Krishna, G. (2014, June 7). It's a Jungle Out There: A former Air Force officer leads a citizens' effort to revive a forest in south Delhi. *The Business Standard*. New Delhi. Retrieved from http://www.business-standard.com/article/specials/it-s-a-jungle-out-there-114060601205_1.html (10.06.14).

Kusenbach, M. (2003). Street Phenomenology: The Go-Along as Ethnographic Research Tool. *Ethnography*, *4*(3), 455–485.

Kvale, S. (1983). The qualitative research interview – A phenomenological and a hermeneutical mode of understanding. *Journal of Phenomenological Psychology*, *14*, 171–196.

Kvale, S. (1996). *Interviews: An introduction to qualitative research interviewing*. Thousand Oaks, CA: Sage.

Lachmund, J. (2004). Knowing the Urban Wasteland: Ecological expertise as local process. In S. Jasanoff & M. Martello (Eds.), *Earthly Politics: Local and global in environmental governance* (pp. 241–262). Cambridge Mas.: MIT Press.

Lakha, S. (2000). The State, Globalisation and Indian Middle Class Identity. In B. H. Chua (Ed.), *Consumption in Asia: Lifestyles and Identities* (pp. 251–274). London and New York: Routledge.

Lalchandani, N. (2010, January 25). A Forest in Decay. *The Times of India*. New Delhi. Retrieved from http://timesofindia.indiatimes.com/city/delhi/A-forest-in-decay/articleshow/5496364.cms (03.06.14).

Lalchandani, N. (2011a, August 30). Nurture, Nature Pull Bhatti Mines out of the Pits. *Times of India*. New Delhi. Retrieved from http://timesofindia.indiatimes.com/city/delhi/Nurturenature-pull-Bhatti-Mines-out-of-the-pits/articleshow/9792225.cms (03.06.14).

Lalchandani, N. (2011b, October 19). Sanjay Van could turn bird sanctuary. *Times of India*. New Delhi. Retrieved from http://timesofindia.indiatimes.com/city/delhi/Sanjay-Van-could-turn-bird-sanctuary/articleshow/10409952.cms (03.06.14).

Lalchandani, N. (2013a, May 22). Forest Department Hamstrung by Severe Staff Shortage. *Times of India*. New Delhi. Retrieved from http://timesofindia.indiatimes.com/city/delhi/Forest-department-hamstrung-by-severe-staff-shortage/articleshow/20184168.cms(03.06.14).

Lalchandani, N. (2013b, July 23). Govt Blamed for Ridge Violations. *Times of India*. New Delhi. Retrieved from http://timesofindia.indiatimes.com/city/delhi/Govt-blamed-for-ridge-violations/articleshow/21261305.cms (03.06.14).

Latour, B. (1986). The power of associations application. In J. Law (Ed.), *Power, Action and Belief. A New Sociology of Knowledge?* (pp. 261–277). London: Routledge.

Latour, B. (1993). *We have never been Modern*. Cambridge Mas.: Harvard University Press.

Latour, B. (2004). *Politics of Nature: How to bring Sciences into Democracy*. Cambridge Mas.: Harvard University Press.

Lawhon, M., Ernstson, H., & Silver, J. (2014). Provincializing Urban Political Ecology: Towards a Situated UPE Through African Urbanism. *Antipode*, *46*(2), 497–516.

Leach, M., & Mearns, R. (Eds.). (1996). *The Lie of the Land: Challenging Received Wisdom on the African Environment*. London: International African Institute.

Lees, L. (2004). Urban geography: discourse analysis and urban research. *Progress in Human Geography*, *28*(1), 101–107.

Lefebvre. (1991). *The production of Space*. Oxford, UK: Blackwell.

Lefebvre. (1996). *Writings on cities*. Cambridge Mas.: Blackwell.

Legg, S. (2006). Postcolonial Developmentalities: From Delhi Improvement Trust to the Delhi Developmnt Authority. In S. Raju, M. S. Kumar, & S. Corbridge (Eds.), *Colonial and Post-Colonial Geographies of India* (pp. 182–204). London: Sage publications.

Lele, S. (2007). A'Defining'Moment for Forests? *Economic and Political Weekly*, *XLII*(25), 2379–2383.

Lele, S., & Norgaard, R. B. (2003). Sustainability and the Scientists Burden. In V. K. Saberwal & M. Rangarajan (Eds.), *Battles over nature: Science and the politics of conservation* (pp. 158–185). New Delhi: Orient Blackswan.

Lemanski, C., & Tawa Lama-Rewal, S. (2013). The "missing middle": Class and urban governance in Delhi's unauthorised colonies. *Transactions of the Institute of British Geographers*, *38*(1), 91–105.

Lemos, M. C., & Agrawal, A. (2006). Environmental Governance. *Annual Review of Environment and Resources*, *31*(1), 297–325.

Litfin, K. T. (1994). *Ozone Discourses: Science and Politics in Global Environmental Corporation*. New York: Columbia University Press.

Little, P., & Painter, M. (1995). Discourse, politics, and the development process: Reflections on Escobar's "Anthropology and the Development Encounter." *American Ethnologist*, *22*(3), 602–609.

Long, A. (1992). Goods, Knowledge and Beer. In *The Battlefields of Knowledge:The Interlocking of Theory and Practise in Social Research and Development* (pp. 147–170). London: Routledge.

Long, N. (1992). From paradigm lost to paradigm regained: The case for actor oriented socialogy of development. In *The Battlefields of Knowledge:The interlocking of theory and practise in social research and development* (pp. 16–46). London: Routledge.

Long, N., & Long, A. (Eds.). (1992). *The Battlefields of Knowledge:The interlocking of theory and practise in social research and development*. London: Routledge.

Lovraj Committee. (1993). *Report of the committee to Recomend the pattern of management of the Delhi Ridge. New Delhi.*

Low, N., Gleeson, B., Elander, I., & Lidskop, L. (2000). *Consuming cities: The urban environment in global economy*. London: Routledge.

Ludwig, D., Mangel, M., & Haddad, B. (2001). Ecology, Conservation and Public Policy. *Annual Review of Ecological Systems*, *32*, 481–517.

Macnaghten, P., & Urry, J. (1998). *Contested Natures*. London: Sage publications.

Madsen, S. T. (1999). Introduction. In *State, Society and the Environment in South Asia* (pp. 1–19). Richmond: Curzon Press.

Mahapatra, D. (2007, July 21). Centre Wants Green Bench Disbanded. *Times of India*. New Delhi. Retrieved from http://timesofindia.indiatimes.com/india/Centre-wants-Green-Bench-disbanded/articleshow/2222042.cms (15.12.14).

Mahapatra, R. (2014, February). The Curious Case of Panchayats in Delhi. *Down To Earth*.New Delhi.

Mailk, I., & Roos, V. (1994, May 26). Asola Sanctuary Grazed into extinction. *The Pioneer*. New Delhi.

Major Colvin. (1883). On the restoration of the ancient canals in the Delhi teritory. *Journal of the Asiatic Society of Bengal*, *15*, 105–127.

Malik, I. (1998). *Static Ridge: Denudation of the Ridge since 1940s, A Vatavaran Report*. New Delhi: Vatavaran

Malik, I., & Agarwal, R. (1993, May 22). The Ridge. *The Pioneer*. New Delhi.

Malik, I., & Agarwal, R. (1994a, January 23). Saving the Ridge. *Hindustan Times*. New Delhi

Malik, I., & Agarwal, R. (1994b, June 9). The Ridge: After notification, comes conservation. *The Pioneer*. New Delhi.

Mandal, S., & Sinha, A. (2008). Vanishing Ridge in Delhi: A conservation Approach. *Architechture+Design: A Journal of Indian Architechture*, *XXV*(5), 124–130.

Mann, M., & Sehrawat, S. (2009). A City with a View: The Afforestation of the Delhi Ridge- 1883 –1913. *Modern Asian Studies*, *43*(2), 543–570.

Manzoor, A. (2014, May 6). Asola Bhatti Demolitions to Resume From May 7. *The Pioneer*. New Delhi. Retrieved from http://www.dailypioneer.com/city/asola-bhatti-demolitions-to-resume-from-may-7.html (07.05.14).

Marcot, B. (1992). Conservation of Indian forests. *Conservation Biology*, *6*(1), 12–16.

Marx, K. (1990). *Capital Volume 1 (Orginially published in 1867)*. London: Penguin Classics.

Masselos, J., & Gupta, N. (2000). *Beato's Delhi 1857, 1997*. New Delhi: Ravi Dayal Publishers.

Massey, D. (1992). Politics and Space/Time. *New Left Review*, (196), 65 – 84.

Mate, M. (2013). Public Interest Litigation and the Transformation of the Supreme Court of India. In D. Kapiszewski, G. Silverstein, & R. A. Kagan (Eds.), *Constitutional Courts: Judicial role in global perspective* (pp. 262–288). New York.

Mathur Committee. (2006). *Report of the K.K. Mathur Committee of Experts set up by the Government of India to look into the issues with regard to the unauthorized development in the form of "farm houses" as well as "unauthorized colonies inhabited by affluent sections" of the society*. New Delhi. Retrieved from http://delhi-masterplan.com/wp-content/uploads/2009/09/mathur-committee-for-farm-house-unauthorised-colonies.pdf (10.04.14).

Mawdsley, E. (2004). India's middle classes and the environment. *Development and Change*, *35*(1), 79–103.

Mawdsley, E., Mehra, D., & Beazley, K. (2009). Nature Lovers, Picnickers and Bourgeois Environmentalism. *Economic and Political Weekly*, *xliv*(11), 49–59.

McCorkel, J. A., & Myers, K. (2003). What Difference Does Difference Make? Position and Privilege in the Field. *Qualitative Sociology*, *26*(2), 119–227.

Meadowcroft, J. (1998). Co-operative management regimes, a way forwad? In P. Glasbergen (Ed.), *Co-operative environmental governance: public-private agreements as policy strategy*. Dordrecht: Kluwer Academic Publishers.

Menon, N. (2010). Introduction. In *Empire and Nation: Selected essays 1985–2005* (Chatterjee., pp. 1–20). Ranikhet: Permanent Black.

Menon, V. (1993, March 23). Delhi's Ridge on the Verge of Destruction. *Times of India*. New Delhi.

Migdal, J. S. (1998). *Strong societies and weak states: State-society relations and state capabilities in the Third World*. Princeton: Princeton University Press.
Miller, J., & Hobbs, R. (2002). Conservation where people live and work. *Conservation Biology*, *16*(2), 330–337.
Misra, B. B. (1961). *The Indian Middle Class: Their growth in Modern Times*. Delhi: Oxford University Press.
Misra, M. (2010). Dreaming of a Blue Yamuna. In *Finding Delhi: Loss and renewal in the megacity* (pp. 71–86). New Delhi: Penguin.
Mitra, A. K. (2003). *Occupational Choices, Networks and Transfers: An Exegesis Based on Micro Data from Delhi Slums*. New Delhi: Manohar Publishers.
MoEF. (2007). *Anual Report, 2006–2007*. Ministry of Environment and Forest. New Delhi: Government of India.
MoEF. (2010). *ETF Evaluation Report*. National Afforestation & Eco-Development Board, Ministry of Environment and Forests. New Delhi: Government of India.
Mohammed, T. (2010). *Writing Labour: Stone quarry workers in Delhi*. Delhi: Oxford University Press.
Mohan, M. (2003). Land Encroachment Mapping Through GIS on the Northern Flank of Aravalli Mountain Hills – The Delhi Ridge. *FIG Working Week 2003 Paris, France, April 13–17, 2003*. Retrieved from http://www.fig.net/pub/fig_2003/TS_17/TS17_4_Mohan.pdf (23.06.14).
Mohapatra, S. (1994, January 8). Govt Stakes Claim to Forest. *Hindustan Times*. New Delhi.
Moore, D. S. (1996). Marxism, Culture and Political ecology. In R. Peet & M. Watts (Eds.), *Liberation Ecologies: Environment, Development,Social Movements* (pp. 125–47). London: Routledge.
Moore, D. S. (1998a). Clear waters and muddied histories: Environmental history and the politics of community in Zimbabwe's Eastern highlands. *Journal of Southern African Studies*, *24*(2), 377–403.
Moore, D. S. (1998b). Subaltern Struggles and the Politics of Place: Remapping Resistance in Zimbabwe's Eastern Highlands. *Cultural Anthropology*, *13*(3), 344–381.
Moore, S. F. (1986). *Social Facts and Fabrications: "Customary" Law on Kilimanjaro, 1880–1980*. New York: Cambridge University Press.
Mortimore, M. J. (1975). Peri-Urban Pressures. In R. P. Moss & J. A. Rathbone (Eds.), *The Population Factor in African Studies* (pp. 188–197). London: University of London Press.
Mueller. (1999). Die Forshungsperspectiven der Third world political ecology am beispiel des Gold und Diamantenbergbaus in Suedosten Venezuela. In 52nd Duetscher geographentag. Hamburg.
Mukherjee, V., & Deenadayalan, Y. (2011, October). Is There a Thorn in Delhi's Green Side? *Tehelka*. Retrieved from http://archive.tehelka.com/story_main50.asp?filename=Ws291011 Environment.asp (10.12.14).
Munshi, S. (2014, May 5). Slum families wait for flats. *Times of India*. New Delhi. Retrieved from http://timesofindia.indiatimes.com/city/delhi/Slum-families-wait-for-flats/articleshow/34655473.cmshttp://timesofindia.indiatimes.com/city/delhi/Slum-families-wait-for-flats/articleshow/34655473.cms (14.05.14).
Mwangi, E., & Swallow, B. (2005). *Invasion of Prosopis juliflora and Local Livelihoods: Case study from the lake Baringo area of Kenya. ICRAF Working Paper – no. 3.* World Agroforestry Centre, Nairobi.
Myrdal, G. (1970). The "Soft State" in Underdeveloped Countries. In P. Streeten (Ed.), *Unfashionable Economics: Essays in honour of Lord Balogh*. London: Weidenfeld and Nicolson.
Nadasdy, P. (2011). Application of Environmental Knowledge the Politics of Constructing Society/Nature. In M. J. Goldman, P. Nadasdy, & M. D. Turner (Eds.), *Knowing Nature: Conversations at the intersection of Political Ecology and Science Studies* (pp. 129–133). Chicago: University of Chicago Press.

Nair, J. (2005). *The Promise of the Metropolis: Bangalore's Twentieth Century*. Delhi: Oxford University Press.

Nandi, J. (2013, October 24). Digitized Maps Show Encroachment in Asola Sanctuary. *Times of India*. New Delhi. Retrieved from http://timesofindia.indiatimes.com/city/delhi/Digitized-maps-show-encroachment-in-Asola-sanctuary/articleshow/24676474.cms (23.05.14).

Nandi, J. (2014a, April 29). Wall of Forest Farmhouse Razed. *Times of India*. New Delhi. Retrieved from http://timesofindia.indiatimes.com/city/delhi/Wall-of-forest-farmhouse-razed/articleshow/34343046.cms (30.04.14).

Nandi, J. (2014b, May 30). Forest chief wants leopards in Asola. *Times of India*. New Delhi. Retrieved from http://timesofindia.indiatimes.com/city/delhi/Forest-chief-wants-leopards-in-Asola/articleshow/35750742.cms (30.05.14).

Nandy, A. (Ed.). (1998). *The Secret Politics of our desires: Innocence, Culpability and Indian Cinema*. Delhi: Oxford University Press.

Narayanan, N. C. (2008). State, Governance and Natural Resource conflicts. In N. C. Narayanan (Ed.), *State, Natural Resource conflicts and challenges to governance* (pp. 15–38). New Delhi: Academic Foundation.

Narayanan, N. C., & Chourey, J. (2007). Environmental Governance: Concept and Contextual Illustration through Three Indian Wetlands Cases. In E. R. N. Gunawardena, B. Gopal, & H. Kottagama (Eds.), *Ecosystems and Integrated Water Resources Management in South Asia.* (pp. 302–337). New Delhi: Routledge.

Nayar, M. (2004, July 7). Malls may add to water woes. *The Hindu*. New Delhi.

Neale, W. C. (1998). Property: Law, Cotton-Pickin' Hands, and Implicit Cultural Imperialism. In R. C. Hunt & A. G. Lanham: (Eds.), *Property in Economic Context* (pp. 47–66). University Press of America: Monographs in Economic Anthropology, No. 14.

Neumann, R. P. (1998). *Imposing Wilderness: Struggles over livelihood and nature preservation in Africa*. London and Berkley: University of California Press.

Neumann, R. P. (2003). The Production of Nature: Colonial Recasting of the African Landscape in Serengeti National Park. In K. S. Zimmerer & B. J. (Eds.), *Political ecology: An Integrative Approach to Geography and Environment-Development Studies*. New York: The Guilford Press.

New Delhi Development Committee. (1939). *New Delhi Development Committee Report*. New Delhi.

Noor, M., Salam, U., & Khan, M. A. (1995). Alleopathic Effects of Prosopis juliflora swartz. *Journal of Arid Environments*, *31*(1), 83–90.

O'Farrell, C. (2005). *Michel Foucault*. London: Sage Publications.

Paavlova, J. (2007). Institutions and environmental governance: A Reconceptualization. *Elogical Economics, 63*(1), 93–103.

Pandit, A. (2007, September 30). Greening of Bhatti Mines runs into dry spell. *Times of India*. New Delhi.

Parker, R. N. (1919). Afforestation of the Ridge at Delhi by Parker. *Indian Forester*, *January*, 21 –28.

Paulson, S., Gezon, L. L., & Watts, M. (2003). Locating the Political in Political Ecology: An introduction. *Human Organization*, *62*(3), 205–217.

Peet, R., & Watts, M. (1993). Introduction: Development theory and environment in an age of market triumphalism. *Economic Geography*, *69*(3), 227–253.

Peet, R., & Watts, M. (1996). *Liberation Ecologies: Environment, Development, Social movements*. London: Routledge.

Peluso, N. (1992). *Rich Forests, Poor People: Resource control and resistence in Java*. Berkley: University of California Press.

Peluso, N. (1993). Coercing conservation? The politics of state resource control. *Global Environmental Change*, *3*(2), 199–218.

Perappadan, B. S. (2006a, January 9). Ridge losing its green cover: Report. *The Hindu*. New Delhi. Retrieved from http://www.hindu.com/2006/01/09/stories/2006010921220300.htm(03.06.13).

Perappadan, B. S. (2006b, July 6). Unsettled life in Bawana resettlement colony. *The Hindu*. New Delhi. Retrieved from http://www.hindu.com/2004/07/06/stories/2004070609450400.htm (14.05.13)

Pieterse, E. (2012). "High Wire Acts: Knowledge imperatives of Southern urbanisms." In *Paper presented to the conference Emergent Cities: Conflicting Claims and the Politics of Informality, University of Uppsala, 9–10 March.*

Pioneer (1994, February 5). A March to Save the Capital's Lungs. Pioneer. New Delhi

Pioneer (1998, June 4). No too bare for bear. Pioneer. New Delhi.

Poffenberger, M. (Ed.). (1990). *Forest management partnerships: Regernerating India's forests*. New Delhi: Ford Foundation.

Prakash, S. (2013). Urban Avifaunal Diversity: An Indicator of Anthropogenic Pressures in Southern Ridge of Delhi, *4*(June), 135–144.

Rademacher, A., & Sivaramakrishnan, K. (Eds.). (2013a). *Ecologies of Urbanism in India: Metropolitan Civility and Sustainability*. Hong Kong: Hong Kong University Press.

Rademacher, A., & Sivaramakrishnan, K. (2013b). Introduction: Ecologies of Urbanism in India. In A. Rademacher & K. Sivaramakrishnan (Eds.), *Ecologies of Urbanism in India: metropolitan civility and sustainability* (pp. 1–41). HongKong: Hong Kong University Press.

Rajamani, L. (2007). Public Interest Environmental Litigation in India: Exploring Issues of Access Participation. Equity, Effectiveness and Sustainablity. *Journal of Environmnetal Law*, *19*(3), 293–321.

Ramanathan, U. (2006). Illegality and the Urban Poor. *Economic and Political Weekly*, *41*(29), 3193–3197.

Rangan, H. (1997). Property vs. Control: The State and Forest Management in the Indian Himalaya. *Development and Change*, *28*(1), 71–94.

Rangan, H. (2000). *Of Myths and Movements: Rewriting Chipko into Himalayan History*. Delhi: Open University Press.

Rangarajan, M. (1996). *Fencing the forest: Conservation and ecological change in India's Central Provinces 1860–1914*. New Delhi: Oxford University Press.

Rangarajan, M. (2001). *India's wildlife history*. New Delhi: Permanent Black.

Rangarajan, M. (2003). The Politics of Ecology: The debate on wildlife and people in India. In V. K. Saberwal & M. Rangarajan (Eds.), *Battles over nature: Science and the politics of conservation* (pp. 189–229). New Delhi: Sangam Press.

Rangarajan, M. (2009). Striving for a balance: Nature, power, science and India's Indira Gandhi, 1917–1984. *Conservation and Society*, *7*(4), 299–312.

Ravindran, K. T. (2000). A state of siege: How four decades of official apathy, corruption and distorted plan implementation have resulted in a royal mess in Delhi. *Frontline*, *17*(25), 116–118.

Reisigl, M., & Wodak, R. (2009). The Discourse Historical Approach. In R. Wodak & M. Meyer (Eds.), *Methods for Critical Discourse Analysis* (2nd ed., pp. 87–121). London: Sage publications.

Relph, E. (1985). Geographical experiences and being in the world. , M. Nijhoff and Hingham, Boston. In D. Seamon & R. Mugerauer (Eds.), *Dwelling, Place and Environment: Towards a Phenomenology of Person and World* (2nd 2000 ed.). Dordrecht: Krieger Publishing Company.

Ribot, J. C. (1995). From Exclusion to Participation: Turning Senegal's Forestry Policy Around. *World Development*, *23*(15), 87–99.

Ribot, J. C., & Peluso, N. (2003). A Theory of Access. *Rural Sociology*, *68*(2), 153–181.

Richmond, R., Rongo, T., Golbuu, Y., Victor, S., Idechong, N., Davins, G., Wolanski, E. (2007). Watersheds and Coral Reefs: Conservation Science, Policy, and Implementation. *BioScience*, *57*(7), 598–607.

Rittel, H. W. J., & Webber, M. M. (1973). Dilemmas in a General Theory of Planning. *Policy Sciences*, *4*, 155–169.

Robbins, P. (2000). The Practical Politics of Knowing : State Environmental Knowledge and Local Political Economy. *Economic Geography*, *76*(2), 126–144.

Robbins, P. (2001). Tracking Invasive Land Covers in India, or Why Our Landscapes Have Never Been Modern. *Annals of the Association of American Geographers*, *91*(4), 637–659.

Robbins, P. (2003a). Fixed Categories in a Portalable Lanscape : The causes and consequeses of land cover categorisation. In K. S. Zimmerer & J. Basset (Eds.), *Political ecology: An integrative approach to geography and environment-development studies* (pp. 181–201). New York: The Guilford Press.

Robbins, P. (2003b). Political ecology in political geography. *Political Geography*, *22*(6), 641–645.

Robbins, P. (2012). *Political Ecology: A Critical Introduction*. Malden, MA: Blackwell.

Robbins, P., & Sharp, J. (2006). Turfgrass subjects: The political ecology of urban monoculture. In N. Heynen, M. Kaika, & E. Swyngedouw (Eds.), *In the Nature of Cities Urban Political Ecology and the Politics of Urban Metabolism* (pp. 110–128). New York: Routledge.

Rohilla, S. (2005). Defining "Peri-urban" – A review. In V. Dupont (Ed.), *Peri-urban dynamics: Population, Habiat and environment on the peripheries of large Indian metropolisie: A review of concepts and general issues. CSH Occassional paper No. 14* (pp. 103–120). New Delhi: Centre de Sciences Humaines.

Rome, A. (2003). "Give Earth a Chance": The Environmental Movement and the Sixties. *The Journal of American History*, *90*(2), 525–554.

Rootes, C. (2003). *Environmental Protest in Western Europe*. Oxford, New York: Oxford University Press.

Rosencranz, A., Boenig, E., & Dutta, B. (2007). The Godavaram Case: The Indian Supreme Court's Breach of Constitutional Boundaries in Managing India's Forests. *ELR News and Analysis*, *37*, 10032–10042.

Rosencranz, A., & Lele, S. (2008). Supreme Court and India's Forests. *Economic and Political Weekly*, *43*(5), 11–14.

Rosenzweig, M. L. (2003a). Reconciliation ecology and the future of species diversity. *Oryx*, *37*(2), 194-205.

Rosenzweig, M. L. (2003b). *Win-win ecology: How Earth's species can survive in the midst of human enterprise*. New York: Open University Press.

Routray, S. (2014). The Postcolonial City and its Displaced Poor: Rethinking "Political Society" in Delhi. *International Journal of Urban and Regional Research*, *38*(6), 2292–2308.

Roy, A. (2002). *City Requiem, Calcutta: Gender and Politics of Poverty (Globalization & Community Series)*. Minneapolis: University of Minnesota Press.

Roy, A. (2009). Civic Governmentality: The Politics of Inclusion in Beirut and Mumbai. *Antipode*, *41*(1), 159–179.

Roy, A. (2011a). Postcolonial Urbanism: Speed, Hysteria, Mass Dreams. In A. Roy & A. Ong (Eds.), *Worlding Cities: Asian Experiments and the Art of Being Global* (pp. 307–335). Oxford: Blackwell Publishing Ltd.

Roy, A. (2011b). The Blockade of the World-class City. In A. Roy & A. Ong (Eds.), *Worlding Cities: Asian Experiments and the Art of Being Global* (pp. 259–278). Oxford: Blackwell Publishing Ltd.

Roy, A. (2011c). Urbanisms, Worlding Practices and The Theory of Planning. *Planning Theory*, *10*(1), 6–15.

Roy, B. (2005, December 27). Inside Delhi's lungs: Illegal roads, temples, DDA shed. *Indian Express*. New Delhi. Retrieved from http://el.doccentre.info/website/DOCPOST/dec05-frombom/TS1-H-ie-inside-delhis-lungs-illegal-roads-temples-dda-shed.pdf (13.07.14)

Roy, D. (1994). Factoid of Dismal Nights. *Development Alternatives Newsletter. Delhi Special*, *4*. Retrieved from http://www.devalt.org/newsletter/may94/of_1.htm (09.13.14)

Roy, D. (2004). From Home to Estate. *Seminar*, *533*. Retrieved from http://www.india-seminar.com/2004/533/533 dunu roy.htm (13.04.14)

Roy, D. (2010). City makers and city breakers. In M. Dayal (Ed.), *Celebrating Delhi* (pp. 143–161). UK: Penguin.

Roy, S. (2006, September 16). NGOs meet to save the Ridge. *Hindustan Times*. New Delhi.

Roychoudhury, S. (2013, December). From a Barren Land to a Lush Green Sanctuary, The Aravalli Biodiversity Park Has Come To Life. *Mail Today*. Retrieved from http://indiatoday.intoday.in/story/aravalli-comes-to-life-delhi-tourism/1/332912.html (10.01.14).

Saberwal, V. K. (1996). Pastoral politics: Gaddi grazing, degradation, and biodiversity conservation in Himachal Pradesh, India. *Conservation Biology*, *10*(3), 741–749.

Saberwal, V. K. (2003). Conservation by state fiat. In V. Saberwal & M. Rangarajan (Eds.), *Battles over nature: Science and the politics of conservation* (pp. 240–263). New Delhi: Sangam Press.

Sahgal, B., & Thapar, V. (1996). The Tiger's Trauma. *Sanctuary Asia*, *16*, 24–43.

Salim, S. (2010, March 16). Illegal Mining in Sanctuary. *Indian Express*. New Delhi.

Samanta, A. D. (2004, September 20). Now a `Green' Encroachment on Delhi Ridge. *The Hindu*. New Delhi. Retrieved from http://www.thehindu.com-/2004/09/20/stories/2004092008750400.htm (13.05.14).

Sandberg, J. (2005). How Do We Justify Knowledge Produced Within Interpretive Approaches? *Organizational Research Methods*, *8*(1), 41–68.

Sarkar De, S. (2006, June). The Ridge is Vital Recharge Area and no construction should be allowed here – Vikram Soni. *Toxics Dispatch Toxics Link*. Retrieved from http://issuu.com/toxicslink/docs/toxics_dispatch_28_june2006 (07.04.14).

Schenk, H. (2005). India's urban fringe. In V. Dupont (Ed.), *Peri-urban dynamics: Population, Habiat and environment on the peripheries of large Indian metropolises: A review of concepts and general issues. CSH Occassional paper No. 14* (pp. 121–141). New Delhi: Centre de Sciences Humaines.

Scott, J. C. (1985). *Weapons of the weak: Everyday forms of peasant resistance*. New Haven: Yale University press.

Scott, J. C. (1998). *Seeing like a state: How certain schemes to improve the human condition have failed.* London: Yale University Press.

Sehgal, R. (2007, January 5). Petitions in SC on Rape of the Ridge. *Asian Age*. New Delhi.

Sengupta, R. (2007). *Delhi Metropolitan:The making of an unlikely city*. New Delhi: Penguin India.

Seth, S. (2006, August 11). Ridge under siege of big money, corrupt systems. *Hindustan Times*. New Delhi.

Sethi, A. (2005). Unsettled Lives. *Frontline*, *22*(25), 34–37.

Seur, H. (1992). The engagement of researcher and local actors in the construction of case studies and research themes: Exploring methods of restudy. In *The Battlefields of Knowledge:The interlocking of theory and practise in social research and development* (pp. 115–146). London: Routledge.

Sharan, A. (2002). Claims On Cleanliness: Environment and Justice in Contemporary Delhi. In *Sarai Reader 02: The Cities of Everyday Life* (pp. 31–37). Delhi: Centre for the Study of Developing Societies.

Sharan, A. (2006). In the city, out of place: Environment and modernity, Delhi 1860s to 1960s. *Economic and Political Weekly*, *41*(47), 4905–4911.

Sharma, A. (2006, March 17). Who will stop this illegal Mall? *Hindustan Times*. New Delhi.

Sharma, R. (2008). Green Courts in India: Strengthening Environmental Governance? *Law, Environment and Development Journal*, *4*(1), 50–71.

Sharma, S. (1990, April). Ecological Task Force: Growing army. *India Today*. Retrieved from http://indiatoday.intoday.in/story/greening-operations-by-the-ecological-task-force-become-a-success/1/315057.html (14.09.14).

Shatkin, G. (2011). Coping with actually existing urbanisms: The real politics of planning in the global era. *Planning Theory*, *10*(1), 79–87.

Sheikh, S., & Banda, S. (2014). *The Thin Line between Legitimate and Illegal: Regularising Unauthorised Colonies in Delhi*. Report of Cities of Delhi project, Centre for Policy Research, New Delhi.

Shekhar, S., Purohit, R. P., & Kaushik, Y. B. (2009). Groundwater Management in NCT Delhi. In *Proceedings 5th Asian Regional Conference of Indian National Committee on Irrigation and Drainage (INCID), December 9–11*. New Delhi. Retrieved from www.indiaenvironmentportal.org.in/files/Kaushik.pdf (13.02.13).

Shokoohy, M., & Shokoohy, N. (2003). Tughlaqabad, Third Interim Report: Gates, Silos, Waterworks and Other Features. *Bulletin of School of African and Oriental Studies*, *66*(1), 14–55.

Shrivastava, K. S. (2011, November). Unhappy Bani. *Down To Earth*. Retrieved from www.downtoearth.org.in/content/unhappy-bani (19.12.14).

Singh, D. (2013a, February 7). Activists join villagers to save Ridge from concrete, plan stir. *Hindustan Times*. New Delhi. Retrieved from http://www.hindustantimes.com/india-news/newdelhi/activists-join-villagers-to-save-ridge-from-concrete-plan-stir/article1-1008332.aspx (03.04.14).

Singh, D. (2013b, July 17). Govt claim to protect city Ridge an Eyewash: NGT. *Hindustan Times*. New Delhi. Retrieved from http://www.hindustantimes.com/newdelhi/govt-claim-to-protect-city-ridge-an-eyewash-ngt/article1-1094126.aspx (24.04.14).

Singh, D. (2013c, September 4). Government assures to demarcate Ridge boundaries by mid-October. *Hindustan Times*. New Delhi. Retrieved frombwww.hindustantimes.com/news-feed/newdelhi/govt-assures-to-demarcate-delhi-ridge-boundaries-by-mid-october/article1-1117370.aspx (03.04.14).

Singh, D. (2013d, September 10). Ridge under threat from own guardians. *Hindustan Times*. New Delhi. Retrieved from http://www.hindustantimes.com/india-news/newdelhi/ridge-under-threat-from-own-guardians/article1-1119856.aspx (03.04.14).

Singh, D. (2014a, March 15). Illegal colony eating into Delhi's lone sanctuary. *Hindustan Times*. New Delhi. Retrieved from http://www.hindustantimes.com/newdelhi/illegal-colony-eating-into-city-s-lone-sanctuary/article1-1195303.aspx (16.03.14).

Singh, D. (2014b, March 20). Delhi: Forest Dept Suspects "Foreign" Nationals Living Illegally in Asola sanctuary. *Hindustan Times*. New Delhi. Retrieved from http://paper.hindustantimes.com/epaper/viewer.aspx?noredirect=true (21.04.14).

Singh, D. (2014c, April 12). Farmhouses rob 400 acres of Delhi's forest. *Hindustan Times*. New Delhi. Retrieved from http://www.downtoearth.org.in/node/23699 (15.04.14).

Singh, D. (2014d, April 22). 37 illegal houses built on Neb Sarai forest land face eviction. *Hindustan Times*. New Delhi. Retrieved from http://paper.hindustantimes.com/epaper/viewer.aspx?noredirect=true (14.04.14).

Singh, D. (2014e, April 23). Demolition drive starts in Delhi's wildlife sanctuary. *Hindustan Times*. New Delhi. Retrieved from http://www.hindustantimes.com/india-news/newdelhi/demolition-drive-starts-in-delhi-s-wildlife-sanctuary/article1-1211258.aspx?htsw0023 (25.04.14).

Singh, D. (2014f, May 18). Govt orders relocation of Sanjay Colony from Ridge. *Hindustan Times*. New Delhi. Retrieved from http://paper.hindustantimes.com/epaper/viewer.aspx?noredirect=true (20.05.14).

Singh, D. (2014g, August 2). VIPs may get protection from monkey menace, but what about rest of Delhi? *Hindustan Times*. New Delhi. Retrieved from www.hindustantimes.com/india-news/newdelhi/vips-may-get-protection-from-monkey-menace-but-what-about-rest-of-delhi/article1-1247384.aspx (04.08.14).

Singh, D. (2014h, December 26). Monkey scam brewing in Delhi: Maneka Gandhi Darpan Singh,. *Hindustan Times*. New Delhi. Retrieved from http://www.hindustantimes.com/india-news/newdelhi/monkey-scam-brewing-in-delhi-maneka-gandhi/article1-1300446.aspx?hts0021 (30.12.14).

Singh, G. (2007, November 6). Stop degrading Delhi Ridge: Jackal roams in the heart of city. *Meri News*. New Delhi. Retrieved from hwww.merinews.com/article/stop-degrading-delhi-ridge-jackal-roams-in-the-heart-of-city/127540.shtml (09.03.13).

Singh, S. (2013, October 21). Dumping Monkeys in Asola Will never End the Simian Menace. *Hindustan Times*. New Delhi. Retrieved from www.hindustantimes.com/shivanisingh/dumping-monkeys-in-asola-will-never-end-the-simian-menace/article1-1137792.aspx (19.01.14).

Singh, U. (2006). *Delhi Ancient History*. New Delhi: Social Science Press.

Sinha, G. N. (Ed.). (2014). *An introduction to the Delhi Ridge*. New Delhi: Department of Forests and Wildlife, Government of National Capitlal Teritory of Delhi.

Sivaramakrishnan, K. (1996). *Forest Politics and Governance in Bengal 1794–1994, PhD. Thesis*. Yale University.

Sivaramakrishnan, K. (1999). *Modern forests: Statemaking and environmental change in colonial eastern India*. Oxford, UK: Oxford University Press.

Smith, N. (1984). *Uneven Development*. Athens, Georgia: University of Georgia Press.

Socolow, R. H. (1976). Failures of Discourse. *Bulletin of the Americn Academy of Arts and Sciences*, *29*(6), 11–32.

Someshwar, S. (1995). *Macro Policies, Local Politics:The official and clandestine process of deforestation inthe Western Ghats of South Kanara, India, Phd. Thesis*. University of California, Los Angeles.

Soni, A. (2000). Urban Conquest of Outer Delhi: Beneficiaries, Intermediaries and Victims. The Case of Mehrauli Countryside. In V. Dupont & E. Tarlo (Eds.), *Delhi: Urban Space and Human Destinies* (pp. 75–96). New Delhi: Manohar-CSH.

Soni, A. (2002). *Everybody Loves Bhatti Mines*. New Delhi: Centre for the Study of Developing Societies.

Soni, A. (2006a, July 15). Tell us where to go. *Tehelka*. New Delhi. Retrieved from http://archive.tehelka.com/story_main18.asp?filename=Cr071506Tell_us.asp (13.02.12).

Soni, A. (2006b, July 22). Use us, don't abuse us please. *Tehelka*. New Delhi Retrieved from http://archive.tehelka.com/story_main18.asp?filename=Cr072206Use_us.asp (13.02.12).

Soni, V. (2007). Three waters – An evaluation of urban groundwater resource in Delhi. *Current Science*, *93*(6).

SPA. (1997). *Interim Report on Artificial Recharge Studies JNU-Sanjay Van-IIT Complex*. School of Planning and Architecture, New Delhi.

Spear, P. (2002). *Delhi A Historical Sketch*. New Delhi: Oxford University Press.

Sriram, J. (2011, July 14). Rule of the Kikar Ends, The Ridge is Born Again. *Indian Expres*. New Delhi. Retrieved from http://archive.indianexpress.com/news/rule-of-the-kikar-ends-the-ridge-is-born-again/817258/ (23.11.14).

Srishti. (1994). *Saving the Delhi Ridge: One year of conservation action*. New Delhi: Srishti.

Srivastava, S. (2009). Urban Spaces, Disney-Divinity and Moral Middle Classes in Delhi. *Economic and Political Weekly*, *44*(26-27), 338–345.

Stokstad, E. (2009). The Famine Fighter's Last Battle. *Science*, *324*(5928), 710–712.

Stott, P., & Sullivan, S. (Eds.). (2000). *Political Ecology: Science, myth and power*. London: Oxford University Press.

Sundaram, R. (2004). Uncanny networks: Pirate, urban and new globalisation. *Economic and Political Weekly*, *39*(1), 64–71.

Sundberg, J. (2003). Strategies for authenticity and space in the Maya Biospheric Reserve, Peten Guatemala. In K. S. Zimmerer & T. J. Bassett (Eds.), *Political ecology: An integrative approach to geography and environment-development studies* (pp. 50–69). New York: The Guilford Press.

Supdtg. Engineer PWD Delhi Province. (1938). *Annual Report on Arboricultural Operations in the Delhi Province 1937–38, Delhi State Archives*. Delhi.

Susman, P., O'Keefe, P., & Wisner, B. (1983). Global Disasters: A Radical Interpretation. In K. Hewitt (Ed.), *Interpretations of Calamity from Human Ecology* (pp. 263–83). London: Allen and Unwin.

Sutton, M., & Anderson, E. N. (2004). *Introduction to Cultural Ecology*. Calafornia: Altamira Press.

Swyngedouw, E. (1996). The City as a Hybrid: On Nature, Society and Cyborg Urbanisation. *Capitalism, Nature, Socialism*, *7*(2), 65–80.

Swyngedouw, E. (2004). *Social Power and the Urbanisation of Water: Flows of Power*. New York: Oxford University Press.

Swyngedouw, E. (2006). Metabolic Urbanisation: The making of cyborg cities. In N. Heynen, M. Kaika, & E. Swyngedouw (Eds.), *In the Nature of Cities Urban Political Ecology and the Politics of Urban Metabolism* (pp. 20–39). New York: Routledge.

Swyngedouw, E., & Heynen, N. (2003). Urban Political Ecology , Justice and the Politics of Scale. *Antipode*, *35*(5), 898–918.

Tawa Lama-Rewal, S. (2007). Neighbourhood Associations and Local Democracy: Delhi Municipal Elections 2007. *Economic and Political Weekly*, *42*(47), 53–60.

TCPD. (2012). *Draft Mangar Develpment Plan 2031*. Town and Country Planning Department, Haryana Government.

Thakuria, N., Joshi, S., & Barik, S. (2003, January). Deep in the Woods. *Down To Earth*, 25–31.

The Economist. (2007, February 15). Delhi's monkey menace: Simian agonistes. Delhi. Retrieved from http://www.economist.com/node/8708756 (03.02.14).

The Gazette of India. (2008). *Extraordinary: Part II, Section 3, Sub-Section II*. New Delhi: Department of Publications, Ministry of Urban Development, Government of India.

The Hindu. (2004, February 23). Foundation of biodiversity park laid. New Delhi. Retrieved from www.thehindu.com/2004/02/23/stories/2004022308690400.htm(13.06.14).

The Hindu. (2006a, July 5). Residents Term Public Hearings by Pollution Control Panel "Farce." New Delhi. Retrieved from www.thehindu.com/todays-paper/tp-national/tp-newdelhi/residents-term-public-hearings-by-pollution-control-panel-quotfarce/article3100683.ece (19.04.14).

The Hindu. (2006b, July 11). Objections to One Sided Hearing. The Hindu. New Delhi. Retrieved from www.thehindu.com/todays-paper/tp-national/tp-newdelhi/objection-to-quotonesided hearingquot/article3103318.ece (12.06.14).

The Hindu. (2006c, August 17). Demand to demolish illegal VK office, malls. The Hindu.New Delhi.

The Hindu. (2007, May 22). Construction of Park on the Ridge Stayed. New Delhi.

The Hindu. (2009, November 8). Only 14 of 295 illegal colonies in South Delhi to be regularised - The Hindu. New Delhi. Retrieved from http://www.thehindu.com/news/cities/Delhi/only-14-of-295-illegal-colonies-in-south-delhi-to-be-regularised/article56548.ece (13.12.13).

The Hindu. (2011, October 20). Sanjay Van a bird sanctuary in the making. New Delhi. Retrieved from http://www.thehindu.com/todays-paper/tp-national/tp-newdelhi/sanjay-van-a-bird-sanctuary-in-the-making/article2553756.ece (16.06.14).

The Pioneer. (1994, February 23). HC admits Petition on Ridge Protection. Pioneer. New Delhi.

Tillman-Healy. (2003). Friendship as Method Qualitative Inquiry. *Qualitative Inquiry*, *9*(5), 729–749.

ToI. (1993, February 18). Illegal mining. *Times of India*. New Delhi.

ToI. (1994a, February 5). March to Press Ridge Plan. *The Times of India*. New Delhi.

ToI. (1994b, March 16). Policy on Green Belt in 6 Months. *The Times of India*. New Delhi.

ToI. (2006, September 17). Medha joins "save Ridge" stir. *Times of India*. New Delhi.

Town Planning Organisation. (1956). *Interim General Plan for Greater Delhi*. New Delhi: Ministry of Health, Government of India.

Toxicslink. (2001). A landfill in the midst of Asola forest. *Toxics Dispatch: A Newsletter of Toxics Link*, *13.10*, 1–2.

Trans. Beveridge, A. S. (1922). *The Babur-nama in English (Memoirs of Babur) Volume 1*. London: Luzac and Co.

Trepl, L. (1996). City and Ecology. *Capitalism, Nature, Socialism*, *7*, 785–795.

Truelove, Y., & Mawdsley, E. (2011). Discourses of Citizenship and Criminality in Clean, Green Delhi. In I. Clark-Deces (Ed.), *A Companion to the Anthropology of India* (pp. 407–425). Malden, MA: Blackwell Publishing Ltd.

Tuan, Y. F. (1997). *Space and Place: The Perspective of Experience*. Minneapolis: University of Minnesota Press.

UNFPA. (2007). *State of the world's population 2007. Unleashing the potential of urban growth*. New York: United Nation Population Fund.

Upadhyay, V. (2001). Forests, People and Courts: Utilizing Legal Space. *Economic and Political Weekly*, *36*(24), 2131–2134.

Utpal, B., & Bhan, I. (2006, June 30). Ridge projects under scanner: MP to MoEF. *Business Standard*. New Delhi. Retrieved from http://www.business-standard.com/article/economy-policy/ridge-projects-under-scanner-mp-to-moef-106063001057_1.html (12.06.14).

Vandergeest, P., & Peluso, N. (1995). Territorialization and state power in Thailand. *Theory and Society*, 385–426.

Varma, P. K. (1998). *The Great Indian Middle Class*. Delhi: Penguin.

Vayda, A. P., & Walters, B. B. (1999). Against Political Ecology. *Human Ecology, 27*(1), 167–179.

Vedeld, T., & Siddham, A. (2002). Livelihoods and Collective Action among Slum Dwellers in a Mega-City (New Delhi). In *IASCP conference 2002 The Commons in the Age of Globalisation*. Retrieved from http://dlc.dlib.indiana.edu/dlc/bitstream/handle/10535/1380/vedldt 120402.pdf?sequence=1 (19.04.14).

Venkatesan. (2007, July 21). Wind up Forest Bench: Centre. *The Hindu*. New Delhi. Retrieved from www.thehindu.com/todays-paper/tp-national/wind-up-forest-bench-centre/article1877079 .ece (13.09.14).

Verba, S., Scholzman, H., & Brady, K. L. (1995). *Voice and Equality: Civic Voluntarism in American Politics*. Cambridge: Harvard University Press.

Verma, G. D. (2003). *Slumming India: A Chronicle of Slums and Their Saviours*. New Delhi: Penguin.

Verma, G. D. (2004a). Dispossession in Lal Khet (in Delhi ridge). *National Common Minimum Programme, Delhi Master Plan Monitor*. Retrieved from http://architexturez.net/doc/az-cf-21764 (02.07.24).

Verma, G. D. (2004b). National Common Minimum Programme Opportunities and imperatives: Delhi Master Plan solutions for housing for the poor, ridge and riverbed. *National Common Minimum Programme, Delhi Master Plan Monitor*. Retrieved from http://architexturez .net/doc/az-cf-21810 (02.07.24).

Verma, G. D. (2004c). NCMP subversion: Capital Case: Mall on Ridge, Park on Poor. *National Common Minimum Programme, Delhi Master Plan Monitor*. Retrieved from http:// architexturez.net/doc/az-cf-21759 (02.07.24).

Verma, G. D. (2004d). NGO-CEC clearance for illegal Malls and Park project. *Architexturez Imprints*. Retrieved from http://architexturez.net/doc/az-cf-22947#footnote1_y9ex5sg (02.07.14).

Verma, G. D. (2004e). Unplanned developments in Ridge / Ridge periphery in Mehrauli-Mahipalpur area (Letter sent to the CEC). *Architexturez South Asia*. Retrieved June 12, 2014, from http://architexturez.net/pst/1653a372-4d4e-4e96-b995-36e0c3369671?qt-related_content_posts=1 (02.07.24).

Véron, R. (2006). Remaking urban environments: The political ecology of air pollution in Delhi. *Environment and Planning A*, *38*(11), 2093–2109.

Verschoor, G. (1992). Identity, networks and space. In *The Battlefields of Knowledge:The interlocking of theory and practise in social research and development* (pp. 171–188). London: Routledge.
Victor, C. (2003, November 29). There are blackbuck in your backyard. *The Tribune*. New Delhi. Retrieved from http://www.tribuneindia.com/2003/20031129/ncr1.htm (16.06.14).
Vij, B. (1999, March 12). Zoo struggles to tackle problem of plenty. *Indian Express*. New Delhi.
Waitt, G. (2005). Doing discourse analysis. In I. Hay (Ed.), *Qualitative research methods in human geography* (pp. 163–191). Oxford, UK: Oxford University Press.
Walker, K. J. (1989). The State in Environmental Management: The Ecological Dimension. *Political Studies, 37*(1), 25–38.
Walker, P. A. (2005). Political ecology: Where is the Ecology? *Progress in Human Geography, 29*(1), 73–82.
Walker, P. A. (2007). Political ecology: Where is the Politics? *Progress in Human Geography, 31*(3), 363–369.
Warren-Rhodes, K., & Koenig, A. (2001). Escalating trends in urban metabolism of Hong Kong:1971-1997. *Ambio, 30*, 429–438.
Watts, M. (1983). On poverty of theory: Natural hazards research in context. In K. Hewitt (Ed.), *Interpretations of Calamity from Human Ecology* (pp. 231–62). London: Allen and Unwin.
Watts, M. (1989). The agrarian question in Africa: Debating the crisis. *Progress in Human Geography, 13*, 1–41.
White, L. G. (1994). Policy Analysis as discourse. *Journal of Policy Analysis and Management, 13*(3), 506–525.
WII. (1994). *Study Report: Management Planning for Asola Bhatti Wildlife Sanctuary*. Dehradun: Wildlife institute of India.
Williams, G., & Mawdsley, E. (2006a). India's Evolving Political Ecologies. In S. Raju, M. S. Kumar, & S. Corbridge (Eds.), *Colonial and Post-Colonial Geographies of India* (pp. 261–278). London: Sage Publications.
Williams, G., & Mawdsley, E. (2006b). Postcolonial environmental justice: Government and governance in India. *Geoforum, 37*(5), 660–670.
Williams, R. (1973). *The Country and the City*. New York: Oxford University Press.
Willig, C. (2001). *Introducing Qualitative Research in Psychology: Adventures in theory and method.* Buckingham: Open University Press.
Wilson, R. K. (1999). "Placing nature": The politics of collaboration and represent ation in the struggle for La Sierra in San Luis, Colorado. *Cultural Geographies, 6*(1), 1–28.
Winkel, G. (2012). Foucault in the forests—A review of the use of "Foucauldian" concepts in forest policy analysis. *Forest Policy and Economics, 16*, 81–92.
Wodak, R., & Meyer, M. (2008). Critical Discourse Analysis: History, Agenda, Theory, and Methodology. In R. Wodak & M. Meyer (Eds.), *Methods for Critical Discourse Analysis* (2nd ed., pp. 1–33). London: Sage Publications.
Wolf, E. (1972). Ownership and Political Ecology. *Anthropological Quarterly, 42*, 201–205.
Wolman, A. (1996). The Metabolism of the Cities. *Scientific American, 213*(3), 178–193.
WWN. (2012). Restoring Sanjay Van: Making of an Urban Aravalli Forest (Handout). New Delhi: Working With Nature, Delhi Development Authority.
Yanow, D. (2000). *Conducting Interpretative policy analysis (Qualitative research methods; volume 47). Qualitative research methods.* Thousand Oaks, CA: Sage.
Yin, R. (2003). *Applications of Case Study Research.* New Delhi: Sage.
Zérah, M.H. (2007). Conflict between green space preservation and housing needs: The case of the Sanjay Gandhi National Park in Mumbai. *Cities, 24*(2), 122–132.
Zérah, M.H., & Landy, F. (2012). Nature and urban citizenship redefined: The case of the National Park in Mumbai. *Geoforum, 46*(1), 25–33.
Zierhoffer, W. (2002). *Gesellschaft:Transformartion, eines problems.Wahrnehmungsgeographische Studien, 20.* Oldenburg: BIS Verlag.

Zimmer, A. (2010). Urban Political Ecology. Theoretical concepts, challenges, and suggested future directions. *Erdkunde*, *64*(4), 343–354.
Zimmer, A. (2012). The Politics of Regularising Delhi's Unauthorised Colonies. *Economic and Political Weekly*, *xlviI*(30), 89–97.
Zimmerer, K. S. (1993). Ecology. In C. Earle & M. Kenter (Eds.), *Concepts in Human Geography.* (pp. 161–188.). Maryland: Lanham.
Zimmerer, K. S. (1994). Human Geography and the "New Ecology": the prospect and promise of integration. *Annals of the Association of American Geographers, 84*(1), 108–125.
Zimmerer, K. S. (2000). The Reworking of Conservation Geographies: Nonequilibrium Landscapes and Nature-Society Hybrids. *Annals of the Association of American Geographers*, *90*(2), 356–369.
Zimmerer, K. S., & Basset, J. (Eds.). (2003). *Political Ecology : An Integrative Approach to Geography and Environment-Development Studies*. New York: The Guilford Press.

ACTS

Delhi Development Act (1957), available at: http://dda.org.in/tendernotices-_docs/march09/DDA_conduct_disciplanary_and%20_appel_Regulation1999.pdf (14.03.14)
Delhi Preservation of Trees Act (1994), available at: www.forest.delhigovt.nic.in/act/chap1.html (15.06.13)
Environment (Protection Act) (1986), available at: http://envfor.nic.in/legis/env/env1.html (04.06.13)
Forest (Conservation) Act (1980), available at: with Amendments Made in 1988 http://wrd.bih.nic.in/guidelines/awadhesh02c.pdf (12.06.14)
National forest Policy (1988), available at: http://envfor.nic.in/sites/default/files/introduction-nfp.pdf (15.06.13)
National Green Tribunal Act (2010), available at: www.moef.nic.in/downloads/public information/NGT-fin.pdf (19.09.14)
The Indian Forest Act (1927), available at: http://envfor.nic.in/legis/forest/forest4.html (04.06.13)
The Indian Wildlife (Protection) Act (1972), available at: http://envfor.nic.in/legis/wildlife/wildlife1.html (23.06.14)

NOTIFICATIONS (IN CHRONOLOGICAL ORDER)

Chief Commissioner of Delhi (1913, 6 December), Notification No. 8734h-R&A.
Chief Commissioner of Delhi (1915, 7 September), Notification No. 5911-R&A.
Chief Commissioner of Delhi (1915, 7 September), Notification No. 5913-R&A
Office of Chief Commissioner, Delhi (1942, 16 September), Notification No. F.F. 14 (122)/41-LSB
Chief Commissioner, Delhi (1944, 10 April), Notification No. F.14(80)/44-L.S.G
Secretary (Development) Delhi Administration (1957), Letter no. F.11.(3)/55-P&D
Under Secretary (Development) Delhi Administration, Delhi (1958, 19 November), Notification No.F.8 (1)/58
Development Commissioner (Delhi) (1965, 15 March), Notification N.F.1.(1)/64-65/SCO
Lt. Governor, Delhi (1968, 30 November), Notification No. D.O.No.Secy/Vac/187/68
Lt. Governor of Delhi (1980, 10th April), Notification No.F.SCO 32(c)
Delhi Administration, Delhi (1986, 9 October), Notification No.F.3(116)/CW/84/897/to 906
Delhi Administration: Delhi (1991, 15 April), Notification No.F2(19)/DCF/90-91/1302-91

Lt. Governor, Delhi (1991, 13 December), Notification no. F.2/DCF/1990-91/5927-5935 dated 13.12.91
Lt. Governor of Delhi (1993, 24 April), Notification No.F.2 (11)/DCF/1990-91
Lt. Governor of National Capital Territory of Delhi (1994, 24 May), Notification Declaring Reserved Forests, Notification No.F.10(42)-1/PA/DCF/93/2012-17(1)
Lt. governor of National Capital Territory of Delhi (1994, 24 May), Notification No.F.10(42)-I/PA/DCF/93/2018-23(ii)
Office of the Development Commissioner, Government of NCT, Delhi (1995, 6 October), Constitution of Ridge management board No. F56(225)/95/Dev./HO/5596
Lt. Governor of National Capital Territory of Delhi (1996, 2 April), Notification Declaring Reserved Forests, Notification No.F.1(29)/PA/DC/95.
Ministry of Environment and Forests (1998, 22 January), Environment Pollution (Prevention and Control) Authority for National Capital Region Notification No. S.O.93 (E)
Ministry of Environment and Forests, Government of India (2002, 3 June), Notification No. 1-1/CEC/2002
Lt. Governor of National Capital Territory of Delhi (2006, 10 May), Department of Environment, Forests and Wildlife, Notification No.F.10(42)-1/PA/DCF/93/II/181-198
Ministry of Urban Development (Delhi Division), Government of India (2007, 5 October), Revised Guidelines 2007 for Regularisation of Unauthorised Colonies in Delhi NO. 0.33011/2/94-DDIIB/Vol XI
The Gazette of India: Extraordinary [Part II-Sec.3 (ii)] (2008, 25 March), Notification Regulations for of Unauthorised Colonies in Delhi (Under Section 57 of DD Act, 1957)
Lieutenant Governor of Delhi (2010, 5 March), Minutes of the Meeting Regarding Regularisation of Unauthorised Colonies and Finalisation of Plan for the Special Area Held at Raj Niwas on 5 March 2010. Retrieved from: http://lgdelhi.nic.in/min_meeting. html (03.02.12)
Ministry of Environment and Forests, GNCTD (2012, 28 February), Notification No. F.8 (118)/PA/ CF/RUC/Pt.IV/7709-7723
GNCT, Urban Development Department (2012, September 4), Unauthorised Colonies Cell order No. F.No.1-33/UC/Policy/21012/549-553. Retrieved from: http://delhi.gov.in/DoIT/DOIT/DOIT_UDD/895uc/895uc1.pdf (13 July 2014)
Ministry of Urban Development (Delhi Division), Government of India (2012, 6, December), List of 40 Unauthorised Colonies whose Provisional Regularization Certificates has been Cancelled out of 127 Retrieved from http://www.delhi.gov.in/wps/wcm/connect/DOIT_UDD/urban+development/unauthorised+colonies+under+the+jurisdiction+of+government+of+nct+of+delhi/prc+cancelled (11 July 2014)
GNCTD (2013, 29 July), Subject: Permission for development work to be carried out in 90+39 more unauthorized colonies including unauthorized colonies affected by forest and ASI Order No.FN 1-33/UC/UD/Policy/07/Part file/1050-1072.

JUDICIAL DOCUMENTS

Citizens for the Preservation of the Quarries and Lakes Wilderness (2004), Application to Centrally Empowered Committee of the Supreme Court, I.A. no. 331
Delhi High Court (2004), Shiv Narayan v/s DDA & Ors., Writ Petition (Civil) No. 8523/2003
Delhi High Court (2007), New Friends Colony Residents vs Union of India (UoI) And Ors. Writ Petition (Civil) No. 2600/2001. Orders dated 14 March, 2007
Delhi High Court (2009), Nav Yuwak Gram Vikas Samiti versus Government OF NCT of Delhi and others Writ petition (Civil) 4362/2007. Orders dated 27 August 2009
Delhi High Court (2010), DDA vs. Kenneth Builders Pvt. Limited. SLP (Civil) No. 35374/2010
Delhi High Court (2011), IA in Almitra Patel vs UOI and Others Writ Petition (Civil) No. 888/1996, Order dated 26.05.2011

Delhi High Court (2011), Ashok Tanwar vs. Union of India Writ Petition (Civil) 3339/2011
Delhi High Court (2011), Freedom Fighters Social Welfare Association and others vs Union of India & others. Orders dated 15 March, 2011
Delhi High Court (2011), Shree Hazur Baba Sadhu Singh Ji Maharaj Trust and others vs Union Of India & others. Orders dated 11 November, 2011,
Supreme Court of India (1995), T.N. Godavarman Thirumulkpad vs Union of India & Ors. Writ Petition (Civil) No. 202 of 1995 2SCC267 No.5
Supreme Court of India (1996), M.C. Mehta Vs. UOI and others Writ Petition (civil) No. 4677/1985. Orders dated 25.01.1996 and 13.03.1996
Supreme Court of India (1996), M.C. Mehta Vs. UOI and others Writ Petition (civil) No. 4677/1985. Orders dated 03.01.1996 and 09.04.1996
Supreme Court of India (1996), M.C. Mehta vs. Union of India and others, I.A.No.18 in Writ Petition (Civil) No.4677/85. Order dated 13.9.1996
Supreme Court of India (1996), T.N. Godavaram vs. Thirumulpad vs. Union of India and others. No.202 of 1995 Order dated 12 December 1996
Supreme Court of India (1997), M.C. Mehta Vs. Union of India and ors. Writ Petition (civil) No. 4677/1985 SLP No. 8960/97
Supreme Court of India (1997), M.C. Mehta Vs. Union of India and ors. Writ Petition (civil) No. 4677/1985. Orders dated 19.08.97
Supreme Court of India (1999), MI builders Pvt. Ltd. Vs Radhey Shyam Sahu and others (1999) 6 SCC 464
Nayar, Kuldip (1996), Letter from Kuldip Nayar to Justice Kuldip Singh, Judge of Supreme Court of India, New Delhi Dated 6th August 1996 (EPCA, 2000, p. 16)
Supreme Court of India (2004), M.C. Mehta vs. Union of India and others Writ Petition (civil) 4677 of 1985. Decision dated 18.03.2004
MPISG (2004), Letter to CEC Subject: With reference to news reports about NGOs petitioning it for declaring Lal Khet area ridge, to draw attention to available statutory protections, PIL being heard by High Court, etc. Dated 04.06.2004
Ridge Bachao Andolan (2006), Review Petition No. INI.A. NO. 1463 of 2006 In Writ Petition (Civil) NO. 202 OF 2005
MoEF (2004), Note titled "Objection of the Ministry of Environment & Forest (MoEF) in the matter of report prepared by Shri Shekhar Singh, Special Invitee, Central Empowered Committee (CEC) in IA No. 331 Regarding the preservation of environment and bio diversity in the Delhi Ridge Area. Dated 26.07.2004
Supreme Court of India (2006), M.C. Mehta vs. UOI and others Writ Petition (civil) No. 4677/1985. Order dated 07.02.2006
Supreme Court of India (2006), T.N. Godavarman Thirumulpad vs Union of India And Ors. Writ Petition (Civil) No. 202 of 1995 Orders Dated 17 October 2006
Supreme court of India (2011), Almitra Patel vs UOI and Ors Writ Petition (Civil) No. 5236/2010. Orders dated 26.05.11
Supreme court of India (2011), Mc. Mehta vs. Union of India and Ors. IA No. 1868 of 2007 in Writ Petition (civil) No.4677 of 1985. Order dated 6.07.2011

WEBSITES

Central Empowered Committee (accessed 20.03.13): http://cecindia.org/aboutcec.php
Dargah Ashiq Allah Nazriya Peer (accessed 24.07.14): http://www.ashialla.com
DDA, Aravalli Biodiversity Park (accessed 20.12.14): https://dda.org.in/greens/biodiv/aravalli-biodiversity-park.html
DDA, Members of Delhi Biodiversity Foundation (accessed 13.06.14): https://dda.org.in/greens/biodiv/delhi-biodiversity-foundation.html

Delhi Master Plan Implementation Support Group (accessed 09.03.14): http://architexturez.net/doc/az-cf-22055
Delhi University, CEMDE (accessed 06.04.14): Biodiversity Parks Programme of DDA www.du.ac.in/du/uploads/rti/Biodiversity_Park_2652011.pdf
Department of Planning Delhi (accessed 03.09.13): delhiplanning.nic.in/TPP2006.pdf
Department of Forests and Wildlife, NCT, Delhi (accessed 23.12.14): hwww.delhi.gov.in/wps/wcm/connect/DOIT_Forest/forest/home/rti/particulars+of+organization
Natural Heritage First Team (accessed 02.03.12): http://naturalheritagefirst.in/the-team/
Sanjay Colony (accessed 12.02.12): http:// www.bhattimines.com

UNPUBLISHED DOCUMENTS

Onkareshwar, K. (2006, 9 August), Press release, Ridge Bachao Andolan
RBA (2006, 16 September), Invitation for Demonstration: You are invited to a last chance to save the urban environment in India, Ridge Bachao Andolan, New Delhi
Singh, Diwan (2005, 20 June), Letter to RWA Front from Diwan Singh, RBA convener
Verma, Gita Diwan (2004, 26 July), Letter to DDA vice-chairman from Gita Diwan Verma

APPENDICES

APPENDIX A: INTERVIEW AND FIELD NOTE CODES

Interview Codes

.No.	Code	Organisation/Designation/Role	Dates of interview
1.	BSLS:1	Local Resident, Bhawar Singh Camp/ PDS shop owner	22.02.13
2.	DDA:1	Director, Landscape Department, DDA	15.01.13
3.	DDA:2	Assistant Director, Horticulture Department, Zone VI, DDA, Sanjay Van	20.02.13
4.	DDA:3	Senior consultant, Landscape Department, DDA	23.03.14
5.	DDA:4	Field Staff, DDA Horticulture Department, Zone VI, Sanjay Van (3 respondents)	14.02.13
6.	DDA:5	Former Vice-Chairman, DDA	03.03.12
7.	ETF:1	Commanding Officer, Eco-Task Force	13.03.13
8.	EV:1	Former Director, Natural Heritage Division, INTACH	13.03.12
9.	EV:2	Srishti member /Currently Founder, Toxics Link	16.03.12
10.	EV:3	Kalpavriksh member	23.01.14
11.	EV:4	Kalpavriksh member	13.02.14
12.	EV:5	Chief Coordinator, Srishti/ Convenor, JNGO for the Ridge/ Founder, Vatavaran	10.03.14
13.	EV:6	Kalpavriksh member/Conservation biologist Ambedkar University	20.02.12
14.	EV:7 EV:7(b)	WWN member/ Delhi Bird Club member	07.02.12 03.3.13
15.	EV:8	Kalpavriksh/ Ridge management Board NGO Member	23.03.12
16.	EV:9	Kalpavriksh member	28.01.12
17.	EV:10	Director Natural Heritage Division, INTACH	09.03.12
18.	EV:11	Founder, Srishti/Founder, Wildlife Institute of India	04.02.14
19.	EV:12	Srishti member	09.02.12
20.	EX:1	Head Scientist CEMDE, Delhi University	09.02.13
21.	EX:2 EX:2 (b)	Scientist In-charge, Aravalli Biodiversity Park, CEMDE	03.02.12 16.01.14
22.	EX:3	Field Biologist, Aravalli Biodiversity Park, CEMDE	03.02.12

.No.	Code	Organisation/Designation/Role	Dates of interview
23.	EX:4	Former CEMDE Scientist/Currently Faculty Human Ecology Department, Ambedkar University	14.04.13
24.	EX:5	BNH-CEE, Environmental Education officer Asola Bhatti Sanctuary	12.02.12
25.	FD:1	Deputy Conservator of Forests (South), Forest Department, GNCTD	09.03.13
26.	FD:2	Additional Principal Chief Secretary and Head of Department, Department of Forests and Wildlife, Government of NCT of Delhi	5.01.13
27.	FD:3	Chief Conservator of Forests and Chief Wildlife Warden, Department of Forests and Wildlife, Government of NCT of Delhi	16.02.13
28.	KPLS:3	Local Resident, Kusumpur/Headman (Pradhan)	22.02.13
32.	KPLS:2	Local Resident, Kusumpur/BJP activist	23.02.13
29.	MEV:1 MEV:1 (b)	CPQLW member	02.02.12 15.03.13
30.	MEV:2	Ridge Bachao Andolan member	12.02.12
31.	MEV:3	Ridge Bachao Andolan member/Lawyer	06.03.13
33.	MoEFD:1	Secretary, Environment and Forest, Government of National Capital Territory of Delhi	5.02.14
34.	MoEF:2	Senior Scientific Officer, Department of Environment, GNCTD	09.03.14
35.	SA:1	Social Activist, Kachra Kamgar Union/ RBA member	13.02.13
36.	SA:2	Social Activist, Kisan Mahasangh/Mahipalpur Village Council leader and AAP party member	21.03.2014
37.	SVEV:1 SVEV:1 (b)	Founder, WWN	04.03.13 03.03.14
38.	SVEV:2	Environmentalist, complied report in Sanjay Van for Srishti/Toxics Link	26.02.12
39.	SVFO	Sanjay Van Field Owners (5 respondents)	15.02.13
40.	SVLS:1	Temple Priest, Sanjay Van	05.02.13
41.	SVLS:2	Nazaria Peer Trust/ President Dargah board	05.02.13
42.	SVLS:3	Local resident, Mehrauli Ward No. 2	08.02.13

Field Notes Codes

S.No.	Codes	Location	Date of notes cited
1.	FN-ABDP	Aravalli Biodiversity Park	20.02.2013
2.	FN-BS	Bhawar Singh Camp	03.03.13
3.	FN-KG FN-KG (b)	Kishangarh	12.02.13 06.03.14
4.	FN-KP FN-KP (b)	Kusumpur Pahadi	04.03.13 14.02.14
5.	FN-SC FN-SC (b)	Sanjay Colony	14.03.13 18.02.14
6.	FN-SGV FN-SGV (b)	Sangam Vihar	10.03.13 20.02.14
7.	FN-SV	Sanjay Van	14.02.13
8.	FN-MRW2 FN-MRW2 (b)	Mehrauli Ward No.2	13.02.13 07.03.14

APPENDIX B: GOOGLE MAP IMAGES OF THE CONSERVATION UNITS

Map 4: Aravalli Biodiversity Park (Source, Google Earth, April 2014)

Map 5: Sanjay Van (Source, Google Earth, April 2014)

Map 6: Asola Bhatti Wildlife Sanctuary (Source, Google Earth, April 2014)

APPENDIX C: PHOTOGRAPHS FROM THE FIELD

Bhawar Singh Camp as seen from Aravalli Biodiversity Park

Visitors' trail in Aravalli BiodiversityPark

Information board in Sanjay Van

A small masoleum in Sanjay Van

One of the lakes in Asola Bhatti Wildlife Sanctuary

Tree plantation in Asola Bhatti Wildlife Sanctuary

A female Nilgai in Asola Bhatti Wildlife Sanctuary

Cows graze in Asola Bhatti Wildlife Sanctuary

Od Women in Sanjay Colony

Birdwatchers chance upon a herd of nilgai in the South Central Ridge

MEGACITIES AND GLOBAL CHANGE / MEGASTÄDTE UND GLOBALER WANDEL

herausgegeben von Frauke Kraas, Martin Coy, Peter Herrle und Volker Kreibich

Franz Steiner Verlag ISSN 2191–7728

7. Kirsten Hackenbroch
The Spatiality of Livelihoods – Negotiations of Access to Public Space in Dhaka, Bangladesh
2013. 396 S. mit 31 z.T. farb. Abb., 10 Tab. und 57 z.T. farb. Fotos, kt.
ISBN 978-3-515-10321-3
8. Anna Lena Bercht
Stresserleben, Emotionen und Coping in Guangzhou, China
Mensch-Umwelt-Transaktionen aus geographischer und psychologischer Perspektive
2013. 445 S. mit 92 Abb., 6 Tab. und 9 Textboxen, kt.
ISBN 978-3-515-10403-6
9. Shahadat Hossain
Contested Water Supply: Claim Making and the Politics of Regulation in Dhaka, Bangladesh
2013. 336 S. mit 25 Abb., 1 Tab., 4 Textboxen und 51 z.T. farb. Fotos, kt.
ISBN 978-3-515-10404-3
10. Pamela Hartmann
Flexible Arbeitskräfte
Eine Situationsanalyse am Beispiel der Elektroindustrie im Perlflussdelta, China
2013. 201 S. mit 18 Abb., 29 Tab., 15 farbigen Fotos und 6 farbigen Karten, kt.
ISBN 978-3-515-10400-5
11. Juliane Welz
Segregation und Integration in Santiago de Chile zwischen Tradition und Umbruch
2014. 258 S. mit 19 Abb. und 41 Tabellen, kt.
ISBN 978-3-515-10467-8
12. Daniel Schiller
An Institutional Perspective on Production and Upgrading
The Electronics Industry in Hong Kong and the Pearl River Delta
2013. 253 S. mit 7 Abb. und 49 Tabellen, kt.
ISBN 978-3-515-10479-1
13. Benjamin Etzold
The Politics of Street Food
Contested Governance and Vulnerabilities in Dhaka's Field of Street Vending
2013. XVIII, 386 S. mit 103 z.T. farb. Abb. und 7 Karten, kt.
ISBN 978-3-515-10619-1
14. Tibor Aßheuer
Klimawandel und Resilienz in Bangladesch
Die Bewältigung von Überschwemmungen in den Slums von Dhaka
2014. 285 S. mit 21 Abb., 9 farb. Abb. und 37 Tab., kt.
ISBN 978-3-515-10786-0
15. Matthias Garschagen
Risky change?
Vulnerability and adaptation between climate change and transformation dynamics in Can Tho City, Vietnam
2014. 432 S. mit 38 Abb., 25 farb. Abb. und 18 Tab., kt.
ISBN 978-3-515-10876-8
16. Sebastian Homm
Global Players – Local Struggles
Spatial Dynamics of Industrialisation and Social Change in Peri-urban Chennai, India
2014. 219 S. mit 15 Abb., 26 Tab. und 10 Farbkarten, kt.
ISBN978-3-515-10877-5
17. Sophie Schramm
Stadt im Fluss
Die Abwasserentsorgung Hanois im Lichte sozialer und räumlicher Transformationen
2014. 278 S. mit 38 Abb., 29 farb. Abb. und 5 Tab., kt.
ISBN978-3-515-10905-5
18. Malte Lech
Institutions, Innovation and Regional Economic Change
2015. 220 S. mit 34 Abb. und 23 Tab., kt.
ISBN978-3-515-11120-1
19. Markus Keck
Navigating Real Markets
The Economic Resilience of Food Wholesale Traders in Dhaka, Bangladesh
2016. XV, 240 S. mit 93 Abb. und 33 Tab., kt.
ISBN978-3-515-11379-3